T0146279

Hydrocarbon Process Safety

A Text for Students and Professionals

Second edition

J.C. Jones
School of Engineering
University of Aberdeen

Whittles Publishing

Published by
Whittles Publishing Ltd.,
Dunbeath,
Caithness KW6 6EG
Scotland, UK
www.whittlespublishing.com

ISBN 978-184995-055-8

Printed by Latimer Trend & Company Ltd., Plymouth

CONTENTS

Preface   xi

I   Background to the oil and gas industry   I

1.1   The importance of hydrocarbons in the modern world   1
1.2   The nature of crude oil and of natural gas   2
1.3   The present-day industry   3
1.4   Some relevant archival material   4
      1.4.1   Oil and gas production   4
      1.4.2   Petrochemicals   6
      1.4.3   Legislative aspects   7
1.5   Units   9
1.6   Concluding remarks   9
      References   9
      Numerical problems   9

2   Hydrocarbon leakage and dispersion   II

2.1   Preamble   11
2.2   Gas leakage through an orifice   11
      2.2.1   Leakage of a single quantity of gas   11
      2.2.2   Allowances for pressure drop during discharge   13
      2.2.3   Allowances for friction   14
2.3   Leakage of a non-flashing liquid: Bernoulli's equation   15
2.4   Two-phase discharge   17
2.5   Dispersion of hydrocarbon once leaked   17
      2.5.1   An empirical approach suitable for risk assessment   17
      2.5.2   More detailed approaches   19
2.6   Dispersion of liquefied natural gas   21
2.7   Detection of leaked hydrocarbon   21

2.8   Background levels of oil in the sea   23
2.9   Concluding remarks   24
      References   24
      Numerical problems   25

3   The combustion behaviour of hydrocarbons   27

3.1   Introduction   27
3.2   Heats of combustion   27
      3.2.1   Adiabatic flame temperatures   29
3.3   Flash points   30
      3.3.1   Introduction   30
      3.3.2   Correlation of flash points of pure organic compounds
              with flammability limits   30
      3.3.3   Calculated flash points of petroleum fractions   32
      3.3.4   Recent developments in the understanding of
              flash points   33
      3.3.5   Flash points in law   34
      3.3.6   Standards for flash points   34
3.4   Thermal radiation and its relevance to flames   36
3.5   Hydrocarbon combustion phenomenology   38
      3.5.1   Preamble   38
      3.5.2   Low-temperature oxidation   38
      3.5.3   Jet fires   39
      3.5.4   Pool fires   41
      3.5.5   Fireballs and BLEVEs   43
      3.5.6   Vapour cloud explosions (v.c.e.s) and flash fires   46
3.6   The use of probit equations in fire and explosions   49
      3.6.1   Introduction   49
      3.6.2   Application to a flash fire   50
      3.6.3   Application to overpressure damage   51
3.7   Concluding remarks and further numerical example   51
      References   52
      Numerical examples   53
      Appendix Hypothetical case study involving dimethyl ether   58

4   Physical operations on hydrocarbons and associated hazards   60

4.1   Introduction   60
4.2   Storage and transportation   60
      4.2.1   Fire loads and case studies   60
      4.2.2   Buncefield, 'the biggest fire in peacetime Europe'   61

4.2.3   Safety measures in storage   61
4.2.4   Effects of solar radiation on storage of hydrocarbons   65
4.2.5   Storage codes   66
4.2.6   HAZOP studies   67
4.2.7   Thermal ignition theory applied to storage and
        pumping of unstable substances   69
4.3   Refining   73
4.3.1   Introduction   73
4.3.2   Accidents at refineries   75
4.3.3   The Marcus Hook and Richmond CA refinery
        accidents   76
4.3.4   Possible process integration in refining   77
4.4   Stirring and mixing   78
4.5   Heat exchange   79
4.5.1   Introduction   79
4.5.2   Hazards with heat exchangers   81
4.6   Refrigeration   81
4.6.1   Introduction   81
4.6.2   The provision of cooling water for plant   81
4.6.3   Accidents due to refrigeration failure   82
4.7   Site layout   82
4.8   Concluding remarks   86
      References   86
      Numerical examples   86

5   Chemical operations on hydrocarbons and
    hydrocarbon derivatives   93

5.1   Introduction   93
5.2   Cracking and hydrocracking   93
5.3   Hydrodesulphurisation and hydrodenitrogenation   96
5.4   Partial oxidation   98
5.5   Chlorination   101
5.6   Gasification   103
5.7   Hydrogenation   105
5.7.1   Introduction   105
5.7.2   Process details   105
5.8   Nitration   106
5.9   Polymerisation   108
5.10 Alkylation   110
5.11 Safety issues relating to catalysis   111
5.12 Concluding remarks   111

References   111
Numerical examples   112

6 Some relevant design principles   117

6.1 Background   117
6.2 Design of pressure vessels   117
   6.2.1 LPG storage   117
   6.2.2 Extension to other hydrocarbons   119
6.3 Pipes   120
   6.3.1 Liquids in pipes   120
6.4 Vessel support   122
6.5 Design features at the scenes of major accidents   124
6.6 Design data   125
   6.6.1 Introduction   125
   6.6.2 Densities   126
   6.6.3 Viscosities   127
   6.6.4 Enthalpies   129
   6.6.5 Vapour pressures   130
   6.6.6 Other quantities relevant to design   131
6.7 Concluding remarks   132
References   132
Numerical problems   132

7 Some relevant measurement principles   137

7.1 Introduction   137
7.2 Flow measurement   137
   7.2.1 The venturi meter and the orifice meter 137
   7.2.2 The weir   139
7.3 Pressure measurement   140
7.4 Temperature measurement   142
   7.4.1 Use of thermocouples   142
   7.4.2 Resistance thermometry   149
   7.4.3 Measurement of cryogenic temperatures   152
7.5 Fire protection of sensitive measurement instruments   153
7.6 Concluding remarks   154
References   154
Numerical examples   156

8 Offshore oil and gas production   159

8.1 Introduction   159

8.2   Some features of an offshore platform   161
8.3   The role of structural components in platform safety   161
8.4   Background to offshore accidents   163
8.5   Measures taken in the event of an initial leak   163
8.6   Background on frequencies and probabilities   165
8.7   Consequence analysis   167
      8.7.1   Jet fires   167
      8.7.2   Pool fires   170
      8.7.3   Fireballs   170
      8.7.4   Smoke   172
8.8   Construction of escalation paths   174
8.9   Offshore accident case studies   176
8.10 Other matters relating to offshore safety   178
8.11 Concluding remarks   180
      References   180
      Numerical questions   182

9   Hazards associated with particular hydrocarbon products   186

9.1   Introduction   186
9.2   Crude oil   186
9.3   Natural gas   187
      9.3.1   Background   187
      9.3.2   Case studies   189
      9.3.3   Liquefied natural gas (LNG)   191
      9.3.4   Compressed natural gas (CNG)   197
9.4   Liquefied petroleum gas (LPG)   197
      9.4.1   Nature of LPG   197
      9.4.2   Examples of risk assessment for LPG transportation 197
      9.4.3   Combustion phenomenology and case studies   200
9.5   Natural gas condensate   201
9.6   Oxygenated hydrocarbons   202
      9.6.1   Introduction   202
      9.6.2   Combustion characteristics   202
9.7   Organic peroxides   206
      9.7.1   Introduction   206
      9.7.2   Case studies and related calculations   206
      References   207
      Numerical problems   208

10 Toxicity hazards 213

10.1 Introduction 213
10.2 Chlorine 213
    10.2.1 Introduction 213
    10.2.2 Threshold limit values and trends in fatality
          through exposure 214
    10.2.3 Chlorine leakage case studies 214
10.3 Ammonia 215
10.4 Hydrogen fluoride 216
    10.4.1 Introduction 216
    10.4.2 Toxicity 216
    10.4.3 A case study 217
10.5 Selected hydrocarbon derivatives 217
    10.5.1 Introduction 217
    10.5.2 Methyl isocyanate: $CH_3NHCOCl$ 217
    10.5.3 Benzene, toluene, xylenes (BTX) 217
    10.5.4 Vinyl chloride, $CH_2{=}CHCl$ 219
    10.5.5 Acrylonitrile ($CH_2{=}CHCN$) 220
    10.5.6 Fully halogenated organic compounds 222
    10.5.7 Toxicity of combustion products in hydrocarbon fires 223
10.6 Control of major accident hazards (COMAH) 226
10.7 Classification and signage 226
10.8 Concluding remarks 228
    References 228
    Numerical problems 228
    Summary of US classification of hazardous substances 231

11 Safe disposal of unwanted hydrocarbon 233

11.1 Flaring 233
    11.1.1 Introduction 233
    11.1.2 Hazards in flaring 233
11.2 Afterburning 235
    11.2.1 Introduction and basic principles 235
    11.2.2 Catalytic afterburning 236
    11.2.3 Heat recovery 237
11.3 Use of adsorbent carbons 238
11.4 Venting 239
11.5 Disposal methods in which the hydrocarbon is utilised 240
    11.5.1 Introduction 240
    11.5.2 Blending with solid waste 240

11.5.3 Gasification 240
11.6 Non-destructive disposal on land 241
11.7 Re-refining 242
11.8 Steam raising 243
References 243
Numerical problems 243

12 Means of obtaining hydrocarbons other than from crude oil and related safety issues 247

12.1 Introduction 247
12.2 Oil from shale 247
12.2.1 Background on shale oil 247
12.2.2 Retorting processes 248
12.3 Hydrocarbons from tar sands 250
12.4 Hydrocarbons from coal 251
12.5 Tight gas, CBM and hydraulic fracture 252
12.6 Concluding remarks 253
References 254
Numerical example 254

Appendix The Canvey and Rijnmond studies 255

Part 1 Introduction 255
Part 2 Background to the study 255
Part 3 Some points from the First Canvey Report (1978) 256
Part 4 Recommendations of the Second Canvey Report 257
Part 5 Concluding remarks on the Canvey study 257
Part 6 The Rijnmond Report (brief) 257
Numerical problems 261

Solutions to numerical examples 262

Transformation of percentages to probits 314
Type K thermocouple tables 315

True/false questions 318

Introduction 318
Questions 318
Answers 326

Index 333

PREFACE

I am pleased that the first edition of this book did well and that there has been demand for a second. Readers will find that the structure of the book is fundamentally unaltered, but that many parts have been expanded and some new topics including tight gas have been added. The first edition was seen as an innovative approach to the subject and the approach has been to a considerable degree vindicated by reviews and purchases.

After nearly 20 years I am leaving Aberdeen for greener pastures in the near future. There will be no difficulty in tracking me down in my new affiliation via the internet, and readers need not hesitate to do so. As always thanks go to Whittles Publishing.

Clifford Jones
Aberdeen, January 2014

Preface to first edition

Hydrocarbon processing is an area in which many new graduates find suitable employment opportunities. The whole range of science and engineering subjects, as well as such disciplines as financial studies and law, are represented in the hydrocarbon industry. The new recruit to the industry soon becomes aware that safety is paramount. This of course is true anywhere but, because of the extreme hazards associated with hydrocarbons, safety issues feature very centrally in the development and implementation of practices and also strongly influence planning and policy-making.

Consequently, graduates who have entered the hydrocarbon industry sometimes feel the need to supplement their basic professional expertise with a suitable formal course on safety, leading to a further qualification. Such courses attract students from diverse backgrounds, and this can of course make for difficulties in the planning of a suitable course content. Over several years of teaching process safety to MSc students at the University of Aberdeen, the author has developed the following approach. The irreducible minimum of common background knowledge which the class is assumed to have goes no further than certain aspects of four subjects at first-year university level: organic chemistry, physical chemistry, heat transfer and fluid flow. Any necessary revision of these will not involve a student in an inordinate amount of effort. The content of the process safety course is then planned appropriately, and each part of it is backed up with a significant number of numerical problems. It is quite a challenge to the lecturer to develop course material which, whilst requiring only the background specified above, is closely focused on industrial realities and will enable the student to return to his or her duties in industry with a truly enhanced knowledge of process safety.

Usually when a branch of applied science or technology previously viewed as 'interdisciplinary' becomes itself the sole topic of a degree course there is a need for carefully developed numerical problems. It is hoped that this book meets this need in relation to hydrocarbon process safety. Of course, the book is by no means entirely numerical problems, but all of the material presented is backed up with such problems and further numerical problems for student use are at the ends of the chapters.

In the book there are frequent references to more advanced literature, including the monumental three-volume tome by the late Professor F.P. Lees. This and other books have been drawn on by the present author a great deal. Lees' text is a comprehensive and highly detailed coverage of the subject written for the benefit of specialists, whereas this book is a synthesis of selected material for use in teaching and private study. Readers (and reviewers!) should therefore be aware that there is no possible basis for comparison of this text with that by Lees.

In concluding this preface I should like to invert the point made previously that to present a course to a group of students with very different backgrounds can be problematic. The positive side to that is that their respective backgrounds are what the students have to contribute to class discussion and project work. I am conscious that my contact over the years with our MSc students, especially those with extensive industrial experience, has helped to advance my own knowledge. Though I should have liked all of the MSc students I have taught at Aberdeen to feel included equally in this acknowledgement, there has to be an Orwellian slant whereby 'some are more equal than others'!

I

BACKGROUND TO THE OIL AND GAS INDUSTRY

I.I The importance of hydrocarbons in the modern world

The world in which we live has become critically dependent upon supplies of liquid hydrocarbons. In historical terms, this is quite a new development. There was, of course, no such dependence in the pre-industrial era. Industrialisation went hand in hand with developments in thermodynamics, in particular steam cycles. In steam appliances the only function of the fuel is to raise the steam which is itself the 'working substance' in a steam cycle such as the Rankine cycle. Heat released in such cycles is 'latent heat', due to phase change, not the heat released by whatever fuel is used, for example coal.

Then came other thermodynamic cycles, notably the Diesel and Otto cycles, in which the working substance is air, a mixture of two 'permanent gases'. No such heat release is possible with these, and it is necessary to introduce something capable of chemical heat release at the appropriate stage of the cycle, that is, a suitable fuel. Diesel himself used powdered coal in his exploratory work, but a fuel vapour is more satisfactory in practical devices. At roughly the same time that these thermodynamic advances were occurring, the oil industry was coming into being; crude oil was being processed into fractions, a technology which was itself the fruit of advances in thermodynamics. So it would appear that quality liquid fuels became available in large quantities at just the right time to be used in the development of internal combustion engines.

But there was more to come. Whereas fractionating is a physical process – separation along a boiling point range – new technologies were developed for chemically modifying oil-derived hydrocarbon stock by cracking or reforming. There were also advances in polymerisation, so that simple hydrocarbon compounds made by cracking petroleum fractions were processed into materials which in many applications replaced natural ones in the manufacture of products including furnishings (plastic instead of wood) and clothing (synthetic fibre materials instead of wool or cotton). In other words, the petrochemical industry had been born, and although the production of fuels from crude oil continued to be very important there was the extra dimension of the supply of hydrocarbons as raw materials for the manufacturing sector.

Long before 'the millennium' oil had become a commodity upon which world trade was strongly dependent, and factors affecting its availability would have far-reaching effects on world finance. The 2005 Gulf Coast hurricanes were such a factor. The erratic behaviour in oil prices which resulted were such as had not been observed since the 1870s when expansion of discovered reserves led to severe dips and surges in the price of a barrel of oil. In late 2007 the price of oil reached $US100 for the first time and that was seen by some as a portent that lack of availability of oil was about to plunge world finances into something comparable to the 1929 Wall St. crash. That of course did not happen. By September 2012 the West Texas Intermediate price was $99.08 per barrel, the Brent price $116.90 per barrel and the OPEC Basket price $114.87 per barrel. In the intervening period prices were on occasion below $50 per barrel.

1.2 The nature of crude oil and of natural gas

There are 103 elements in the Periodic Table. One of them, carbon, has an entire branch of chemistry to itself, organic chemistry. The uniqueness of carbon which enables it to form so many more compounds than other elements is that it combines with itself to form carbon structures which, when complemented with the hydrogen atoms needed to satisfy valency requirements, constitute hydrocarbons. The following are examples of hydrocarbon series of compounds:

$$\text{alkanes e.g., propane, } CH_3CH_2CH_3$$

$$\text{alkenes e.g., propene, } CH_3CH=CH_2$$

$$\text{aromatics e.g., benzene, toluene.}$$

All hydrocarbons, on reacting with oxygen, are very powerful heat releasers. Variations between hydrocarbon molecules in the number of carbon atoms and in their skeletal arrangement, and the possibility of double or triple bonds as well as single ones, mean that the number of possible hydrocarbon compounds even in the range up to about C_{20} is gigantic. Crude oil therefore consists of a huge number of organic compounds, and to characterise a crude on the basis of the amount of one particular constituent hydrocarbon is impossible. Crudes can however be characterised according to

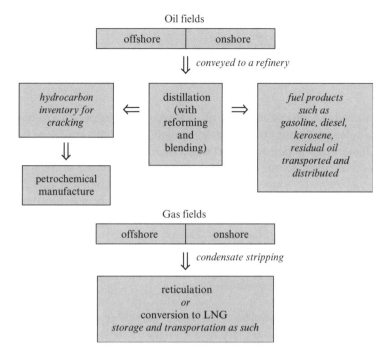

Figure 1.1 *Structure of the modern oil and gas industry.*

a preponderance of hydrocarbon types, there being paraffin (alkane)-base, naphthene (cycloalkane)-base, aromatic-base and mixed-base.

The dominant constituent of natural gas is methane (CH_4), the simplest hydrocarbon of all. There might also be small amounts of higher hydrocarbons, e.g., ethane, propane or propene, as well as inerts such as nitrogen and carbon dioxide.

Hydrocarbons are fire hazards wherever they are present in significant quantities, e.g., offshore installations, onshore processing such as refining, the petrochemicals industry and fuel transportation. Liquid hydrocarbons, pure compounds or petroleum fractions, always have calorific values in the neighbourhood of $45MJkg^{-1}$. This is a high exothermicity: liquid explosives such as DINA (dinitroxydiethylnitramine) have exothermicities of less than a tenth of this.

1.3 The present-day industry

This is the background against which the present-day hydrocarbon industry has to be understood. Events marking the beginning of the conventional oil industry are the discovery of crude oil in Pennsylvania by E. Drake in 1859 and the formation of the Standard Oil Company by J.D. Rockefeller in 1870.

Figure 1.1 shows in outline how the present-day industry is structured. Crude oil production began in the late nineteenth century and was entirely onshore until 1945, when offshore activity began off the Gulf Coast. In Chapter 8 offshore installations will be discussed, in particular from the point of view of the unique hazards which exist at such installations. Oil has to be refined by fractionating, and the products are used partly as fuels and partly as feedstock for the petrochemical industry. Gas, which may be onshore or offshore and often occurs with crude oil, can be reticulated to domestic and commercial users or converted to liquefied natural gas, a cryogenic substance which can be transported in large quantities by land or sea. Condensate is usually taken to be synonymous with natural gas liquids (NGL) and has become increasingly important over the last few years, partly because of its presence in shale gas, a.k.a. tight gas, which means gas from formations of very low permeability. Access to such gas by hydraulic fracture has led to environmental and safety concerns.

The most comprehensive study of hydrocarbon hazards to date is the Canvey study. Reference to this is made in a number of subsequent chapters, and this is appropriate in that the study, though carried out over 30 years ago, continues to be widely cited.

1.4 Some relevant archival material

1.4.1 Oil and gas production

Mention has been made previously of the events marking the creation of the oil industry. It is helpful for a professional in the present-day industry to have an idea of how the industry developed and expanded. Accordingly the next few paragraphs will describe in outline what the author sees as some of the milestones in the hydrocarbon industry.

Of course, long before Drake's discovery, in fact from the beginning of recorded history, there was an awareness of the existence of oil and its flammability. There had unquestionably been, in the ancient world, usage not only of oil but also of natural gas in certain places including China. Such usage was only ever of local significance and oil and gas were not traded between countries. Moving closer to the present time, a map of Pennsylvania published in 1748 – over a century before Drake's discovery – shows the oil fields clearly marked as such. Natural gas occurring in England had been brought to the notice of the Royal Society of London in the seventeenth century. In the mid-nineteenth century, there was importation into England of oil brought from Burma. This was later distilled and the distillation process was the subject of a succession of patents.

Figures for the breakdown of energy usage in the USA into coal, oil and gas for the period 1829–1924 are given by Haslam and Russell [1]. Only coals of various rank – anthracite and bituminous – appear until the entry for 1869. 'Domestic oil', i.e., oil produced in the USA, makes its first appearance in the row of the table for 1869 as having contributed 25 trillions of BTU, compared with 879 in the same units for coal. Natural gas makes its first appearance in the row for 1889, as having contributed 268 trillions of BTU; for the same year the contributions of coal and oil, respectively, in the same units, were 3746 and 211. By 1924 the figures were coal 15030, oil (domestic and imported) 4782 and gas 1161 trillions of BTU. These figures give a perspective on the growth of fuel use of hydrocarbons. It should be remembered that processing of shale to make oil predates by about 20 years the manufacture of fuels from crude oil. In Europe there was significant shale oil processing by 1850, though the first discovery of shale in the USA was not until much later[1].

A century ago no less than today, world events were reflected in trends in oil supply. Though until now our discussion has focused mainly on North America, Russia was also a very significant producer of oil in the early days of the industry. In the year 1901 Russian oil production exceeded that of the USA. The Russian production rate dropped dramatically between 1917 and the early 1920s, a result of the Revolution. World War I also had a marked effect. Although world oil production increased between 1914 and 1918 it did so only modestly, and growth was much more rapid from 1919 onwards.

J.D. Rockefeller has been selected for special mention as a pioneer of the American industry. In Britain, William Knox D'Arcy, founder of the British Petroleum Company (BP), originally called the Anglo–Persian Oil Company, perhaps deserves equivalent acknowledgement. In 1901 he began searching for oil in Persia, and his early endeavours were not without severe disappointments. It is recorded [2] that at the first exploration site the conditions under which D'Arcy's men worked were quite atrocious, and that only very small amounts of crude oil were found. It was not until seven years after D'Arcy's exploration began, by which time he had accepted help from the Burmah Oil Company, that crude oil in commercially usable quantities was discovered.

By the mid-1920s total world oil production was 10^9 US barrels, with the USA production accounting for about 70% of that[2]. By 1950 [3] USA production was roughly 2×10^9 US barrels per year and world production was about 3×10^9 US barrels per year. Two factors in particular had contributed to the increase: the growth of the automobile industry and the development of markets for petroleum distillates as chemical feedstocks. Natural gas utilisation also increased significantly in the second quarter of the twentieth century, providing 5% of the total energy requirements of the USA in 1924 and 15% in 1950.

The second half of the twentieth century saw many more advances including a proliferation of offshore oil and gas production facilities. At the same time there developed an awareness of the finiteness of hydrocarbon resources, creating a need to use what we have wisely and to plan for the future by investigating alternative means of making liquid fuels. These have often led to very fruitful research programmes, for example into flash pyrolysis of high-volatile coals to yield liquids which can be engineered into the equivalents of petroleum fractions.

With the exception of the mention of use of natural gas in China in ancient times, oil from Russia and the importation of oil from Burma in the nineteenth century, our discussion until now has been concerned with what might loosely be called the western world. There are of course highly industrialised nations in the Far East and it is instructive to conclude this section with a summary of the development of fuel utilisation in those[3].

By the mid-twentieth century there was onshore oil production in Malaya (as it then was), China and Burma. The shortage of indigenous fuels has been an important issue with Japan throughout her modern history. Even in the opening years of the twenty-first century Japanese oil fields supplied only a very small proportion of the total energy requirement of the country. Between about 1955 and 1970 refining capacity in Japan increased hugely but this simply involved importing more crude. There has however been some Japanese offshore activity – close to Honshu Island – since about 1970. Again, this only provides a small fraction of the total requirement. By the end of the twentieth century there was continuing activity in China, though by 1993 that country had become a net importer of oil. There is also major oil production in Vietnam and a growing offshore industry in Thailand which is more productive of gas than of oil. The previously very productive onshore oil field at Sarawak was in decline by about 1970.

1.4.2 *Petrochemicals*[4]

Serious investigation into using petroleum products as chemical feedstocks, instead of simply as fuels, began in the early 1920s. The previous feedstocks were all coal derived and were obtained more from steelworks than from the chemical industry because of the practice of preparing coke on site for manufacturing steel. This sometimes made availability uncertain. Also, the visionaries of the 1920s sensed that petroleum was so superior to coal tars as a chemical feedstock that there was the potential to revolutionise the organic chemistry industry. Essential in effecting such a revolution was the development of a means to break down long-chain hydrocarbons present in petroleum fractions to simple olefins. Olefin production by cracking finally became an industrial reality in about 1930. Afterwards pipelines from refineries to factories, sometimes hundreds of kilometres in length, were provided in order to keep the organic chemical industry supplied with feedstock. Many chemical companies found it worth their while to relocate to the Gulf Coast.

The USA organic chemistry industry was therefore stimulated and fortified by the ready availability of feedstock. In Germany, which in wartime had depended greatly on chemicals from coal, there was the disadvantage that all of the refineries were on the northern coast with no pipeline structure to convey hydrocarbons to the industrial centres. In order to maintain competitiveness with the USA, the German industry began to prepare its own petrochemical feedstocks from crude oil. This contrasted with the USA situation where cracking was of previously refined material. In 1953 small-scale experiments were performed by researchers at Hoechst in which crude oil was passed along a heated tube charged with hot pebbles. Contact of the oil vapours with the pebbles leads to the formation of cracked material with a good yield of olefins. Scale-up soon followed, and in February 1956 a unit capable of producing 10^4 tons per annum of ethylene by cracking of crude oil commenced operations. Significant amounts of methane were produced as a by-product and this was largely diverted to methyl chloride manufacture. A present-day cracker for ethylene production from naphtha can have a productivity up to ten times that of the 1956 Hoechst facility.

1.4.3 Legislative aspects

This is a vast and extremely complex topic, a detailed account of which is within neither the scope of this book nor the area of specialism of the author. Nevertheless, the topic is so central to the functioning of all aspects of the hydrocarbon industry that no engineer or technologist in that industry would want to be without at least a background knowledge. Such knowledge will normally be acquired as necessary in the discharge of day-to-day professional duties in association with others. Knowledge so obtained will usually relate to the contemporary scene and will be enhanced if supplemented with some key facts relating to the development of workplace safety legislation. That is the purpose of this section.

The hydrocarbon processing industry is subject to workplace legislation in the broad sense as well as to much legislation relating specifically to the industry. As described in the introductory part of the text by Lees [4], the proliferation of factories in the nineteenth century was accompanied by exploitative practices a description of which one can find, for example, in certain of the writings of Charles Dickens. In Britain a succession of 'Factory Acts' was passed, the first in 1802, and such an Act in 1833 brought the Factory Inspectorate into being. The first of the series of Acts to be concerned with a specific safety issue was that of 1844, which imposed a duty upon employers to install and maintain guards around flywheels and other rapidly moving devices. By the end of the nineteenth century the earlier Acts had been subsumed into a single Factory and Workshop Act. By that time employer liability was established in law and hazards associated with particular operations had been identified and special regulations imposed.

What distinguishes chemical hazards from other sorts of industrial hazard such as falling objects or faulty electrical installation is that chemicals, whether they ignite and burn or simply escape, have the potential to harm the public as well as the occupants of the site of the hazard. The Robens Report of 1972 recommended the creation of a specialised branch of the Factory Inspectorate which would be concerned with industrial hazards which can affect the public. At first this was concerned only with fire and explosion hazards, but shortly afterwards toxic hazards also came within its remit.

Legislation specific to hydrocarbons has been accompanied by the development of practices and standards for incorporation into legislation. Perhaps the most notable is the flash point as an indicator of storage and transportation safety of hydrocarbon liquids. The *Encyclopaedia Brittanica* for 1911 states that, in Britain at that time, only a hydrocarbon liquid with a flash point of 73°F (23°C) or higher could be stored without special precautions. Other countries had similar requirements. For example, in France at that time the flash point was set, for the same purpose, as 35°C, in Germany at 21°C and in Russia at 28°C. In the USA it was State rather than Federal legislation which applied. Pennsylvania and New York independently set the value at 100°F (38°C). Of course there were (and are) different methods of determining the flash point, in particular 'closed cup' and 'open cup' methods[5]. Each piece of relevant legislation specified a method and some also prescribed a detailed procedure, drawn up by one of the standards bodies such as the ASTM, which must be followed if a flash point measured and documented was subsequently to stand up in law.

The first UK Act of Parliament concerned specifically with offshore safety (Lees, *op. cit.*) was passed in 1971. The Piper Alpha disaster of 1988 (outlined in Chapter 8) led to a change of approach towards monitoring offshore safety; in particular, after Piper Alpha less emphasis was placed on site inspections and more on the auditing of management practices and this is developed in considerable detail in Chapter 8. Other legislative and regulatory material is given at suitable places in the later chapters.

1.5 Units

In the world of oil and gas supply, non-SI units abound especially in the US and literature therefrom. The barrel – 0.159 cubic metres – is as non-SI as one could imagine but it is inconceivable that it will be replaced. Similarly, natural gas is sold on a heat basis, not a quantity basis, and pricing is in $ per million BTU. The conversion factor to SI is helpful: 1 BTU = 1055 joules.

1.6 Concluding remarks

This chapter gives a concise background on the hydrocarbon industry. It is intended that a reader will retain the gist of it in his or her mind as a framework within which the quite detailed contents of the subsequent chapters will be assimilated.

References

[1] Haslam R.T., Russell J.A. *Fuels and their Combustion*, First Edition. McGraw-Hill, New York (1926)
[2] *Our Industry Petroleum*, Third Edition. British Petroleum, London (1958)
[3] Johnson A.J., Auth G.H. *Fuels and Combustion Handbook*, First Edition. McGraw-Hill, New York (1951)
[4] Lees F.P. *Loss Prevention in the Process Industries*, Second Edition. Butterworth-Heinemann, Oxford (1996)
[5] http://www.indexmundi.com/energy.aspx?product=gas&graph=production

Numerical problems

1. Reference [1] gives a value of 1099 trillion BTU for the energy available from oil produced in the USA in 1909. Reference [2] gives a figure of 28 million tons for US oil production the following year, 1910. Examine these data for order-of-magnitude agreement using the conversion factor given in the main text.
2. A widely cited rule-of-thumb is that the world uses a cubic mile of crude oil per year. In 2011, crude oil production internationally was 26.6 billion barrels. How close is this to a cubic mile?
3. Natural gas is sold on a heat basis, per million BTU. Pure methane has a calorific value of 889 kJ mol^{-1}. What volume of methane, referred to 1 bar pressure and 15°C, would yield one million BTU of heat on burning? Use the conversion factor in the main text and express your answer in cubic metres.
4. Annual world natural gas production is about 10^{14} cubic feet [5] referred to 1 bar pressure and 15°C. How many cubic miles is this? What are the respective amounts of heat releasable by this quantity of gas and by a cubic mile of oil?
5. Some reserves of natural gas contain significant amounts of ethane. If a particular natural gas is composed of methane and ethane, no other hydrocarbons and no inerts, and has a calorific value of 1150 BTU per cubic foot what is the proportion of ethane, volume or molar basis?

End notes

[1] It was fully reported in the March 1874 issue of *Scientific American*, having actually occurred about five years earlier.

[2] 1 US barrel = 159 litre.

[3] Most of the information in the remainder of this section has been taken from successive editions of the *International Petroleum Encyclopedia* Pennwell Publishers, Tulsa, Oklahoma.

[4] Most of the information in this section is taken from *A Century of Chemistry* E. Baumler, Econ Verlag, Dusseldorf (1968).

[5] More information on flash points is given in Chapter 3.

2

HYDROCARBON LEAKAGE AND DISPERSION

2.1 Preamble

An offshore platform is uniquely hazardous in that persons are miles out to sea and surrounded by huge quantities of powerfully combustible material. In the event of leakage, emergency responses come into operation: these include isolation of inventory at either side of the leak, to limit the quantity able to escape. Offshore, pipes and vessels may contain gas, crude oil or both. At refineries there will be, as well as the influx crude oil, distillation products of widely differing boiling range. Wherever hydrocarbons are processed, stored, transported or utilised there is the possibility of leakage and subsequent ignition. Ways of describing such leakage according to the principles of physics and fluid mechanics will be covered in this chapter. In computer software for offshore risk assessment (ORA)[6] leakage has to be incorporated by suitable equations, and much recent development work in the modelling of hydrocarbon leakage has been in the context of such software. This chapter will trace some of these developments and will therefore be largely, but not exclusively, offshore focused.

2.2 Gas leakage through an orifice

2.2.1 Leakage of a single quantity of gas

This can be described by:

$$Q = C_d A P \sqrt{(M\gamma / RT)[2/(\gamma+1)]^{(\gamma+1)/(\gamma-1)}}$$

Eq. 2.1

where Q = mass flow rate of gas ($kg\,s^{-1}$)
C_d = coefficient of discharge
A = discharge area (m^2)

P = upstream pressure (N m^{-2})
M = molecular weight (kg mol^{-1})
R = gas constant = 8.314 J K^{-1} mol^{-1}
T = gas temperature (K)
γ = ratio of principal specific heats

Equation 2.1 is currently used in ORA, for example to represent natural gas leakage through a small, accidentally created orifice. The coefficient of discharge arises from friction, and has value 1.0 if friction is negligible. Correlations or graphical charts in the literature can be used to determine C_d from quantities including the Reynolds number and the ratio of orifice diameter to the vessel dimensions. The coefficient is often hard-coded as 1.0 in software for ORA. Some programs calculate it, and such refined values are usually in the range of 0.7–0.8.

The equation is a classical one[7], and can be found for example in any edition of *Perry's Chemical Engineers' Handbook* [1]. It is based on steady, critical flow by a perfect gas. By *critical* is meant that the flow rate is independent of the downstream pressure, depending only on the upstream pressure. The criterion for critical flow [1] is:

$$P_{down}/P_{up} = [2/(\gamma + 1)]^{\gamma/\gamma-1}$$

For methane at 20°C, γ = 1.307, from which the critical pressure ratio 0.545 that is:

$$P_{internal}/P_{external} = 1.8$$

At an offshore platform crude oil and gas often enter the platform together via the wellhead and 'Christmas tree' (see Chapter 8), and have to be separated before piping ashore. At the separator, the gas pressure might be as low as 1.1 bar absolute. Having regard to the fact that the external (downstream) pressure will be 1.0 bar, leakage at the separator will not display critical flow. By contrast, in a natural gas pipe containing the separated gas pressures are tens of bar, possibly as high as 100 bar. Here the requirement for critical flow is met. A simple application of the equation is in the shaded area below.

Consider natural gas inventory at 100 bar abs., 15°C. A hole of 1 cm^2 is created in the pipe containing this gas. Assuming a coefficient of discharge of unity, what will be the rate of release at 15°C?

Solution
$A = 10^{-4}$ m^2, $P = 10^7$ Pa, $\gamma = 1.3$, $T = 288$ K, $M = 0.016$ kg mol^{-1},

$$R = 8.314 \text{ J K}^{-1}\text{mol}^{-1}$$

$$\Downarrow \text{Eq. 2.1}$$

$$Q = 1.7 \, \text{kg s}^{-1}$$

If the gas so released, on encountering atmospheric air, ignites the result will be a jet fire. As we shall see in Chapter 3, a value for the gas discharge rate enables the length of the jet fire to be estimated. This length can be compared with the distance from the position of the leak to the nearest hydrocarbon-containing pipe or vessel, and a judgement made as to whether there will be flame impingement on to the pipe or vessel.

2.2.2 Allowances for pressure drop during discharge

The treatment in the previous section oversimplifies matters by using a single value of the pressure upstream of the orifice; clearly this pressure will drop as transfer of the pipe contents to the outside takes its course. The equation for gas discharge in the form in which we have already met it, viz.:

$$Q = C_d A P \sqrt{(M\gamma/RT)[2/(\gamma+1)]^{(\gamma+1)/(\gamma-1)}} \qquad \text{Eq. 2.1}$$

can be improved upon by putting:

$$P = P(t), \text{ given by:}$$

$$P(t) = P_{\text{initial}} \exp\{-(Kt/MV)RT\} \qquad \text{Eq. 2.2}$$

where $K = C_d A \sqrt{(M\gamma/RT)[2/(\gamma+1)]^{(\gamma+1)/(\gamma-1)}}$

and V is the enclosure (pipe or vessel) volume. In the shaded area below, the calculation carried out previously is redone with a time-dependent pressure.

Reconsidering the above example, where the release rate of methane was calculated as $1.7 \, \text{kg s}^{-1}$, what will be the release rate after one hour if the gas which can exit the leak occupies a volume of $500 \, \text{m}^3$?

Solution

Substituting into the expression above, K (SI units) $= 1.7 \times 10^{-7}$
after 1 h, $P/P_{\text{initial}} = 0.83$, $P(t) = 0.83 \times 10^7 \, \text{N m}^{-2}$

$$\Downarrow$$

$$Q = 1.4 \, \text{kg s}^{-1}$$

The more advanced model predicts only a 17% drop in the discharge rate after one hour and in ORA one is seldom interested in the value after as long as an hour. Depending on the volume of gas able to leak, the simpler treatment using a single pressure may well suffice.

2.2.3 Allowances for friction

In the worked examples so far, the coefficient of discharge has been taken to be unity. For escape through an orifice this is easier to justify than for escape (planned or accidental) from the open end of a long pipe. Here, friction can give rise to pressure drops. Energy balance for a gas flowing isothermally along a hori- zontal pipe, incorporating kinetic energy, flow work and friction, furnishes [1]:

$$\left(p_1^2 - p_2^2\right) = 4FLG^2RT/DM \qquad \text{Eq. 2.3}$$

where p_1, p_2 = the pressures at the ends of a section of pipe of length L, position 1 being upstream of position 2

F = Fanning friction factor
G = mass flow rate $(\text{kg}\,\text{m}^{-2}\text{s}^{-1})$
T = temperature (K)
D = pipe diameter (m)
M = molecular weight of the gas $(\text{kg}\,\text{mol}^{-1})$

A related calculation is summarised in the shaded area below.

At an offshore platform, natural gas at 10 bar pressure enters a circular pipe of diameter 1 m at a rate of $30\,\text{kg}\,\text{s}^{-1}$. This gas travels 200 km along this pipe to an onshore terminal. Taking the process to be isothermal at 280 K, find the pressure on exit. Use a value of 0.003 for the Fanning friction factor.

Solution

$$G = 30/(\pi \times 0.5^2)\,\text{kg}\,\text{m}^{-2}\text{s}^{-1} = 38.2\,\text{kg}\,\text{m}^{-2}\text{s}^{-1}$$
$$p_1^2 - p_2^2 = 5.1 \times 10^{11}\,\text{Pa}^2 \Rightarrow p_2 = 7 \times 10^5\,\text{Pa (7 bar)}$$

The Fanning friction factor can be estimated from the Reynolds number via the Reynolds analogy [2], which links degree of turbulence with friction. For laminar flow in smooth tubes, the form of the Reynolds analogy which applies is:

$$\text{Re} \times F = 16$$

where Re = Reynolds number and F = Fanning friction factor. Below is a calculation in which this is utilised.

Methane flows at 300 K, 1.5 bar pressure along a tube of 10 cm diameter. What is the maximum flow speed (ms^{-1}) consistent with laminar flow? What is the Fanning friction factor corresponding to this condition? The dynamic viscosity (usual symbol μ) of methane at this temperature has a value of $1.12 \times 10^{-5} kg\,m^{-1}\,s^{-1}$.

Solution

$$Re = ud\sigma/\mu \text{ where } u = \text{speed of flow } (m\,s^{-1})$$
$$d = \text{tube diameter}$$
$$\sigma = \text{fluid density } (kg\,m^{-3})$$

Now $\sigma = M \times P/RT$, where $M = \text{molecular weight } (kg\,mol^{-1})$,
$P = \text{pressure } (N\,m^{-2}) \text{ and } T = \text{temperature } (K)$.

$$\Downarrow$$

$$\sigma = 0.96\,kg\,m^{-3}$$

Transition from laminar to turbulent, in a tube, at $Re = 2500$ [3], hence the maximum speed is given by:

$$u_{max} = 2500 \times 1.12 \times 10^{-5}/(0.1 \times 0.96)\,m\,s^{-1}$$
$$= 0.29\,m\,s^{-1}$$
$$F = 16/2500 = 0.0064$$

2.3 Leakage of a non-flashing liquid: Bernoulli's equation

Imagine a tank of crude oil initially full which develops a small leak $H(m)$ below the top as shown in Figure 2.1. The top is at atmospheric pressure. We can obtain an expression for the velocity of the fluid on exit by means of Bernoulli's equation. Let the pressure at position 1 be P_1, that at position 2 be P_2. Also let the respective velocities of the fluid be v_1 and v_2, and note that these are orthogonal; at position 1 the liquid is flowing vertically and at position 2 horizontally.

Energy balance on the arrangement in Figure 2.1 gives:

$$z_1 g + v_1^2/2 + P_1/\sigma = z_2 g + v_2^2/2 + P_2/\sigma \qquad\qquad \text{Eq. 2.4}$$

where g is the acceleration due to gravity. Since the vessel diameter is very much larger than the leak diameter, v_1 is negligible in comparison with v_2. Also, $z_1 = z_2$, therefore:

$$0.5\,(v_2^2 - v_1^2) = (1/\sigma)(P_1 - P_2) \implies v_2 = \sqrt{[2\,(1/\sigma)(P_1 - P_2)]}$$

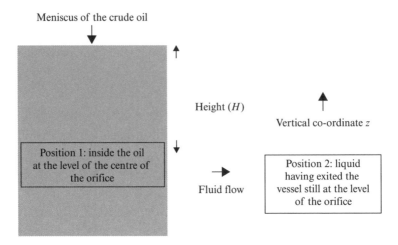

Figure 2.1 *Leakage of crude oil from a tank.*

Now at position 2 the pressure is atmospheric. At position 1 the pressure exceeds atmospheric by $\sigma g H$, where H(m) is as defined in Figure 2.1, from which:

$$v_2 = \sqrt{[2gH]}$$

These ideas will be illustrated in a worked example below.

A large tank is used to store crude oil. Part way down the vertical wall of the tank, 6.5 m from the head of liquid, is a circular orifice. This is used occasionally to tap off small samples for analysis, otherwise it is shut off. The tank is at the centre of a circular area of radius 50 m with a 'bund' (dike) around the circumference to prevent any leaked oil from spreading. If there is accidental discharge of oil through the leak, calculate how long it will take for the oil to reach the dike:

(a)   if it is assumed that the exit velocity (v_{exit}) is maintained
(b)   if it is assumed (This expression is utilised in the Canvey study (see Appendix) that the oil front travels with an average velocity

$v_{exit}/\sqrt{2}$

Solution

(a) $v_{exit} = \sqrt{2gH} = 11.3 \text{ m s}^{-1}$

Time taken to travel 50 m = 4.4 s

(b) $v_{\text{average}} = 11.3 / 1.414 \text{ m s}^{-1} = 8.0 \text{ m s}^{-1}$

Time taken to travel 50 m = 6.3 s

A very similar expression to that for the velocity of liquid exiting an orifice, as in the above example, applies to the 'slumping' of a catastrophically failed vertical column of liquid. This was used in the Canvey study (see Appendix). The equation for slumping of a catastrophically leaked liquid through total containment failure is:

$$\text{velocity of liquid} = \sqrt{(gh)} \text{ m s}^{-1}$$

where h = height of the column of liquid in the container before discharge. For $h = 10$ m this gives a velocity of 10 m s^{-1} (≈ 20 m.p.h.).

2.4 Two-phase discharge

An equation applicable to two-phase flow through an orifice is [4]:

$$Q = C_d A \sqrt{2\sigma_m (P - P_c)} \qquad \text{Eq. 2.5}$$

where Q = discharge rate (kg s^{-1})
P_c = critical orifice pressure
P = upstream pressure
σ_m = 'mean density' of the two-phase hydrocarbon C_d,
A as previously defined

The critical pressure is, in ORA software, usually taken to be 60% of the upstream pressure. Two-phase hydrocarbon inventory can be characterised for the purposes of flow calculations by taking a sample initially at the upstream pressure and rapidly reducing the pressure ('flashing'), which of course simulates leakage. Weight fractions and densities of the two phases after flashing provide a route to a calculation of σ_m.

2.5 Dispersion of hydrocarbon once leaked

2.5.1 An empirical approach suitable for risk assessment

Most hydrocarbon gases and vapours are denser than air, though an important exception is methane. Imagine that a quantity of hydrocarbon gas or vapour is accidentally released, for example during chemical processing (as happened at the Flixborough accident in the UK, in the 1970s: the hydrocarbon was cyclohexane and the process partial oxidation). Once a quantity is leaked, it might ignite or it might disperse without igniting. During dispersion it becomes diluted, eventually to a concentration below the lower flammability limit (LFL). Clearly, beyond this stage the substance cannot ignite. It is therefore useful to be able to estimate the time required for a specified quantity of a particular hydrocarbon substance to disperse sufficiently for the concentration to drop below the LFL.

The dispersion of a hydrocarbon cloud in the atmosphere is a complex process. Disciplines relevant to its analysis include fluid mechanics, and very advanced computer modelling of hydrocarbon dispersion has been carried out. For the purposes of risk assessment, once a hydrocarbon has been released it is helpful to know the time taken for dispersion to below the LFL as a function of wind conditions. This is readily incorporated into software, and enables predictions to be made as to how far away from the position of leak a possible ignition source such as a flare is rendered harmless.

Leakage may be 'catastrophic' as at Flixborough or continuous. Examination [5] of the results of independently reported field tests for slow, continuous flow have suggested that correlations of the type:

$$d = d(0)e^{-ku} \qquad \text{Eq. 2.6}$$

where d is the distance travelled before dilution to sub-flammable concentrations at wind speed u, $d(0)$ is the distance travelled before dilution to sub-flammable concentrations in still air ($u = 0$) and k is a constant, are applicable. In the shaded area below this is applied to propane.

For propane at $2\text{--}4\,\mathrm{m}^3\mathrm{s}^{-1}$ continuous discharge rate, field tests fit the expression:

$$d = 620\,e^{-0.116u}$$

with values of d in m and u in $\mathrm{m\,s}^{-1}$. What distance is required for dilution to sub-flammable proportions: (a) in still air and (b) in a wind of speed $10\,\mathrm{m\,s}^{-1}$?

Solution
By examination of the above expression, the drift distance for dilution to below the LFL in still air is 620 m

For a wind speed of $10\,\mathrm{m\,s}^{-1}$, $d = 194\,\mathrm{m}$

2.5.2 More detailed approaches

2.5.2.1 Passive dispersion

A widely used model for gas dispersion is the Pasquil model. This is for passive release, that is, where the gas density is comparable to that of air so that gravity effects are small therefore the primary effect is movement of molecules from an area of high concentration to an area of low concentration. It is also for non-instantaneous release, though the release duration might be short. For short releases, values of the lateral 'spread' θ (degrees) as a function of the drift distance 'd' is given for six classifications of weather conditions. For example, on a sunny cloudless day the lateral displacement for an initially almost vertical leak will, after a drift of 100 m, be 60°, decreasing eventually to about 20°. Figure 2.2 shows the behaviour of leaked gas as predicted by the model whereby as dispersion takes its course the envelope of gas becomes more squat and forms a smaller angle with the horizontal. The height h is the height at which the concentration is one tenth that at the ground level, and the line making the angle θ with the horizontal is the distance from the centre of the envelope at which the concentration is one-tenth the axial value. Pasquil's treatment leads to correlations for the vertical spread h (see Figure 2.2) as a function of drift distance for the six weather conditions. These ideas are best illustrated by means of a suitable numerical example.

A leakage of ethane occurs which is initially almost vertical. The wind speed is low ($\approx 1 \, m \, s^{-1}$) and conditions cloudless and sunny. At what distance from the source will the height of the gas envelope be 100 m? If the day is overcast with stronger winds, what will the distance be?

Solution
First note that as ethane is only slightly more dense than air, a passive dispersion treatment is appropriate. Lees [6] gives correlations for d and h according to the six weather conditions specified in Pasquil's treatment. The first of the weather conditions here is termed by Pasquil category A, and the correlation is:

$$\log_{10}h = 2.95 + 2.19\log_{10}d + 0.723(\log_{10}d)^2$$

with h in m and d in km. Substituting $h = 100$ m and rearranging gives:

$$-0.434 = \log_{10}d + 0.330(\log_{10}d)^2$$

Trial-and-error solution gives:

$$d = 0.30 \, km$$

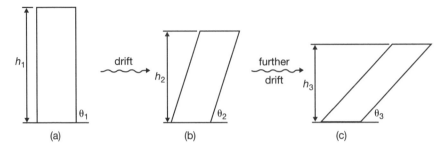

Figure 2.2 *Representation of passive dispersion of gas on the Pasquil model: (a) initial leak; (b) and (c) at successively longer times.*

The second set of weather conditions are Pasquil category D, for which:

$$\log_{10}h = 1.85 + 0.835 \log_{10}d - 0.0097(\log_{10}d)^2$$

Substituting $h = 100\,\text{m}$ and rearranging gives:

$$0.18 = \log_{10}d - 0.0116(\log_{10}d)^2$$

Trial-and-error solution gives:

$$d = 1.5\,\text{km}$$

and intuitively a longer drift distance is expected in the second case because of the stronger wind conditions.

There are other approaches to passive dispersion including an extension of the Pasquil model, called the Pasquil–Gifford model [6]. There is also a relatively simple Gaussian model, though this has found greater application to the dispersion of effluent from stacks than to leaked hydrocarbon.

2.5.2.2 Dense gas dispersion

Gases which, on entry into air, are subject to initial slumping through gravity are termed 'dense gases'. These are observably distinct from buoyant releases by their tendency to slump, spread and then 'passively disperse' when leaked as a single quantity. There are several advanced models for these, a selection of which we shall examine. One of these is by van Ulden, and models a single gas release according to the phenomenology outlined above: slumping, spread and passive dispersion. The model takes the initial quantity of 'instantaneously released' gas to be a cylinder of height H_o and radius R_o, and by means of an equation for mixing incorporating an 'entrainment coefficient' gives a value for the rate of increase of radius as:

$$dR/dt = (c_E/R)\sqrt{(\sigma_o - \sigma_a)gV_o / \pi\sigma_o}$$ Eq. 2.7

where R = radius at time t

c_E = a constant, termed a 'slumping constant', having a value close to unity verified in many experimental and field tests

σ_o = density of the cloud at time zero ($kg\,m^{-3}$)

σ_a = density of air ($kg\,m^{-3}$)

V_o = initial volume (m^3)

This is of course integrable by elementary means to give:

$$R^2 - R_o^2 = 2c_E t\sqrt{(\sigma_o - \sigma_a)gV_o / \pi\sigma_o}$$ Eq. 2.8

The equation for dR/dt above is equivalent to one for the velocity of the vapour front at time t (hence radius R). Van Ulden concluded that as a cloud develops a stage is reached where the vertical component of the mixing, gravitationally driven, becomes negligible in comparison with the radial mixing, whereupon the dense gas starts to behave like a buoyant one. This leads to stage on the graph of R against t (or equivalently, of R against distance from the site of the original leak) beyond which the slope declines markedly. Van Ulden confirms this experimentally, with a halogenated organic compound as the leaked substance, in trials conducted at his Netherlands laboratory. Figure 2.3 shows these trends schematically.

2.6 Dispersion of liquefied natural gas

It is well known that although methane is only just over half as dense as air, LNG when leaked displays dense gas dispersion as would, for example, propane which is 50% more dense than air. The reason LNG disperses as a dense gas is that droplets become suspended in the gas and these have the effect of raising the density. It has been shown [7] that a monodisperse system of LNG droplets of radius $0.2\,\mu m$ would have a molar mass of $10^6\,g$, making the LNG a 'transient macromolecule'.

2.7 Detection of leaked hydrocarbon

Semiconductors – n-type and p-type – are often applied to hydrocarbon leakage detection [8]. These include the tin (IV) semiconductor, which can be used to sense methane. Zinc oxide and silicon carbide are similarly used, and the latter is able to distinguish one hydrocarbon from another. Techniques other

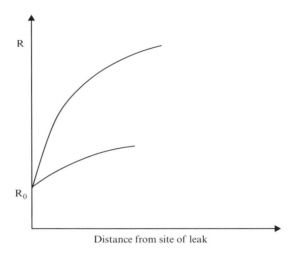

Figure 2.3 *Van Ulden's results for leakage of a halogenated organic compound. Upper curve: behaviour expected on the basis of van Ulden's model and closely matched by experiment. Lower curve: behaviour expected on the basis of a simple Gaussian model.*

than semiconductors include ultrasound, and this finds wide application in gas sensing at offshore installations. There are also infra-red devices for detecting leaked hydrocarbon.

The catalytic gas sensor or pellistor (pelletised resistor) can be used to detect methane amongst other gases. Such a device uses catalytic combustion of a flammable gas as the principle of operation. It comprises a platinum heating coil inside a ceramic pellet the surface of which is coated with a catalyst. Contact of a flammable gas with the heated catalyst surface causes heating of the assembly and the rise in electrical resistance is detected and forms the basis of the signal. In practice two resistors, identical except for the presence of the catalyst on one and its absence from the other, are incorporated into the sensor, the difference in resistance of the two being compared. This cancels out any effects other than those due to the flammable gas being detected. Different gases and vapours can be distinguished from each other by different responses from a pellistor. A related calculation follows.

The table below is taken from information provided by a manufacturer of a particular pellistor. If it is calibrated for, and used in, measuring methane, response 1 means that the methane is at its lower flammability limit (LFL). If it is calibrated for methane and used to measure propane a response of 0.65 indicates that the propane is at its LFL. Response up to the LFL is approximately linear.

The pellistor when used to measure acetone vapour gives a response of 0.5. What is the concentration of acetone, expressed as a proportion of the lower flammability limit?

Solution: At the LFL of acetone the response would be 0.7. A response of 0.5 therefore signifies that the acetone is at a proportion 0.5/0.7 = 0.71 of its LFL.

Combustible gas/ vapour exposure	Relative response when sensor is calibrated on		
Hydrogen	2.2	1.7	1.1
Methane	2.0	1.5	1.0
Propane	1.3	1.0	0.55
n-Butane	1.2	0.9	0.5
n-Pentane	1.0	0.75	0.5
n-Hexane	0.9	0.7	0.45
n-Octane	0.8	0.6	0.4
Methanol	2.3	1.75	1.15
Ethanol	1.6	1.2	0.8
Isopropyl alcohol	1.4	1.05	0.7
Acetone	1.4	1.05	0.7
Ammonia	2.6	2.0	1.3
Toluene	0.7	0.5	0.35
Gasoline (unleaded)	1.2	0.9	0.5

Laser devices are available which detect only methane at ranges up to 100m. The principle of operation is attenuation of a reflected laser beam by the absorptive methane molecules. Such a device therefore detects the methane without contacting it.

2.8 Background levels of oil in the sea

The theme of the chapter changes here from dispersion of gases in the atmosphere to dispersion of liquid hydrocarbon in the sea. A major source of leakage of oil into the sea is tankers which receive oil from a production platform or a terminal. It is arguable that globally oil tankers contribute more to discharge of oil into the sea than do offshore production facilities. This is not due simply to leakage but also to the periodic cleaning of the tank interiors. In this regard let it be remembered that a 'supertanker' holds of the order of a million barrels and travels, in a single voyage, thousands of kilometres. Broadly speaking one expects oil levels in the sea to be p.p.m. or a few tens thereof and 1 p.p.m. is $1\,cm^3$ of oil in $1\,m^3$ of water.

The author has written in detail of the Gulf Coast oil spill in 2010 [9]. 'Time zero' was 21:49 on 20th April 2010. Because of the success of 'Top Hat 10' oil discharge into the sea ceased at 15:25 on 15th July. There was much uncertainty and conflict of information over the leakage rate. Published estimates of the daily release rate over the duration of the spill range from 3–5 million barrels.

2.9 Concluding remarks

This chapter, though written in such a way that the reader is continually reminded of the realities of hydrocarbon technology, is primarily concerned with fluid leakage *per se* so that where it is encountered in subsequent chapters he or she will have the necessary background. The next chapter is concerned with hydrocarbon combustion and this also has to some extent the nature of background material. The chapter does however, draw on the contents of the present chapter in its discussion of combustion phenomenology in hydrocarbon accidents. The 'fluid flow' and 'combustion' themes are brought even closer together later in the book when risk assessment is discussed.

References

[1] Perry R.H., Green D. *Perry's Chemical Engineers' Handbook*, 6th edition (or any available edition), McGraw-Hill, New York (1984)

[2] Jones J.C. *The Principles of Thermal Sciences and their Application to Engineering*, Whittles Publishing, Caithness, UK (2000)

[3] Holman J.P. *Heat Transfer*, SI metric edition. McGraw-Hill, New York (1989)

[4] User Manual. PLATO© Software for Offshore Quantitative Risk Assessment, Version 1.3. Four Elements Ltd., London (1994)

[5] Jones J.C. On the estimation of drift distances required for the dilution of leaked hydrocarbon clouds *Journal of Fire Sciences* **19** 97–99 (2001)

[6] Less F.P. *Loss Prevention in the Process Industries*, 2nd edition. Butterworth-Heinemann, Oxford (1996)

[7] Jones J.C. LNG Chemistry, *LNG Industry,* Autumn 2012 pp. 111–112

[8] Jones J.C. *Dictionary of Fire Protection Engineering*, Whittles Publishing, Caithness, UK (2010)

[9] Jones J.C. *The 2010 Gulf Coast Oil Spill*, Ventus Publishing, Fredericksberg (2010) This electronic book can be downloaded free of charge by going to the BookBoon web site.

Numerical problems

1. At a (natural) gas-bearing pipe at an offshore installation, the internal pressure is 95 bar absolute, the temperature 288 K. A leak develops and in the follow-up it is estimated, from the recorded responses of methane detectors at the platform, that methane entered the atmosphere at $7.5 \, m^3 s^{-1}$ referred to the conditions outside the vessel, which were 288 K, 1 bar pressure. Use the single quantity of gas model to estimate what the size (area) of the leak must have been. Take the coefficient of discharge to be 1.

2. A leak develops in a pipe bearing natural gas at 10 bar pressure, 288 K. The leak area is $0.5 \, cm^2$. Emergency shutdown valves come into operation and isolate a total inventory volume of $150 \, m^3$ which incorporates the leak. Neglecting friction, what will be the exit rate of gas: (a) initially, (b) after 15 min.

3. A large tank of crude oil develops a small leak 7 m below the meniscus of the oil. Starting with Bernoulli's equation and explaining each step of your working, estimate the speed of the oil as it exits the leak. *From MSc (Safety and Reliability) Examination, University of Aberdeen, 2000.*

4. When liquefied natural gas (LNG)[8] was allowed to evaporate under conditions of measured wind speed, the following results were obtained:

Wind speed/$m \, s^{-1}$	Distance for dilution to below LFL/m
4.8	110
3.9	150

Calculate the distance for dilution to sub-flammable levels in still air and at a wind speed of $3 \, m \, s^{-1}$.

5. It is possible to extend the means described in the main text of estimating drift distances to the lower flammability limit using results cited by Lees [6] for petrol spilt on the ground. In a wind of speed of $0.9 \, m \, s^{-1}$ the distance was 6 m whereas when the spill was deliberately protected from the wind by having it adjacent to a low wall the distance was 9.1 m. Fit these data to an equation of the form of equation 2.6.

6. Consider a quantity of butane leaked into air, with an initial radius of the leak 15 m, and initial height 4 m. On the basis of the van Ulden treatment calculate the radius after: (a) 10 s, (b) 3 min.

7. Return to the calculation in the main text which considers leakage of ethane and the distance from the source at which will the height of the gas envelope will be 100 m. Recalculate the drift distance for the leakage at night with a moderate wind speed. These are Pasquil category F weather conditions, and the applicable correlation (symbols as in the main text) is:

$$\log_{10} h = 1.48 + 0.656 \log_{10} d - 0.122 (\log_{10} d)^2.$$

End notes

[6] What is referred to in this text as 'Offshore Risk Assessment' (ORA), should be understood as meaning Quantitative Risk Assessment (QRA) applied to offshore operations. Reference 4 uses, in its title, the term 'Offshore Quantitative Risk Assessment', and for convenience the word quantitative is 'taken as read' in this text where the term ORA is used. See also: Morris M., Miles A., Cooper J., Quantification of escalation effects in offshore quantitative risk assessment, *Journal of Loss Prevention in the Process Industries*, **7** 337–344 (1994).

[7] It is to be found in what is perhaps the first ever standard text on chemical engineering, that by Walker W.H., Lewis W.K. and MacAdams W.H., all at MIT, published by McGraw-Hill in 1923. On the connection between MIT and the development of chemical engineering education, *see*: Jones J.C., On the origins of the chemical engineering profession, *Chemistry in Australia*, **67** 2 (2000).

[8] The nature, properties and uses of LNG are discussed in Chapter 9.

3

THE COMBUSTION BEHAVIOUR OF HYDROCARBONS

3.1 Introduction

In order to be capable of displaying burning, a chemically reacting system needs to be able to release heat to the extent of MJ or tens of MJ per kg of substance reacted. If this pre-requisite is fulfilled, the precise type of combustion behaviour which occurs, e.g., explosive or steady, depends upon other factors including the effectiveness of mixing of fuel and oxidant and, very importantly, heat transfer from reacting system to surroundings.

3.2 Heats of combustion

Table 3.1 gives some heats of combustion (calorific values) for common substances including petroleum fractions.

The calorific values of petroleum products are therefore high, twice that for dried wood. The value for natural gas is higher still. Note that the calorific values are heat released per kg of fuel; a full mass balance on the combustion would require that the weight of oxidant – almost always air in practical applications – was also accounted for. Heats of combustion of alkanes (methane, ethane, propane, butane, pentane etc.) on a molar basis obviously increase with carbon number. On a kg basis, methane is about $55\,MJ\,kg^{-1}$ (as we have seen), and this drops gradually to about $45\,MJ\,kg^{-1}$ for octane and higher. $45\,MJ\,kg^{-1}$ is of course the heat value of petroleum distillates, and this also applies to pure alkanes beyond about C_8. The reason why the lower hydrocarbons have higher values is their greater proportion of C—H as opposed to C—C bonds. Biodiesels have been included in the table because of their increasing importance in the moves towards carbon-neutral fuels.

A mass balance on the burning of a petroleum fuel in air, which also predicts emissions, follows. Octane C_8H_{18} (molar mass 0.114 kg) is used as a representative compound.

Table 3.1 *Heats of combustion of common materials*

Material	Heat of combustion/MJ kg^{-1}
Wood	≈17 if 'seasoned', ≈20 if dry
Natural rubber	≈44
Bituminous coal	In the approximate range 20–35, depending on the moisture and mineral contents
Lignite	≈8 in the bed-moist state, rising to ≈20 when air dried, depending on the mineral content
Paper and cardboard	15–17
All liquid petroleum fractions (gasoline, naphtha, kerosine, diesel and residual fuel oil)	43–46
Natural gas	≈55*, depending on the precise composition
Biodiesels	≈37

* It is more conventional to express calorific values of gases in MJ m^{-3}, the volume being referred to 1 bar pressure and 288 K. The calorific value of natural gas so expressed is 37 MJ m^{-3}.

Element	kg per mol fuel	mol per mol fuel	Stoichiometry	Product
C	0.096	0.096/0.012 = 8	$C + O_2$ ↓ CO_2	CO_2 8 mol
H	0.018	0.018/0.002 = 9 (expressed as H_2)	$H_2 + 0.5O_2$ ↓ H_2O	H_2O 9 mol

From the third and fourth columns of the table:
$$O_2 \text{ requirement} = (8 + 4.5) \text{ mol} = 12.5 \text{ mol}$$

For every mole of oxygen in the combustion, 79/21 = 3.76 moles of nitrogen involved

Accompanying nitrogen = (3.76 × 12.5) mol = 47 mol

Composition of the post-combustion gas per kg of the fuel burnt.

N_2	47 mol
CO_2	8 mol
H_2O	9 mol

Allowing for condensation of the H_2O, the composition of the gas is 14.5% carbon dioxide and 85.5% nitrogen.

The calculation is for stoichiometric conditions. If, as will often be the case, there is excess air there will be some excess oxygen and correspondingly more nitrogen in the post-combustion gas.

Even if the water condenses it will not, under equilibrium conditions, be fully free of vapour. There will be the pressure of water vapour corresponding to whatever temperature the liquid water is at. If there is condensation to that degree, then to within about 20% a barrel of petroleum fuel burnt results in a barrel of water in the environment.

3.2.1 Adiabatic flame temperatures

Once a hydrocarbon/air mixture has ignited, the flame temperature attained depends upon the extent of heat transfer from flame to surroundings. Clearly the maximum temperature a flame can reach is that in the limit where burning is adiabatic, there being no heat transfer to the outside so that all of the heat of combustion is retained as sensible heat in the reaction products. To estimate adiabatic flame temperatures for hydrocarbon compound combustion is not difficult, and an example is given below.

We attempt to estimate the adiabatic flame temperature of propane, given the following data [1, 2]:

$$\text{net heat of combustion of propane} = 2045\,kJ\,mol^{-1}$$
$$\text{specific heat of water vapour} = 43\,J\,°C^{-1}mol^{-1}$$
$$\text{specific heat of carbon dioxide} = 57\,J\,°C^{-1}mol^{-1}$$
$$\text{specific heat of nitrogen} = 32\,J\,°C^{-1}mol^{-1}$$

Propane burns according to:
$$C_3H_8 + 5O_2 + (18.8N_2) \rightarrow 3CO_2 + 4H_2O + (18.8N_2)$$
and the net heat of combustion applies where the water remains in the vapour phase: the gross value applies if the water condenses to liquid and in so doing releases its heat of vaporisation. When a mole of propane is burnt, the reaction products clearly have a heat capacity:
$$[(3 \times 57) + (4 \times 43) + (18.8 \times 32)]\,J\,°C^{-1} = 945\,J\,°C^{-1}$$

If conditions are adiabatic, the temperature rise is:
$$(2\,045\,000/945)°C = 2165°C$$

This gives a final temperature, for a starting temperature of 25°C, of 2140°C.

The above calculation could be refined in a number of ways. For example, the specific heats used are those at 1200°C, whereas a more accurate treatment would incorporate these as a function of temperature. Nevertheless, the value obtained above is in fair agreement with the literature value of 2250 K (1977°C) [3] and the important point which a reader should appreciate is that adiabatic flame temperatures for hydrocarbon reactions in air are always in the neighbourhood of 2000°C. Of course, excess fuel or excess air will have the effect of lowering this temperature because of their effect on the heat capacity of the post-combustion gas, to which they add. If the working above is repeated for conditions of 20% excess fuel, the calculated adiabatic flame temperature lowers to about 2100K. As has already been pointed out, this is an upper limit on actual flame temperatures, but a useful one in that it is often approached quite closely. This will become clearer in later sections of this chapter.

3.3 Flash points

3.3.1 Introduction

If it is desired to store or transport a hydrocarbon liquid – a single compound such as benzene or a complex blend such as a petroleum fraction – judgement as to the safety is made on the basis of the flash point. This is the minimum temperature of bulk liquid at which there will be a flash if a flame is brought into contact with the equilibrium vapour, and can be determined by the open- or closed-cup methods according to national standards. As described in Chapter 1, flash points have been in use for a very long time [4] and continue, in the twenty-first century, to be the criterion by which particular liquids are assessed for fire safety. Table 3.2 gives the flash points of some liquids.

The table contains values for hydrocarbon compounds and oxygenated hydrocarbon compounds as well as ranges for diesel and coal tar.

3.3.2 Correlation of flash points of pure organic compounds with flammability limits

The flash point was developed as an empirical means of classifying liquids according to their ignition hazards: the higher the flash point the better. Griffiths and Barnard [6] have pointed out that at least for pure hydrocarbon compounds the flash point can be calculated on the basis that it is the temperature at which the equilibrium vapour pressure of the subject liquid (calculable from the Clausius–Clapeyron equation) is such that the vapour concentration in the air contacting the liquid corresponds to the lower flammability limit. By this means they obtained a good calculated value – in close agreement with experiment – for the flash point of toluene, and this

Table 3.2 *Flash points of selected organic liquids. Taken from [5], all closed-cup values*

Liquid	Flash point/°C
Benzene, C_6H_6	−11
Toluene, $C_6H_5CH_3$	4
n-octane	13
Methanol, CH_3OH	12
Ethanol, C_2H_5OH	13
Acetone, CH_3COCH_3	−18
North American diesel fuels	82–166
Coal tars, by-product of coking	90–135

was followed by an equally satisfactory calculation of the flash point of benzene [7]. For the purposes of these calculations, the flammability limit can be taken to correspond to the proportion of fuel in the fuel–air mixture at half-stoichiometric. The method is outlined below.

Consider a hydrocarbon of 'generalised' molecular formula C_nH_m:

$$C_nH_m + (n + m/4)O_2 + 3.76\ (n + m/4)N_2 \rightarrow$$
$$nCO_2 + (m/2)H_2O + 3.76\ (n + m/4)N_2$$

The stoichiometric proportion ϕ of the hydrocarbon vapour is given by:

$$\phi = 1/\{1 + 4.76(n + m/4)\}$$

The proportion ϕ^* at *half*-stoichiometric is given by:

$$\phi^* = \phi/2 = 0.5/\{1 + 4.76(n + m/4)\}$$

and this is an estimate of the proportion at the flash point. Let us apply this to one or two simple hydrocarbons, starting with hexane (boiling point 69°C) which burns according to:

$$C_6H_{14} + 9.5O_2\ (+ 35.7N_2) \rightarrow 6CO_2 + 7H_2O\ (+ 35.7N_2)$$

The proportion ϕ^* required for there to be a hazard when the vapour mixes with air is therefore:

$$\phi^* = 0.5/\{1 + 4.76(6 + 14/4)\} = 0.0108$$

In a total pressure of 1 atmosphere (10^5 Pa) there will therefore be a hazard if the pressure of hexane is 1080 Pa or greater. The equilibrium vapour pressure of hexane at 15°C is several times this, hence hexane vapour is an ignition hazard at ordinary storage temperatures. Consider now dodecane, $C_{12}H_{26}$, boiling point 216°C, which burns according to:

$$C_{12}H_{26} + 18.5O_2 (+ 69.6N_2) \rightarrow 12CO_2 + 13H_2O (+ 69.6N_2)$$

The proportion ϕ^* required for there to be a hazard when the dodecane vapour mixes with air is therefore:

$$\phi^* = 0.5/\{1 + 4.76(12 + 26/4)\} = 0.0056 \ (0.56\%)$$

In a total pressure of 1 atmosphere this would require a vapour pressure of the hydrocarbon of 560 Pa. Dodecane would have such a vapour pressure at about 70°C, so its vapour is not a hazard at ordinary storage temperatures.

3.3.3 Calculated flash points of petroleum fractions

3.3.3.1 Correlation with vapour pressures

It has been pointed out that the above approach can, with a few very moderate approximations, be extended to any petroleum fraction for which the Reid vapour pressure (RVP) is known [8]. In extending the calculation of flash points for pure organic compounds to petroleum fractions, the difficulties are twofold: the fact that such a fraction does not have a single boiling point, only a boiling range, and the fact that the Clausius–Clapeyron equation only applies to pure compounds. These difficulties are addressed in recent work [8]. Instead of a boiling point the RVP is used to obtain the constant of integration. There can arguably be no single value for the vapour pressure of a petroleum fraction at any one temperature, since evaporation from liquid to vapour phase causes composition changes to the liquid and the extent of this will depend on the extent of evaporation and, therefore, on the volume of the space into which the vapour is evaporating. However, the RVP is determined under standardised conditions and is viewed in the industry as 'the vapour pressure' of the fraction. As for the second difficulty – the non-applicability of the Clausius–Clapeyron equation – over a limited temperature range an equation of that form can be taken to apply on an empirical basis. This requires very thoughtful choice of a value for the latent heat of vaporisation at temperatures well below the boiling point, and such are available in thermodynamic tables. In this way, good values for the flash points of petroleum fractions have been calculated [8].

The difficulty of non-applicability of the Clausius–Clapeyron equation to petroleum fractions also applies to biodiesels, composed of fatty acids of typically C_{18}. These, because of their carbon neutrality, are becoming increasingly prevalent and knowledge of their flash points is important. Attempt to calculate the flash points from the Clausius–Clapeyron equation have however been strongly deprecated [9]. With biodiesels there is no 'operational' quantity corresponding to the RVP for petroleum fractions which can be used in flash point estimations on the basis of flammability limits and stoichiometry.

3.3.3.2 The Factory Mutual equation

An equation originally proposed by fire researchers at Factory Mutual is:

$$t_F = 0.683t_B - 71.7 \hspace{2cm} \text{Eq. 3.1}$$

where t_F is the closed-cup flash point (°C) and t_B the initial boiling point (°C). The concept of 'initial boiling point' relates to a blend of hydrocarbon compounds such as would constitute a petroleum fraction. If the equation is to be applied to pure compounds, t_B is the normal boiling point. Investigations by the present author [10] suggest that this equation gives good agreement with experimental values for all classes of hydrocarbon compounds: alkane, alkene, alkyne and aromatic. There are corresponding correlations for families of hydrocarbon derivatives, e.g., alcohols and ethers. A calculation of the flash point of cyclohexane from the Factory Mutual equation is given below.

For cyclohexane:
$$t_B = 81°C \implies t_F = -16°C$$

Experimental closed-cup flash point of cyclohexane [5] $= -18°C$

3.3.4 Recent developments in the understanding of flash points

Flash points are expected to continue to be the basis of assessing particular liquids for fire safety in storage and transportation. The hydrocarbon specialist in the twenty-first century needs to be aware that the advances in understanding of the flash point according to the principles of ignition limits and stoichiometry [11–16] have revealed uncertainties in the widely accepted experimentally deduced flash points for some substances. For example, it has been shown by calculations based only on vapour pressures and stoichiometry [11] that the flash point of dimethyl ether given in several authoritative sources[9] cannot possibly be correct. That for benzoic acid is also very uncertain [12, 13]. The reader is encouraged to have not a sceptical but a mildly critical attitude towards published flash point data.

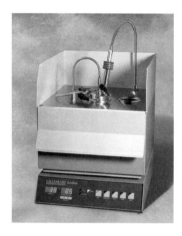

Figure 3.1 *Closed-cup flash point apparatus. Reproduced courtesy of Sanyo-Gallenkamp, Loughborough, UK.*

The *fire point* of a liquid fuel is a few degrees higher than the flash point and is the temperature of the bulk liquid at which a flame is sustained when a pilot flame contacts the vapour. It can, of course, only be determined in an open-cup arrangement. It often occurs in the range 1.2–1.5 times the stoichiometric proportion of vapour. Flash points and fire points are often quite challenging to determine accurately in spite of the apparent simplicity. This is particularly so with the more basic designs of flash point apparatus. A modern and quite advanced flash point apparatus is shown in Figure 3.1. This has the features of programmability of the temperature and a sensor to detect the flash instead of reliance on visual observation.

3.3.5 Flash points in law

If a flash point is invoked in litigation it is not usually sufficient to declare that a flash point was 'determined by the ASTM (or whatever) standard'. A formal report needs to bear the signature of an official authorised by the standards body who will have undergone training (see following section). Additionally to ASTM (USA, HQ in Philadelphia PA) there are British, European and Japanese standards' bodies for flash points. ASTM accreditation of a UK organisation is possible via United Kingdom Accreditation Services (UKAS). As an appendix to this chapter a hypothetical but totally realistic litigation case study relating to flash points is analysed.

3.3.6 Standards for flash points

We first note the role of the standards body ISO, which was founded in 1946–7 by representatives of 25 countries. Its HQ is in Geneva. It is the world's largest organisation for the development of standards but, as will be noted later, by no means the oldest. Its standards appertain to products, services and practices and exceed in number 19000. In fact ISO functions as a network of national standards bodies which include, amongst many others, American National Standards Institute (ANSI), Standards Australia (formerly Standards Association of Australia), European Standards (EN) and the British Standards Institution (BSI). There are over 100 countries which are full members of ISO, that is, there is a standards body within each of the countries which is part of the network referred to. Nominees of those bodies are eligible for membership of ISO technical committees. BSI is represented on 727 such committees, ANSI on 619, Standards Australia on 439.

Sometimes a standard issued by ISO and that by the standards body representing a member country are one and the same standard, for example:

BS EN ISO 13736:2008 Determination of flash point, Abel closed-cup method

This provides a method for the determination of the closed-cup flash point of liquids. It is for those with a flash point between 30.0 °C and 70.0 °C. There is also:

ISO 2592 IP 36 Determination of flash and fire points, Cleveland open cup method

IP denotes Institute of Petroleum, a US body which is not the representative of the US at ISO. However such bodies can unite their standards with those of ISO via (in this case) ANSI. It is by no means essential that a standard issued by a standards body also have an ISO equivalent. ASTM – American Society of Testing and Materials – long predates ISO and has been involved in standards appertaining to petroleum for over a century. ASTM standards for flash points are seen as being authoritative. Calibration liquids available for particular standards, e.g. a liquid with flash point of 'certified value 115 °C' can be purchased for use with:

UKAS /BS EN ISO /IEC 17025   Flash Point Reference Standard, Cleveland Open Cup

where IEC denotes International Electrotechnical Commission, founded in 1906 jointly by UK and US representatives, therefore well predating ISO. There will be a return in the discussion to this particular standard calibration liquid. Flash point standards are available for particular types of flammable liquid, e.g. ISO 3679:2004, the scope of which includes paints and varnishes. Its upper limit is 110 °C. There is also ISO 10156:1996 for paint and lacquer thinners.

Returning to the matter of the 115°C liquid flash point standard, we examine acenaphthalene a.k.a. acenaphthene, $C_{12}H_{10}$ which burns according to:

$$C_{12}H_{10} + 14.5O_2 (+ 54.5N_2) \rightarrow 12CO_2 + 5H_2O + (54.5N_2)$$

Its structural formula is shown below.

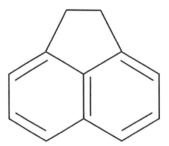

Structure of acenaphthalene

According to the half-stoichiometric rule previously studied, its flash point would be at whatever temperature corresponded to the equilibrium vapour pressure:

$$[10^5 \times 0.5/(1 + 14.5 + 54.5)] \text{ Pa} = 714 \text{ Pa. } (5.4 \text{ mmHg})$$

From Perry's, vapour pressure of acenaphthalene at 114.8°C is 5 mm Hg! This means that this compound matches the specifications of the standard calibration liquid referred to earlier. No claim that it is the standard can of course be made: that is proprietary information.

Examples of case studies where flash points have been invoked are provided in Table 3.3. The examples are all taken from reports published online by the US Chemicals Safety Board (CSB), HQ in Washington.

3.4 Thermal radiation and its relevance to flames

This section will not comprise an introductory sketch of radiation heat transfer *per se* – obtainable from numerous sources elsewhere – but a summary of how quite simple radiation calculations can give good predictions of flame temperatures. In later sections this will be applied to various sorts of combustion behaviour encountered in hydrocarbon accidents, including the jet fire.

Whereas radiation from a solid, e.g., a heating element, is a surface effect that from a gas is a volumetric effect. Consider the combustion of methane:

$$CH_4 + 2O_2 + 7.52N_2 \rightarrow CO_2 + 2H_2O + 7.52N_2$$

The post-combustion gas is 2:1 water vapour. Carbon dioxide and water are 'participating gases', whereas N_2 and O_2 are transparent to thermal radiation. Hydrocarbons are also strongly participating. Post-combustion products alone at say 1500 K will have an emissivity of only about 0.05–0.07. In a turbulent flame such as a jet fire unburnt fuel is mixed with the combustion products and this has the effect of dramatically increasing the emissivity, and this might be aided by the formation of carbon particles as intermediates. The flame is then said to be 'optically thick'. In fact, an estimate of radiative emission from a flame can be obtained by assuming it to be a black body at a single 'flame temperature' in the following way.

Table 3.3 *Case studies involving flash points*

Location and date	Nature of the accident	Details and reference
Houston TX, 2004	Explosion in a polyethylene wax processing facility	Flash point of the wax determined by ASTMD93-02a which is entitled: 'Standard Test Methods for Flash-Point by Pensky-Martens Closed Cup Tester' http://www.csb.gov/assets/document/Marcus_Report.pdf
Wichita KS, 2007	Explosion in a tank of solvent, initiated by static	Flash point of the solvent reported as being14°C http://www.csb.gov/assets/document/CSB_Study_Barton_Final.pdf
de Moines IA, 2008	Fire originating in a 300 gallon tank of ethyl acetate	Flash point of ethyl acetate noted as being –4°C http://www.csb.gov/assets/document/Barton_Case_Study_-_9_18_2008.pdf
Paterson NJ, 1998	Explosion during dye manufacture. One reactant 2-ethylhexylamine	Flash point of 2-ethylhexylamine noted as being 52°C http://www.csb.gov/assets/document/Morton_Report.pdf
Georgetown CO, 2007	Cleaning at a hydroelectric plant with methyl ethyl ketone	Flash point of methyl ethyl ketone at the scene not expected to comply with the literature value because the fire began in a space measurably below atmospheric pressure http://www.csb.gov/assets/document/Xcel_Energy_Report_Final.pdf

The form of the Stefan–Boltzmann law which applies is:

$$q/A = \varepsilon\sigma\{T^4 - T_o^4\}$$

Eq. 3.2

where q/A = radiative flux (W m^{-2})

σ = Stefan–Boltzmann constant = 5.7×10^{-8} W m^{-2} K^{-4}

ε = emissivity = 1 for a black body

T_o = temperature of the surroundings

e.g., a methane jet fire (see section 3.6.3.) at 1300 K, surroundings at 298 K:

$$q/A = 162 \text{ k W m}^{-2}$$

This is the heat transferred by radiation which is, of course, less than the total heat released since radiation is not the only mode of heat transfer: there is convection and also loss of heat on dispersion of the burnt gas. The proportion of heat transferred by radiation is typically about 0.2. Treatment of the flame as a black body at a single temperature is sometimes referred to as the 'solid flame model'.

3.5 Hydrocarbon combustion phenomenology

3.5.1 Preamble

We shall be concerned with types of combustion behaviour observed in hydrocarbon accidents: the jet fire, the pool fire, the fireball, the vapour cloud explosion and the flash fire. These will be given coverage adequate to equip a reader to perform basic calculations, such as might be required in an accident follow-up or for risk assessment, in respect of each.

A hydrocarbon safety professional, particularly one with a background in chemistry or chemical engineering rather than say mechanical engineering or physics, needs to be aware that there are certain features of hydrocarbon oxidation which do not normally come within the province of process safety but which are of great intrinsic interest and have been the subject of laboratory investigations from the nineteenth century to the present time. An outline of these follows.

3.5.2 Low-temperature oxidation

In all combustion phenomena of interest in process safety there is ignition, attainment of temperatures in excess of 1000°C and virtually complete conversion of the hydrocarbon to carbon dioxide and water. In laboratory reactors, batch or continuous, hydrocarbon oxidation under certain conditions of reactor temperature, total pressure and fuel:oxygen ratio enters a regime where there is no ignition, but instead 'low-temperature oxidation'. There is incomplete reaction, therefore there is unburnt fuel in the post-reaction gas, and two principal types of phenomenology. One is 'slow combustion', where the mixture of fuel and oxidant develops (and, in a continuous reactor, maintains) a temperature excess up to a few tens of degrees due to the chemical reaction. The other is 'cool flame' behaviour, where multiple flames each with an amplitude of about 100 K occur, and again conversion to products is incomplete. Cool flames are oscillatory and can, in a continuous reactor, be sustained indefinitely. Cool flames were known to Sir Humphry Davy.

Very many hydrocarbons and oxygenated hydrocarbons have, over the last 60 or so years, been examined for cool flame behaviour. In the 1970s and 1980s, acetaldehyde and propane were studied particularly closely. Modelling of such behaviour requires two coupled differential equations: one for heat balance and hence temperature (T), and one for the concentration of a chain-branching intermediate (x), which is often a peroxide. Neither a thermal nor a kinetic model singly can represent cool flame behaviour, hence the importance of the 'Unified Theory', which brings the two things together and enables oscillatory behaviour to be represented as a limit cycle in the x–T phase plane. Many interesting variants on simple oscillatory behaviour have been reported experimentally, including temperature histories such that cool flames and full ignitions alternate [17] and abrupt changes in both the oscillation amplitude and frequency in response to a very small adjustment in the reacting conditions [18].

Apart from recognition of their possible role in engine knock [3], cool flames have received relatively little attention in the context of *applied* combustion science. This is undoubtedly at least partly because, as already pointed out, the types of combustion behaviour with which fuel technologists and fire protection experts are concerned are not in the low-temperature oxidation regime, but in the 'full ignition' regime. Over the last 25–30 years huge amounts of reliable experimental data on cool flames for various organic compounds, and related modelling, have found their way into the research literature. There is scope for further thought as to how all of this might profitably be incorporated into hydrocarbon safety practice. For example, is the thermal behaviour leading to any of the 'full ignition' phenomena of interest preceded by cool flame behaviour, as has certainly been observed in laboratory experiments? If so, is there scope for enhanced safety by suppression of cool flames using a suitable chemical additive? Here is possibly a fruitful area of future R&D.

3.5.3 Jet fires

A jet fire is the combustion behaviour expected when a gas or two-phase hydrocarbon leaks from a small orifice and ignites. In the previous chapter it was shown how to calculate the rate of discharge from the upstream pressure, and this in turn can be used to make simple estimations of jet fire length.

A jet fire is 'non-premixed': fuel exits the orifice as pure fuel and all contact with oxidant (air) is subsequent to this. It is not a diffusion flame. In such, fuel–air contacting is purely by diffusion, whereas in a jet fire flame there is also mixing due to the momentum with which the fuel exits the orifice. There are, in the literature, many correlations for jet fire length, including:

$$L(m) = 18.5 \, (Q/\mathrm{kg\,s^{-1}})^{0.41} \qquad\qquad \text{Eq. 3.3}$$

where L is the length and Q is the discharge rate. An application is in the following shaded area.

Consider a 1 cm^2 hole in a gas-bearing pipe or vessel containing methane at 30 bar. The ambient temperature is 15°C. Using equations 2.1 and 3.3, and taking the coefficient of discharge to have value unity, calculate the length of the resulting jet flame if there is ignition.

Solution

$$Q = C_d AP\sqrt{(M\gamma/RT)[2/(\gamma+1)]^{(\gamma+1)/(\gamma-1)}}$$

$$\Downarrow$$

$$Q = 0.52\,kg\,s^{-1}$$

$$\Downarrow$$

$$L = 14.1\,m$$

In principle there has to be an orientation dependence of the jet flame length, but in ORA a single equation such as the above is often taken to apply to all orientations. The orientation influences the flame properties via convection coefficients, diffusion (with its gravity dependence) and radiation view factors. Horizontal flames might tend to become 'banana shaped' through buoyancy effects, whilst not differing markedly in length from a vertical one receiving the same influx of fuel gas. In two-phase flow, it is possible for the liquid to 'rain out' at the orifice exit, and burn separately as a pool fire.

In modelling of the radiation field around a turbulent jet flame, the flame can be represented as a point source of energy half way along the axis. In our previous numerical example, therefore, it would be 7 m from the leak orifice and 7 m from the flame tip. Incident flux q' ($W\,m^{-2}$) a distance D from this point source given by:

$$q' = FQ/4\pi D^2 \qquad\qquad \text{Eq. 3.4}$$

where F = proportion of the heat transferred by radiation ($\approx$0.2 if the fuel is methane) and Q = total rate of energy release by the combustion (W).

At distances very close to the flame the model breaks down. Under such circumstances the 'black body' flux is taken to apply for the purposes of simple combustion calculations. Predictions of injuries to persons can be made, e.g., skin burns require $10\,kW\,m^{-2}$ for about 10 s or $100\,kW\,m^{-2}$ for 1 s. These ideas are illustrated in the following example.

A pipe bearing natural gas at an internal pressure of 70 bar, temperature 288 K, develops a leak of area $2\,cm^2$. Calculate:

The leak rate, if the coefficient of discharge = 0.8

The length of the resulting jet flame

The total rate of release of energy by the flame

How close to the flame a person would need to be to experience skin burns through 10 s of exposure

Calorific value of methane $= 55.6\,MJ\,kg^{-1}$

Solution

For methane, $\gamma = 1.3$, $M = 0.016\,kg\,mol^{-1}$

also, $C_d = 0.8$, $R = 8.314\,J\,K^{-1}mol^{-1}$, $T = 288\,K$

$$\Downarrow$$

$Q = 1.9\,kg\,s^{-1}$ flame length $= 24\,m$

Total rate of heat release $=$
$$1.9\,kg\,s^{-1} \times 55.6 \times 10^6\,J\,kg^{-1} = 1.06 \times 10^8\,W\ (106\,MW)$$

Now a distance D from the flame: $q' = FQ/4\pi D^2$ (equation 3.4)

For skin burns through 10 s of exposure
$$q' = 10\,kW\,m^{-2}$$

$$10\,kW\,m^{-2} = 106000\,kW \times 0.2/[(4\pi D^2)\,m^2]$$
$$\Downarrow$$
$$D = 13\,m$$

3.5.4 Pool fires

A pool fire occurs when flammable liquid is spilt and ignites. Pool fires are laminar only up to about 10 cm diameter; larger ones are turbulent. The flame front contains a high proportion of evaporated fuel vapour at a relatively high temperature, making the flame front emissive. Radiative heat transfer from the flame front to the pool surface promotes further evaporation. A pool fire displays a steady burning rate until fuel is close to being depleted, and there are two means of estimating the steady mass loss rate per unit pool area.

Assigning this the symbol m':

$$m' = 0.1\,kg\,m^{-2}s^{-1} \text{ for any pool fire}$$

This is used in ORA, where the only liquid fuel of interest is crude oil. Alternatively:

$$m' = 10^{-3}\,\frac{\text{heat of combustion}}{\text{heat of vaporisation}}\,kg\,m^{-2}\,s^{-1}$$

e.g., benzene: heat of combustion $41859 \, kJ \, kg^{-1}$
heat of vaporisation $436 \, kJ \, kg^{-1}$

$$m' = 10^{-3} \times \{41859/436\} \; kg \, m^{-2} s^{-1} = 0.096 \, kg \, m^{-2} s^{-1}$$

There are also several experimentally based correlations in the literature for pool fire height, e.g.:

$$H/D = 42 \left\{ m'/\left[\sigma_a \sqrt{gD} \right] \right\}^{0.61} \qquad \text{Eq. 3.5}$$

where D(m) = pool diameter and H(m) = pool height, σ_a = density of air at ambient temperature (kg m^3), g = acceleration due to gravity ($9.81 \, m \, s^{-2}$). This applies to a pool fire in still air; a modification is required if there is wind.

Thermal radiation from a pool fire is represented similarly to that for a jet fire. The fire is treated as a point source, and:

$$q_{rad}(x) = Q_{rad}/4\pi x^2 \qquad \text{Eq. 3.6}$$

where $q_{rad}(x)$ = radiative flux (W m^{-2}) a distance x from the source
$\qquad Q_{rad}$ = total radiation rate (W)

$$Q_{rad} = \lambda mQ \qquad \text{Eq. 3.7}$$

where Q = total heat-release rate (W)
$\qquad m$ = mass burning rate (kg s^{-1})
$\qquad \lambda$ = fraction of total combustion heat transferred as radiation

Example

A tank containing gasoline is surrounded by a circular dike of diameter 10 m. The gasoline leaks and occupies the area bounded by the dike. If there is ignition, calculate:

(a) the total radiative flux from the flame
(b) the flame temperature
(c) the radiative flux a person standing 15 m from the circumference will experience

Use a value of 0.4 for λ, a value of $0.1 \, kg \, m^{-2} \, s^{-1}$ for m', a value of $1.17 \, kg \, m^{-3}$ for the density of air at ambient temperature and a value of $45 \, MJ \, kg^{-1}$ for the calorific value of the fuel.

Solution

$$m = m' \{\pi r^2\} = 7.9 \, kg \, s^{-1}$$

Total rate of radiation $(Q_{rad}) = 0.4 \times 7.9 \times 45 \times 10^6 \, W$

$$= 1.4 \times 10^8 \, W$$

Height given by equation 3.5: $H/D = 42 \left\{ m' / \left[\sigma_a \sqrt{gD} \right] \right\}^{0.61}$

$$\Downarrow$$

$$H/D = 2.3, \ H = 23 \, m$$

The 'area' can be approximated by a cylinder of height H and diameter D, whereupon:

radiative flux $= [1.4 \times 10^8/(2\pi \times 5 \times 23)] \times 10^{-3} \, kW \, m^{-2}$

$$= 194 \, kW \, m^{-2}$$

approximating to a black body, temperature =
$\{194000/\text{Stefan–Boltzmann constant}\}^{1/4} = 1360 \, K$

The flux at a distance of 15 m is clearly:

$$q_{rad}(x = 15 \, m) = Q_{rad}/4\pi x^2 = 1.4 \times 10^8/(225 \times 4\pi) = 50 \, kW \, m^{-2}$$

3.5.5 Fireballs and BLEVEs

'Fireball' means rapid combustion of a 'catastrophically leaked' quantity of fuel, with a duration usually of the order of seconds. A BLEVE (boiling liquid expanding vapour explosion) is a particular sort of fireball, involving a substance which is gas at room temperature but which is stored at such temperatures as a liquid under its own highly superatmospheric vapour pressure. The usual examples of substances displaying BLEVE behaviour are LPG (primary constituent propane) and vinyl chloride monomer (VCM). A BLEVE is actually a physical, not a chemical, explosion and can involve non-flammable substances. When a pressure cooker (in which the fluid is of course water) blows up, that is a BLEVE. If the substance having undergone such an explosion is flammable and ignition follows, a fireball is the result and the 'BLEVE-fireball' is the most precise description. This is often simply called a BLEVE.

Liquefied natural gas (LNG) is a cryogen and though it might burn as a fireball it does not BLEVE in the sense described in the final sentence of the previous paragraph. Physical explosions are possible with LNG, in which case BLEVE behaviour in that sense might be considered in a follow-up. BLEVEs are not expected at offshore installations; the susceptible materials are not to be found there. Fireballs are expected in the event of sudden release of gas or two-phase inventory. Generic equations and correlations for 'fireballs' are often applied to BLEVEs.

An experimental correlation for maximum fireball diameter is:

$$D(m) = 5.25[M(kg)]^{0.314} \qquad \text{Eq. 3.8}$$

where M = quantity released.

There are several variants on this, some used in ORA. A correlation for duration of a fireball is:

$$t_p(s) = 2.8(V/m^3)^{1/6}$$

where t_p = duration and V = vapour volume leaked, referred to ambient conditions (1 bar, 288 K).

---

e.g., for 100 tonne (10^5 kg) methane, using equation 3.8:

$$D(m) = 5.25[10^5]^{0.314} \text{ m} = 195 \text{ m}$$

$$V = \{1 \times 10^5/0.016\}/(P/RT) \text{ m}^3 = 149652 \text{ m}^3$$

$$t_p = 20 \text{ s}$$

Fireballs pose a significant radiation hazard. The proportion of the combustion heat radiated from the fireball is about 0.2. Calculations are complicated by the fact that a fireball, unlike a jet fire or pool fire, cannot be treated as being steady. A recent [19] simplified thermal treatment of fireballs uses a previously reported experimental plot of radiation flux against a suitably defined dimensionless time [20] as a 'template', in order to calculate peak fireball temperature for particular amounts of propane so burnt. This will be explained in the following outline.

The dimensionless time is:

$$t' = g^{0.5}t/V^{1/6}$$

where t = time since ignition (s), g and V are as previously defined. The *maximum* heat release occurs at about $t' = 6$. Now the plot of radiative flux against dimensionless this time for a propane fireball [20] can, to a fair approximation, be drawn as an isosceles triangle with its apex at the time corresponding to $t' = 6$ and the baseline spanning t' values of 0 and 12. The plot so re-drawn is reproduced as Figure 3.2. The dashed line is the triangle and the solid line shows approximately where and how the true plot deviates from it. The maximum is shown as a cusp, whereas of course it is a smooth maximum in the original. The calculation in the shaded area below builds upon these ideas. It is based on information given in [20].

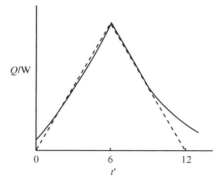

Figure 3.2 *Plot of radiative flux against dimensionless time for a propane fireball, approximated to an isosceles triangle. Reproduced from [19] with the permission of Journal of Fire Sciences, PA.*

A quantity of 10 tonne of propane previously stored as liquid under pressure escapes and forms a fireball. Calculate the maximum flux and the corresponding temperature. Use a value of 0.2 for λ, the proportion of the combustion heat transferred by radiation. The calorific value of propane is $50\,MJ\,kg^{-1}$.

The peak temperature of the fireball is measured as 1561 K. Comment on any discrepancy.

Solution
Referring to Figure 3.2, the total area of the isosceles triangle represents the total heat radiated for the duration of the fireball, which is 20% of the total heat released, i.e., the area is:

$$0.2 \times 50 \times 10^6 J\,kg^{-1} \times 10 \times 10^3 kg = 1 \times 10^{11} J$$

Recalling the definition of t':

$$t' = g^{0.5} t / V^{1/6}$$

$$V = (10 \times 10^3/0.044)/(P/R\,T) = 5437\,m^3$$

$$t' = 0 \text{ corresponds to } t = 0$$
$$t' = 6 \text{ corresponds to } t = 8\,s$$
$$t' = 12 \text{ corresponds to } t = 16\,s$$
$$10^{11}\,J = 0.5 \times 16 \times Q_{max}$$

where Q_{max} is the maximum radiative heat-release rate

$$\Downarrow$$
$$Q_{max} = 1.25 \times 10^{10}\,\text{W}$$
diameter given by: $D(\text{m}) = 5.25[M(\text{kg})]^{0.314}$
$$\Downarrow$$

$$D = 95\,\text{m}$$
surface area $= 4\pi(95/2)^2\,\text{m}^2 = 28357\,\text{m}^2$

flux $= (1.25 \times 10^{10}/28357)\text{W m}^{-2} = 440\,\text{kW m}^{-2}$

this flux corresponds to a black body temperature of 1668 K

Discrepancy?

Assumption that the emissivity of the flame is 1

Assumption that the maximum heat-release rate in Figure 3.2 corresponds to the maximum diameter

Approximation of the fireball to a sphere

The difference between the measured and calculated values is just over 100 K. Perhaps it should be pointed out that actual measurement is likely to have comparable uncertainty. A noble metal (type R or S) thermocouple has a range extending to these temperatures, but an intrinsic uncertainty of about 2% of the reading, i.e. ±30 K. More seriously, heat transfer effects can cause the thermocouple tip temperature to be significantly different from that of the gas in which it stands[10]. The agreement between measured and calculated values is therefore very reasonable indeed. Finally, though spherical geometry was assumed in the calculation, the reader should note that there is limited evidence from photography that a cylinder is at least as good an approximation to the fireball shape at the stage where the maximum flux occurs.

3.5.6 Vapour cloud explosions (v.c.e.s) and flash fires

One or other of these will occur when a quantity of hydrocarbon is leaked and forms a cloud. A vapour cloud explosion (v.c.e.) is accompanied by over-pressure: a flash fire is not. A leakage may lead to one or other, depending upon the quantity released and, importantly, the presence of obstacles such as buildings which will promote turbulence and therefore v.c.e. behaviour. It is also possible for a single leak to lead to both. For example, at the Flixborough (1974) accident the leaked cloud of cyclohexane divided into two: one part displayed v.c.e. behaviour, the other flash fire behaviour. The 2005 refinery accident in Texas City displayed v.c.e. behaviour only.

Both can be fatal. In a v.c.e., there are two classes of casualty: those affected by the heat and those by the overpressure. This tends to be manifest as large differences between fatal and non-fatal injuries. In a flash fire all death and injury are due to the heat. There is a very large number of well documented case histories from about the time of World War II to the present. These have been the basis of identification of trends as to whether a v.c.e. or a flash fire occurs, according to conditions.

There are two types of propagation of gaseous combustion: deflagration, which is subsonic and may or may not create overpressure, and detonation which is supersonic and is always accompanied by a large overpressure. 'High explosives' like TNT display detonation. Hydrocarbons *can* be made to detonate in the laboratory, e.g., under rapid compression. Hydrocarbon accidents in industry are almost always deflagrations. The blast energy accompanying a hydrocarbon explosion is usually taken to be about 5% of the total energy released on combustion. As stated above, the propagation rate of a detonation exceeds the speed of sound under the conditions of the burnt gas, and there is always overpressure. The overpressure will be greatest at the centre, declining at either side. A deflagration, especially one accelerated by turbulence, might have overpressures equivalent to those of a detonation some distance from the centre but will not even approach the detonation value at the centre. Hence pressure profiles for the two will be different at the centre but can merge further out.

The way of quantifying the blast damage done in a hydrocarbon explosion is as 'TNT equivalence': how much TNT would have caused the damage which the hydrocarbon did. An understanding of this requires first an appreciation of the apparent anomaly that 'There is more energy in a candle than in a stick of TNT of the same weight'. This statement is perfectly true. Hydrocarbon wax has a heat of combustion of around $40\,\mathrm{MJ\,kg^{-1}}$, TNT a heat of combustion of $15\,\mathrm{MJ\,kg^{-1}}$. Yet materials such as TNT and dynamite, whilst having smaller heats of combustion than hydrocarbons, obviously have greater blast potential ('brisance'). This is because about 28% of the combustion energy of TNT becomes blast energy, or about $4.2\,\mathrm{MJ}$ of blast energy per kg TNT reacted. By contrast when a hydrocarbon vapour ignites only something of the order of 5% of the heat of combustion becomes blast energy, or 2 to $2.5\,\mathrm{MJ}$ of blast energy per kg. This can be understood on the basis that pre-mixedness promotes high burning velocities. In TNT and other high explosives, the 'oxidant' is intra-molecular, previously bonded to the organic structure of the compound. This is the ultimate in pre-mixedness, and the result is rapid propagation and high blast potential[11]. A related calculation is given below.

At Ludwigshafen, Germany, in 1948 (see Table 3.3), there was an accident involving leakage of 30 ton of dimethyl ether (heat of combustion 31 MJ kg^{-1}). Estimate the blast energy. What weight of TNT would have produced an effect equivalent to that caused by the dimethyl ether at Ludwigshafen?

Solution
As a simplification we equate tons to tonnes.

Dimethyl ether, 30 tonne = 30000 kg capable of releasing 9.3×10^{11} J of energy, 4.7×10^{10} J of blast energy
4.7×10^{10} J = 47000 MJ requiring 47000/4.2 kg of TNT, or 11190 kg, approx. 11 tonne
i.e., the 'TNT equivalence' is 11 tonne.

Mention has already been made of the vast amount of information which has been gleaned from records of vapour cloud explosions and flash fires. Table 3.3 gives an outline of some such incidents in chronological order. In this and other tables giving information on case studies there is no attempt to be comprehensive, rather to give a representative selection from over the years. Details of recent hydrocarbon accidents are available electronically[12] on www.nfpa.org.

Remarkably high overpressures were evident in an explosion in 1999 at a plant in Pennsylvania using (not manufacturing) hydroxylamine, and a brief mention of it here in spite of the absence of a hydrocarbon is not inappropriate. There were five fatalities. Hydroxylamine (NH_2OH) is a white powder at room temperature, and is used in the manufacture of products including pharmaceuticals and paints. Hydroxylamine reacts with oxygen according to:

$$NH_2OH + 0.25O_2 \rightarrow 0.5N_2 + 1.5H_2O$$

and the exothermicity is 12 MJ kg^{-1}. Press reports stated that 'several hundred pounds' of hydroxylamine exploded and also that a wall 100 yards away was cracked. In unpublished calculations by the author this was taken to mean that the overpressure 100 yards away was 0.1 bar and on that basis, using the 'multi-energy' approach originating with TNO in Holland, an estimate of exactly one tonne for the amount of hydroxylamine which had reacted was obtained.

3.6 The use of probit equations in fire and explosions

3.6.1 Introduction

Clearly, if retrospective examination of the consequences of any event such as a fire or explosion provides quantitative information as to the extent of the consequence as a function of the magnitude of the causative factor, and this can be fitted to an equation, such an equation will have subsequent predictive

Table 3.4 *v.c.e. and flash fire case studies*

Location and date	Details
Ludwigshafen, Germany, 1943	Leaked substance a mixture of butadiene (80%) and butene (20%), 16.5 ton. Rupture of a tankcar containing this inventory. v.c.e. behaviour. 57 deaths, 439 injuries. Extensive damage to buildings and plant
Ludwigshafen, Germany, 1948*	Leaked substance dimethyl ether, 30 ton approx. Rupture of a tank car. v.c.e. behaviour. 207 deaths, over 3000 injuries. Extensive damage
Flixborough, England, 1974	Leakage of cyclohexane through a failed temporary connection between two reactors. 28 deaths, over 80 injuries
Decatur, Illinois USA, 1974	Release of isobutane (63 tonne) from a crashed rail car at a freight yard. 7 deaths, 33 injuries
Ufa, USSR, 1989	Leak from an LPG pipeline situated parallel to a railway track. Formation of a vapour cloud subsequently supplied with air and raised in turbulence by two trains passing each other in opposite directions. Explosion, both trains affected. Estimated 462 deaths, 706 injuries
Near Houston, Texas, USA, 1989	Leakage of a mixture of hydrocarbons and hydrogen at a polyethylene plant. v.c.e. 22 deaths
Nagothane, India, 1990	Leak in a pipeline conveying ethane and propane to a cracking facility. Formation of a vapour cloud which drifted before igniting some distance away. 31 deaths and extensive site damage
Bangkok, Thailand, 1990	Road tanker carrying LPG involved in a traffic accident. Release of 5 tonne of LPG. Flash fire and explosion. 68 dead, 100 injured
Morgantown NC, 2006	Polymerisation of acrylic monomers in a 1500 gallon reactor. Loss of containment and explosion. One fatality, two serious non-fatal injuries.

* In both the Ludwigshafen accidents, overfilling and expansion of the tank car contents by solar flux are believed to have played a part. Effects of solar heating on stored hydrocarbon are considered in Chapter 4.

use. In our first example, the consequence of interest is deaths by burning and the causative factor will be radiative flux. Some background on probit equations is first required.

Let the causative factor be expressed in terms of a quantity x and the percentage of a population exposed to this factor, applied to an extent x, affected, be P. Clearly an expression of the form:

$$P = a + bx \qquad \text{Eq. 3.9}$$

can be obtained from retrospective examination. As fully explained by Lees [20], there are good reasons for substituting for P a quantity known as the probit, usual symbol Y. The reader is referred to Lees for a full discussion. Equation 3.9 is thereby transformed to one of the form:

$$Y = k_1 + k_2 \ln x \qquad \text{Eq. 3.10}$$

and, Y having been calculated, a transition back to percentage is possible by means of a standard table (e.g., Lees *op. cit.*, p. 9/73) correlating probits with percentages. The parameters k_1 and k_2 are obtained by fitting data from actual occurrences.

3.6.2 Application to a flash fire

For a flash fire [21], on the basis of retrospective examination of fatal examples, in the probit equation $k_1 = -14.9$, $k_2 = +2.56$ and x has the form:

$$t_e I^{4/3}/10^4$$

where t_e is the flash fire duration (s) and I the radiative flux (W m^{-2}). An application follows.

A quantity of hydrocarbon leaks and drifts for some distance, thereby forming a less rich mixture with air, before igniting. A flash fire ensues which radiates as a black body at 1150 K, and burns out after 5 s. If 30 persons are close enough to experience radiation flux at potentially lethal levels, how many fatalities will there be?

Solution
If the flash fire radiates as a black body at 1150 K:

$$I = 5.7 \times 10^{-8} \times 1150^4 \, \text{W m}^{-2} = 1.0 \times 10^5 \, \text{W m}^{-2} \Rightarrow x = 2321$$

Substituting into the probit equation: $Y = 4.94$, equivalent [20] to a percentage of 48.

The number of fatalities is therefore: $0.48 \times 30 = 14$ to the nearest whole number.

3.6.3 Application to overpressure damage

Various forms of the probit equation for overpressure damage exist according to the precise nature of the 'consequence', which may be glass damage, structural damage or specific types of human injury such as eardrum rupture. This is dealt with in one of the appended numerical problems.

Probit equations apply not only to fire and explosions but also to toxicity. This is taken up in Chapter 10.

3.7 Concluding remarks and further numerical example

The types of combustion behaviour expected in hydrocarbon accidents have been outlined and simple equations for them presented and applied. Of course, sight must not be lost of the fact that all of these behaviour types have been, and are, the subject of very advanced experimental work and modelling. The treatments presented herein are, by comparison, very basic, having been developed into simple, transparent equations for day-to-day calculations in engineering practice. By the same token, the development of flame and fire models for use in process safety is ongoing and treatments less basic than those herein are available to the process safety engineer in the early twenty-first century. These do not replace or supersede the more basic approaches but rather provide, where it is necessary, a finer screening in risk assessment.

A further calculation, which brings together several of the ideas from this and the previous chapter, is given below by way of conclusion to the chapter.

A vessel of methane contains 5 tonne of the gas at 5 bar pressure, 288 K. The surface area of the vessel is $785\,m^2$. A fire begins at a nearby part of the plant, and the vessel of methane receives uniform flux of $5\,kW\,m^{-2}$. The vessel is certified to withstand pressures up to 8 bar.

(a) How long will it take for the contents to attain this pressure?
(b) If once the vessel gets to 8 bar it breaks open, how big will the resulting fireball be?
(c) The vessel is fitted with a safety valve which is set to open at 8 bar. Its opening is equivalent to the creation of an orifice if $1\,cm^2$ diameter with discharge coefficient unity. Use the 'single quantity of an ideal gas' model to predict the flame height if gas so discharged from the vessel ignites.

Use a value of $2250\,J\,kg^{-1}\,K^{-1}$ for the specific heat of methane.

Solution

(a) For a fixed mass of gas at constant volume:

$$P_1/T_1 = P_2/T_2 \Rightarrow T_2 = 461 \text{ K}$$

Heat required to raise the temperature of the gas =

$$5 \times 10^3 \times 2250 \times (461 - 288)\text{J} = 1.95 \times 10^9 \text{ J}$$

Time required = $1.95 \times 10^9/(5000 \times 785)\text{ s} = 497\text{ s } (8.3 \text{ min})$

(b) $$D(\text{m}) = 5.25[M(\text{kg})]^{0.314} = 76\text{ m}$$

i.e., the fireball will, at its maximum size, approximate to a sphere 76 m in diameter.

(c)

$$Q = C_\text{d} A P \sqrt{(M\gamma/RT)\left[2/(\gamma+1)\right]^{(\gamma+1)/(\gamma-1)}} \qquad \text{Eq. 2.1}$$

$$\Downarrow$$

$$Q = 0.14 \text{ kg s}^{-1}$$

$$L(\text{m}) = 18.5 \, (Q/\text{kg s}^{-1})^{0.41} = 8.3 \text{ m}$$

A reader having mastered the text thus far will be comfortable with calculations at this sort of level.

References

[1] Aylward G.H., Findlay T.J.V. *SI Chemical Data*, Second Edition. John Wiley (1974)

[2] Perry R.H., Green D. *Perry's Chemical Engineers' Handbook*, any available edition. McGraw-Hill, New York.

[3] Barnard J.A., Bradley J.N. *Flame and Combustion*, Second Edition. Chapman and Hall, London (1985)

[4] Haslam R.T., Russell J.A. *Fuels and their Combustion*, First Edition. McGraw-Hill, New York (1926)

[5] Kanury A.M. 'Ignition of liquid fuels' in *Handbook of Fire Protection Engineering*, Second Edition. SFPE, Boston, MA (1995)

[6] Griffiths J.F., Barnard J.N. *Flame and Combustion*, Third Edition. Blackie, London (1995)

[7] Jones J.C. *Topics in Environmental and Safety Aspects of Combustion Technology*, First Edition. Whittles Publishing, Caithness, UK (1997)

[8] Jones J.C. 'Reid vapour pressure as a route to calculating the flash points of petroleum fractions' *Journal of Fire Sciences* **16** 222–229 (1998)

[9] Jones J.C. 'Possible difficulties with calculation of vapour pressures of biodiesel fuels' *Fuel* **84** 1721 (2005)

[10] Jones J.C. 'An examination of flash points as an indicator of fire hazards with flammable liquids' Proceedings of the 33rd International Conference on Fire Safety, Sissonville, West Virginia (2001)

[11] Jones J.C. Uncertainties in the flash point of dimethyl ether *Journal of Loss Prevention in the Process Industries* **14** 429–430 (2001)

[12] Jones J.C. On the flash point of benzoic acid *Journal of Fire Sciences* **19** 177–180 (2001)

[13] Godefroy J., Fowler G. and Jones J.C. On the flash point of benzoic acid. Part 2. Experimental measurements *Journal of Fire Sciences* **19** 306–308 (2001)

[14] Jones J.C. Further information on the hazards of n-hexane *Journal of Chemical Education* **78** 1593 (2001)

[15] Jones J.C., Godefroy J. 'Anomalies in the flash points of four common organic compounds' *Journal of Loss Prevention in the Process Industries* – in press

[16] Godefroy J., Jones J.C. A reappraisal of the flash point of formic acid *Journal of Loss Prevention in the Process Industries* **15** 245–247 (2002)

[17] Gray P., Griffiths J.F., Hasko S.M., Lignola P.G. Oscillatory ignitions and cool flames accompanying the non-isothermal oxidation of acetaldehyde in a well-stirred flow reactor *Proceedings of the Royal Society of London* A **374** 313 (1981)

[18] Gray B.F., Jones J.C. The heat-release rates and cool flames of acetaldehyde oxidation in a continuous stirred tank reactor *Combustion and Flame* **57** 3–14 (1984)

[19] Jones J.C. A simplified thermal treatment of fireballs *Journal of Fire Sciences* **19** 100–105 (2001)

[20] Mudan K.S., Croce P.A. 'Fire Hazard Calculations for Large Open Hydrocarbon Fires' in *Handbook of Fire Protection Engineering* Second Edition. SFPE, Boston, MA (1995)

[21] Lees F.P. *Loss Prevention in the Process Industries* Second Edition. Butterworth-Heinemann, Oxford (1996)

Numerical examples

1. The condensation of 1 kg steam at its normal boiling point releases 2257 kJ. How much gasoline, if burnt in air, would release this amount of heat?

Diesel, in his exploratory work on what later became known as the Diesel engine, actually used powdered coal as the fuel. How much coal would be required to supply 2257 kJ?

Make estimates of your own for the calorific values of gasoline and powdered coal.

2. meta-Xylene (C_8H_{10}) burns according to:

$$C_8H_{10} + 10.5O_2\ (+39.5N_2) \rightarrow 8CO_2 + 5H_2O\ (+39.5N_2)$$

and its vapour pressure at 28°C is 1316 N m^{-2}. Deduce, giving your reasoning in full, whether meta-Xylene stored at this temperature would be above or below its flash point. *From the MSc (Safety and Reliability) Examination, University of Aberdeen, 2000.*

3. A vessel of meta-Xylene stands in a storage area at 288 K. The vessel is vented so that the total pressure is always 1 bar. A fire starts nearby and the vessel and contents are heated. The latent heat of vaporisation of the compound is 43 kJ mol^{-1} and its normal boiling point is 412 K. Calculate the initial vapour pressure of the m-Xylene (i.e., the vapour pressure at 288 K). Using the stoichiometric information given in the previous question, determine the temperature at which there will be a sustained flame if an ignition source contacts the vapour/air mixture in the ullage space. Use the rule that the fire point occurs at 1.5 times stoichiometric.

4. Styrene (C_8H_8) is the monomer of polystyrene and burns according to:

$$C_8H_8 + 10O_2 (+37.6N_2) \rightarrow 8CO_2 + 4H_2O (+37.6N_2)$$

At a polystyrene manufacturing plant there is leakage of styrene at 18°C. Perform a calculation to determine whether styrene is above or below its flash point at this temperature, given that its vapour pressure at 18°C is 658 Pa. *From MEng/BEng Level 4 Examination, University of Aberdeen, 1998.*

5. Cyclohexane burns in air according to:

$$C_6H_{12} + 9O_2 (+33.8N_2) \rightarrow 6CO_2 + 6H_2O (+33.8N_2)$$

The heat of combustion is 3658 kJ mol^{-1}. Calculate the stoichiometric adiabatic flame temperature for cyclohexane. Use the following values of the specific heats:
water vapour $= 43$ J °C^{-1} mol^{-1}
carbon dioxide $= 57$ J °C^{-1} mol^{-1}
nitrogen $= 32$ J °C^{-1} mol^{-1}

6. Return to the problem in the main text, a methane jet fire 14 m long receiving 0.52 kg s^{-1} of methane. Generally the diameter of such a flame is about 1/100th of the length, hence take the diameter to be about 15 cm. If the energy radiated is 20% of the total energy released, estimate the flame temperature. Approximate the flame to the curved surface area of a cylinder 14 m in length and 0.15 m diameter.

7. What is the total energy released by combustion of two-phase (approximately 50–50, by weight) hydrocarbon released at an offshore installation at a rate of 5 kg s^{-1}? Make a reasoned estimate of your own for the calorific value.

8. In a fire at an offshore installation a tank of crude oil is, during escalation[13], damaged and the oil leaks onto a solid floor and forms an approximately circular pool of 5 m diameter which promptly ignites. The surrounding area approximates to a large enclosure which, as a result of the earlier stages of the fire, has heated to 900 K. Calculate the temperature of the pool fire. Use a value for λ of 0.25 and make an estimate of your own for the calorific value. Density of air at ambient temperature = 1.2 kg m^{-3}.

9. A quantity of 45 tonne of propane previously stored as liquid under pressure escapes and forms a fireball. Utilising the 'template' provided [as Figure 3.1] and making a reasonable estimate of your own for the proportion (symbol λ) of the combustion heat transferred by radiation, calculate the maximum flux and the corresponding temperature. *From MEng/BEng Level 4 Examination, University of Aberdeen, 2001.*

10. A BLEVE involving LPG radiates as a black body at 2000 K. Calculate the resulting heat flux from the BLEVE to the surroundings.

11. In an outdoor demonstration for trainee fire fighters, a small container of LPG is torched by a gas flame until the container fails, the contents are suddenly released and there is a fireball. If the container contained 5 kg, what would be the maximum flux from the fireball and its temperature. Make estimates of your own for quantities required.

12. At Decatur, Illinois in 1974 there was a fatal accident involving release of 63 tonne of isobutane from a crashed rail car, as reported in Table 3.3. A follow-up concluded:
'… the TNT equivalence at Decatur lay between 20 and 125 tonne'.
On this basis find the yield – proportion of the combustion energy manifest as blast – at Decatur. Take the mid-range value of the TNT equivalence and a use a value of 50 MJ kg^{-1} for the heat value of isobutane.

13. For eardrum rupture the coefficients in the probit equation are:

$$k_2 = 1.93, k_1 = -15.6$$

and x is the peak overpressure in N m^{-2}. Imagine that 150 persons experience, as a result of a confined explosion, an overpressure of 0.2 bar. Estimate how many will suffer eardrum rupture.

14. Repeat the calculation on fireballs in section 3.6.5. of the main text using this alternative equation for fireball diameter:

$$D = 7.71 V^{1/3}$$

where D and V are as defined previously.

15. The organic chemical n-butyl alcohol (C_4H_9OH) burns in air according to:

$$C_4H_9OH + 6O_2 + 22.6N_2 \rightarrow 4CO_2 + 5H_2O + 22.6N_2$$

A quantity of n-butyl alcohol is stored at a site where the ambient temperature is 30°C. The vapour pressure of n-butyl alcohol at this temperature is 1316 Pa. Deduce by calculation whether the n-butyl alcohol is above or below its flash point at this temperature.

16. A jet fire burns natural gas at a rate of $2 \, kg \, s^{-1}$ and has a length of 25 m. Calculate the rate at which it will radiate heat to surroundings at 300 K and also estimate its temperature. Make reasonable estimates of your own for the calorific value, the proportion of total combustion heat radiated and the diameter of the jet fire. Express the jet fire temperature in degrees centigrade.

17. A petroleum distillate is stored in a spherical container of 2 m diameter. Its outside surface has become quite tarnished through formation, over time, of a thin layer of metal oxide. A type K thermocouple is immersed in the liquid which the container holds and at daybreak this records a temperature of 15°C. It subsequently receives solar flux of $1400 \, W \, m^{-2}$ for 10 hours in the day. For the contents to reach 60°C would cause incipient boiling which has to be avoided, consequently if the temperature of the bulk liquid rises to this value the thermocouple output activates an alarm.
Determine:
(i) How long it will take for the contents of a vessel which is four-fifths full to reach 60°C. State any assumption made in your calculation. Use a value for the specific heat of the liquid of $2000 \, J \, kg^{-1} \, K^{-1}$. Make an estimate of your own for the liquid density.
(ii) To what value the absorptivity of the vessel surface must be reduced, by painting white or by polishing, to ensure that the contents do not reach 60°C during daylight hours.
(iii) The minimum voltage in the thermocouple measurement circuit to which the alarm must be set to respond. Take it that the circuit contains an 'intrumentational cold junction' at 0°C and use thermocouple tables. *From the MSc Process Safety Examination, University of Aberdeen, 2002.*
18. At a hydrocarbon facility at Pembroke near the Welsh coast in 1994 there was an accident as a result of which 20 tonnes of hydrocarbon liquid escaped and ignited. What was the TNT equivalence?

19. It is a rule of thumb that in a fire inside an enclosure flashover – transition from a localised fire to one involving the entire enclosure – occurs at a heat release of about 1MW. Beyond flashover anything combustible by way of floor coverings and wooden structural members will become alight. If there is leakage of petroleum liquid into such an enclosure and ignition resulting in a pool fire, how big will the 'pool' need to be for flashover?

20. A container for natural gas under pressure at an indoor site is cylindrical and supported at its curved surface by two 'saddles' (see Chapter 6). A case must be made that if the contents ignited there would not be flashover. The wall thickness of the container and the time for extinguishment measures to respond are such that it is expected that in the event of ignition the wall surface temperature will not exceed 800°C. If the diameter of the container is 1 m, how long can the vessel be consistent with this requirement? Approximate the container when heated to a black body and use a value of 300°C for the temperature of the surroundings.

For how long could the amount of gas in the cylinder, the length of which you have calculated, supply a flame of heat release 0.1 MW if the initial pressure within the container is 100 bar? Heat of combustion of methane = 889 kJ mol^{-1}.

21. There was an explosion at an oil terminal at Avonmouth, England in 1951 in which two young employees died. The site contained quantity of liquid hydrocarbon of the order of 70,000 tonnes. Using this figure, what would have been the TNT equivalence?

22. The atomic bomb which was dropped on Hiroshima in 1945 had a TNT equivalence of 15 kilotonnes. What quantity of natural gas would, on explosion, give this TNT equivalence? Express the answer in cubic metres at 1 bar 15°C.

23. A supertanker is an oil tanker which can hold a million barrels or more of crude oil. The supertanker was developed for the benefit of post-war Japan, where huge amounts of oil were needed and none was available domestically. Imagine that a supertanker holding 1 million barrels were to explode catastrophically. What would be the TNT equivalence?

24. The compounds 1,8-cineole and terpinen-4-ol occur in Ecualyptus oil. They have different structural formulae but the same molecular formula $C_{10}H_{18}O$ and burn in air according to:

$$C_{10}H_{18}O + 14O_2 \ (+ \ 52.6 \ N_2) \rightarrow 10CO_2 + 9H_2O \ (+ \ 52.6 \ N_2)$$

Properties are as follows.

1,8 cineole: normal boiling point 176°C (449K), heat of vaporisation 44.2 kJ mol^{-1}

terpinen-4-ol: the normal boiling point 212°C (485K), he heat of vaporisation is 51.8 kJ mol^{-1}.

Calculate the flash point of each on the half-stoichiometric rule. These can be compared with the literature values given in the solution.

Endnotes

[9] Following the work described [11], corrections to the flash point of dimethyl ether have started to appear in reference tables. The Fifth Edition of the SI Chemical Data Book (John Wiley), published in February 2002, cites [11] and gives a corrected flash point of dimethyl ether accordingly.

[10] Performance limitations of thermocouples are discussed in Chapter 7.

[11] For a further discussion of this point see: Jones J.C. 'Demonstrations with nitrocellulose: possible further pedagogic value' *Journal of Chemical Education* **78** 1596 (2001).

[12] The author owes this information to Dr. M.M. Hirschler.

[13] The term 'escalation' as it relates to offshore hazards is explained more fully in Chapter 8.

Appendix to Chapter 3

Hypothetical case study involving dimethyl ether

At a chemical storage site there is a fire involving dimethyl ether. Authoritative sources, including editions of the *Handbook of the Society of Fire Protection Engineers* and of the *Handbook of Physics and Chemistry* (CRC Press) give the closed-cup flashpoint of dimethyl ether as –41°C. The company responsible assert that they were had consulted the literature, were aware of the flash point and therefore had organised the storage conditions accordingly. They therefore claim that all regulations were met.

BUT:

It is known from the article. 'Uncertainties in the flash point of dimethyl ether' (Jones J.C., (2001) *Journal of Loss Prevention in the Process Industries* **14** 429–430) that the true flash point is –83°C and that the widely accepted value is too high by about 40°. An expert witness could of course produce this publication in court. Ten or twelve years, the time since the article cited above first appeared, might not be long enough for its contents to filter through to the standard reference works many of which are only lightly revised at intervals of several years. A court might then rule that an organisation could not defend itself by proving that flash point tables had been consulted. It might take the view that the company had a duty to keep itself abreast of recent developments and that if readers of the article cited above knew that the accepted published flash point of dimethyl ether is too low they ought to have done! The author would not care to predict which way the judgement would go in court.

4

PHYSICAL OPERATIONS ON HYDROCARBONS AND ASSOCIATED HAZARDS

4.1 Introduction

In the category of 'physical operation' we shall include storage and transportation as well as such processes as refining and heat exchange. Many severe accidents have resulted from hydrocarbon material which was previously being merely stored or transported, not being processed or utilised in any way.

4.2 Storage and transportation

4.2.1 Fire loads and case studies

The fire load of a room or outdoor storage space is the averaged weight of combustibles per square foot of floor area. In this sense a combustible material is one with a calorific value of 16–19 MJ kg^{-1}, the range of values for seasoned wood. Materials of interest in this book have a higher calorific value than this, and in storage practice this is allowed for by adjusting the weights. Residential premises have fire loadings typically in the 7–8 lb ft^{-2} range, whilst industrial premises, depending on the nature of the business conducted, may have 25 to 30 lb ft^{-2}. A related numerical example follows.

A particular manufacturer of gases makes available for sale propane in cylinders of 32 cm diameter with flat bases which, on initial supply, contain 15 kg of propane. What will be the contribution made by one such cylinder to the area in which it is standing?

Calorific value of propane = 50 MJ kg^{-1}.

Area of the cylinder base = $\pi \times (0.16)^2$ m^2 = 0.080 m^2 = 0.86 ft^2

Fire load = $(15 \times 2.205 \text{ lb}/0.86 \text{ ft}^2) \times (50/17) \text{ lb ft}^{-2} = 113 \text{ lb ft}^{-2}$

scaling for the higher calorific value of propane than that of the benchmark material

This is an order of magnitude higher than the fire load in storage of combustible materials other then hydrocarbons, and indicates very clearly the need for effective precautionary measures.

Table 4.1 gives an outline of a few accidents in storage, one of which cost 508 lives (three times the death toll at Piper Alpha). The information is taken chiefly from [1].

Returning to the matter of fire loads, we consider a road or rail tanker for gasoline, the cylindrical tank of which is 8 m in diameter and 30 m long. By simple mensuration, the quantity of gasoline of density 750 kg m^{-3} which it will hold is 11000 tonne. On the basis of the projected area, that is, the area of a rectangle of 8 m by 30 m, the fire load of the space below the tank is:

$$1100000 \text{ kg} \times 2.205 \text{ lb kg}^{-1}/(240 \text{ m}^2 \times 10.8 \text{ ft}^2 \text{ m}^{-2}) \times 45/17 = 25000 \text{ lb ft}^{-2}$$

exceeding the value given above for 'commercial premises' by three orders of magnitude.

4.2.2 Buncefield, 'the biggest fire in peacetime Europe'

Buncefield, where a serious accident occurred early on Sunday 11th December 2005, is a major facility for storage and distribution of hydrocarbons. It receives refined products from several places. Gasoline and fuel oil from Canvey Island are transferred by pipeline to Buncefield to places including the airports of London. Buncefield is a distribution centre and hydrocarbons which pass through the facility are already refined and ready for their intended uses. The total capacity of the Buncefield storage facility is about 2×10^5 tonne. There were 43 injuries due to the 2010 accident at Buncefield, none of them fatal.

4.2.3 Safety measures in storage

When a hydrocarbon is stored, close attention must be paid to its vapour pressure not only at the storage temperature but also at higher temperatures. This is because if the vessel becomes heated by a nearby fire it will burst if the vapour pressure of its contents consequently increases beyond a certain value. To prevent this, safety release valves are fitted which are programmed to start releasing vapour once the bulk reaches a certain temperature. Slow gradual release through such a valve is less dangerous than sudden catastrophic failure

Table 4.1 *Hydrocarbon accidents originating from storage or transport*

Location and date	Details	Cause/lessons learnt
Avonmouth, UK, 1951	Diesel ('gas oil') being transferred from a ship to a stationary storage facility. 2 deaths	The diesel mixed with petrol from a leaking adjoining partition of the tank. Explosion of the vapour
Louisiana, USA, 1967	Isobutane explosion and fire spread. 7 fatalities, 13 injuries	Failure at an alkylation unit
Brazil, 1984	Leak of petrol from a pipeline passing through a 'shanty town'. 508 dead	Residential properties illegally built on land belonging to the petroleum company, which nevertheless provided essential services. 45 minutes for the fire brigade to arrive
Greece, 1986	Oil spillage at an oil terminal ignited by hot work	Escalation largely due to previous spillages which had not been cleared up
France, 1987	Escalating fires and explosions at a depot having 76 tanks containing various petroleum products. 2 deaths, 8 injuries	Most tanks old and without a rupture seam at the roof, with the result that they tended to burst at the base with total release of contents
U.K., circa 1990	Collision of a road tanker bearing toluene. Minor injuries, significant damage	Overturning of the tanker, spillage of toluene
U.K., 1992	Road tanker containing petrol in collision with a small vehicle. Burn injuries to 6 persons, 70 persons treated for smoke inhalation and/or shock. Evacuation of nearby buildings	Fire due to escaping petrol, but no explosion
Dronka, Egypt, 1994	Leakage of aviation fuel and diesel from storage tanks during a rainstorm. >410 deaths	Ignition of the fuel by lightning and propagation towards a centre of population

of the vessel. These ideas are brought out in the context of a numerical example in the shaded area below.

A quantity of n-decane is stored in a closed and evacuated vessel, the ullage space being occupied solely by the compound's own vapour. There is a safety valve which must be set to open when the pressure reaches 5 bar, and this is controlled not by a pressure measurement but by a temperature measurement in the bulk liquid. The normal boiling point is 447 K and the heat of vaporisation 39 kJ mol^{-1}. At what temperature reading must the valve be programmed to respond? The bulk temperature is measured by means of an immersed type K thermocouple with an instrumentational cold junction at 0°C. At what voltage must the valve be set to respond? It will be necessary to consult thermocouple tables.

Solution
To calculate the temperature at which the hydrocarbon has a vapour pressure of 5 bar we use the Clausius–Clapeyron equation [2] (briefly encountered in Chapter 3):

$$\frac{d(\ln P)}{d(1/T)} = -\frac{\Delta H_{vap}}{R}$$

where P = pressure (Pa)
T = temperature (K)
ΔH_{vap} = heat of vaporisation (J mol^{-1})
R = gas constant (8.314 J mol^{-1} K^{-1})

Integrating with the normal boiling point (where, by definition, the vapour pressure is 1 bar) as one limit:

$$\ln 5 = -(\Delta H_{vap}/R) \{(1/T) - (1/447)\} \Rightarrow T = 528 \text{ K } (255°C)$$

From standard thermocouple tables e.g. [3], e.m.f. = 10.4 mV

Whilst most hydrocarbons are stored at atmospheric pressure some, notably LPG and VCM (vinyl chloride monomer), are stored under their own highly superatmospheric vapour pressure. These have critical temperatures such that liquefaction at room temperature is possible and have to be held in containers capable of withstanding the vapour pressures involved. The example in the shaded area below brings out some of these ideas.

With reference to the installations at Canvey Island at the time of the reports (see Appendix), the Occidental plant contained two identical spherical containers for propane storage at ordinary temperatures (say 25°C) each designed to hold 750 tonne. The diameter of each sphere we take to be 15 m. If such a sphere contains its nominal payload of 750 tonne, calculate:

(a) the vapour pressure of the propane
(b) the volume of the ullage space
(c) the weight of propane in the vapour phase, and the proportion of the propane in the vapour phase.

Data required:
    latent heat of vaporisation of propane = 18.5 kJ mol^{-1}
    density of liquid propane at 25°C = 493 kg m^{-3}
    normal boiling point of propane = −42°C (231 K)

Solution
(a) using the Clausius–Clapeyron equation:

$$\mathrm{d}\ln P / \mathrm{d}T = \Delta H / RT^2$$

$$\Downarrow$$

$$\int_{P^o}^{P} \mathrm{d}\ln P = (\Delta H / R)\int_{T_b}^{T} 1/T^2 \, \mathrm{d}T$$

where P^o denotes atmospheric pressure, T_b the normal boiling point. Integrating:

$$\ln(P/P^o) = -(\Delta H/R)\{(1/T) - (1/T_b)\} = 2.17$$

$$P = P^o e^{2.17} = 9\,\mathrm{bar}$$

(b) volume occupied by the liquid propane = $750 \times 10^3/493\,\mathrm{m}^3$

$$= 1521\,\mathrm{m}^3$$

volume of the sphere = $(4/3)\pi \times (7.5)^3 = 1767\,\mathrm{m}^3$
volume of the ullage space = $246\,\mathrm{m}^3$

(c) pressure of propane in this volume = 9 bar. Using the ideal gas equation:

$$PV = nRT, n = PV/RT = (9 \times 10^5 \times 246)/(8.314 \times 298)$$
$$= 8.9 \times 10^4\,\mathrm{mol} \equiv 3.9\,\mathrm{tonne}$$

proportion in the vapour phase = 0.5%

Another issue in storage and transportation, perhaps particularly in the transition to the former to the latter, is static electricity, a possible ignition source. When hydrocarbon is offloaded from a road tanker to a stationary tank via a manifold of pipes, it is important that all the way along the path taken by the flowing hydrocarbon the electrical potential is the same. That is why electrical earthing is very evident in such situations.

The mechanism of static electricity generation is simple: removal of electrons as one layer of hydrocarbon liquid moves against another, entirely analogously to elementary experiments in physics, using the gold leaf electroscope which gives a response when a piece of ebonite rubbed with a rag is applied to it. Clearly, anything which increases the electrical conductivity of a hydrocarbon liquid or of a container holding it will mitigate hazards from static [4]. To that end there are anti-static paints which can be applied to hydrocarbon storage vessels and to surfaces with which such vessels are in electrical continuity. Alternatively a chemical agent called a fuel conductivity improver can be added, in a small proportion, to the hydrocarbon liquid itself. This enables an incipient ignition source within the bulk of the liquid to be conducted to the vessel wall and from there to earth. Biodiesels are composed of much more polar molecules than are petroleum fractions, and this gives biodiesels better electrical conductivity and therefore a safety advantage in these terms.

4.2.4 Effects of solar radiation on storage of hydrocarbons

The sun radiates as a black body at about 5800 K. A minuscule proportion of this radiation ($\approx 10^{-12}$) reaches the earth. In fact 'solar flux' experienced by an object on the earth's surface is roughly 1400 W m^{-2} and is of course periodic. Solar flux is an issue in hydrocarbon storage, a point that was touched on in Chapter 3 when the Ludwigshafen accidents were under discussion.

As the present author has recently pointed out elsewhere [5], there is only limited truth in the identification of black body behaviour in thermal radiation with black colour. So whilst it is often useful to paint the outside of a storage tank white to minimise its absorption of radiation, to do so without due thought might avail nothing; some white paints and pigments have very high emissivities. There are suitable paints for making the surface of a vessel less absorptive. One of the appended numerical examples is concerned with effects of solar radiation on a hydrocarbon vessel.

Storage practices appertaining to particular hydrocarbon products are dealt with in a later chapter.

4.2.5 Storage codes

An example of such a 'code' is that issued by the Institute of Petroleum (IP) for liquefied petroleum gas (LPG) that any two adjacent LPG tanks be separated by one quarter the sum of the diameters of the tanks. This is followed up in the numerical example below.

At a refinery there are LPG tanks of two sizes, 500 tonne and 1000 tonne. If each is spherical and 80% full when bearing the nominal payload calculate the separation required according to the IP code above.

$$\text{For the 500 tonne tank, volume} = (500000 \text{ kg}/550 \text{ kg m}^{-3}) \text{ m}^3$$
$$= 0.8 \times (4/3)\pi r^3 \text{ where } r \text{ is the tank radius}$$
$$\downarrow$$
$$r = 6.5 \text{ m diameter} = 13.0 \text{ m}$$

$$\text{For the 10 tonne tank, volume} = (1000000 \text{ kg}/550 \text{ kg m}^{-3}) \text{ m}^3$$
$$= 0.8 \times (4/3)\pi r^3 \text{ where } r \text{ is the tank radius}$$
$$\downarrow$$
$$r = 8.2\text{m diameter} = 16.4 \text{ m}$$
$$\text{Minimum separation} = 0.25(13.0 + 16.4) \text{ m} = 7.4 \text{ m}$$

The IP codes previously referred to for LPG also require that adjacent tanks containing liquids with flash points in the range 21 to 55°C are separated by 15 m between the tops of the tanks. For spherical tanks or cylindrical tanks with parallel axes, the distance between the curved surfaces of the tanks is therefore $(15 - 2r)$ m.

Imagine that it is required to store naphtha, flash point < 55°C, in two equivalent horizontal cylindrical tanks with parallel axes and centres 5 m apart. What length (L) must the cylinders be if 20000 barrels of the naphtha are to be so stored?

$$\text{Each tank must hold 10000 barrels} = 10000 \times 0.159 \text{ m}^3 = 1590 \text{ m}^3$$
$$\text{Now } (15 - 2r) = 5 \text{ m} \rightarrow r = 5 \text{ m}$$

$$\pi \times 5^2 L = 1590 \text{ m}^3 \rightarrow L = 20 \text{ m}$$

Two of the appended numerical examples continue this theme.

4.2.6 HAZOP studies

Just as in an subsequent chapter a transportation example will be used to illustrate the principles of ALARP (As Low As Reasonably Practicable), so in this chapter simple movement of hydrocarbon during processing will be used to elucidate principles of HAZOP (HAZard and OPerability studies). This is a means of hazard identification, and utilises certain keywords in relation to an operation which is the subject of the HAZOP analysis. These are summarised as Table 4.2.

Table 4.2   *Keywords in HAZOP*

None/Not done – negation of intent
More (than) – quantitative increase
Less (than) – quantitative decrease
As well as – qualitative increase
Part of – qualitative decrease
Reverse – logical opposite of intention
Other than – complete substitution

A light-hearted, but pedagogically sound, application with which the author has occasionally entertained his students at Aberdeen is given as Figure 4.1. It relates to preparation of a cup of black coffee. The following points should be noted. First, one *instruction* (here to prepare a cup of black coffee) involves more than one *operation*, each of which can be analysed according to HAZOP.

In this example we consider the single operation of transferring boiling water from jug to cup. Also, a HAZOP study is only concerned with identification of hazards, but in the example below the effects of the various deviations on production are given in parenthesis. Finally, it is possible for one or more of the deviations to be inapplicable, as in this case is 'part of'.

Principles learnt from this tongue-in-cheek example can now be carried through to a truly realistic application: transfer of an organic liquid from a batch reactor to a tank. A HAZOP on this is given as Figure 4.2.

Note that the contents of the right-hand column of a HAZOP study such as that above have the nature of suggestions or even debating points, and might be enlarged or modified in the light of further information such as the flash point of the liquid and its toxicity.

HAZOP can be used in more than one way, and its application to a diagram process – a 'flow sheet' – is referred to as coarse HAZOP. The next

The operations are:

Add coffee to cup.
Heat water.
Remove heated jug from power supply.

Transfer water from heated jug to cup. ⇐ HAZOP on this

Stir.

Deviation	Comments
None/not done	*Water not poured at all.* No hazard. (No coffee)
More (than)	*Too much water transferred from jug.* Spillage. Possible scalding. Possible water damage to surroundings. (Coffee too weak)
Less (than)	*Too little water transferred from jug to cup.* No hazard. (Less than a full yield of coffee, and too strong)
As well as	*Absent-minded addition of milk as well.* No hazard. ('Product' other than that required)
Part of	Not applicable
Reverse	*Addition of the coffee granules to the water in the jug.* Possible scalding or steam burn. Possible contamination of jug surface (Poor coffee)
Other than	*Operator with a shaky hand pours the water but misses the cup.* Possible scalding. Possible water damage to the surroundings.

Figure 4.1 *HAZOP on making a cup of black coffee.*

stage, termed 'full HAZOP', uses a piping and instrumentation diagram, containing more detailed information than the flow sheet and requiring application of the HAZOP key words to many of the features.

Tens of thousands of HAZOP studies have been applied. The handful chosen for Table 4.3 vary in the scale of the operation to which HAZOP is applied, from cavern storage of a gas to an undertaking formidable in its

Deviation	Meaning and consequences
None/not done	*Liquid not transferred at all.* No hazard.
More (than)	*More liquid transferred than the tank can hold.* Flammability and possibly toxicity hazard through spillage.
Less (than)	*Room to spare in the tank after transfer of liquid.* Possible vapour explosion hazard if the space also contains air.
As well as	*Solid residue from the reactor transferred as well* *as the liquid.* Possible hazard if the residue is capable of explosive decomposition (as some organic solids are).
Part of	Not applicable if a single liquid is being transferred.
Reverse	*Transfer of liquid from tank back to reactor.* Possible hazard if this is accompanied by depressurisation and consequent air ingress.
Other than	*Liquid spilt on the floor instead of being transferred* *to the tank.* Flammability and possibly toxicity hazard.

Figure 4.2 *HAZOP on transfer of an organic liquid from a batch reactor to a tank.*

complexity involving subsea engineering. It is the variation of scale of processes to which HAZOP can be applied that is seen as one of its major strengths and the reason for its wide adoption and acceptance.

4.2.7 Thermal ignition theory applied to storage and pumping of unstable substances

4.2.7.1 Preamble

We have already seen that some substances are capable of very powerful and dangerous heat release by decomposition or, conceptually equivalently, by 'combustion' with intramolecular oxygen as the oxidant. These include certain substances which are used as cetane enhancers for compression ignition engines. Storage of these lends itself to analysis by traditional thermal ignition theory, in particular Frank–Kamenetskii (FK) theory. The reader is referred to one of the many basic texts which cover fundamental ignition theory (e.g., [6]) for background on what follows.

Table 4.3 *Examples of HAZOP*

Location	Process, selected details and reference
India	Removal of propane and butane from natural gas for use as LPG
	Each deviation identified. 'More than' taken to mean presence of components other than those expected from the gas composition. A missing expected component classified under 'part of'
	http://www.letsconserve.org/Hazop-Gas%20Processing%20Complex.pdf
Germany	Cavern storage of natural gas
	'More than' includes air entry to the 'cavern' 'Less of' includes drop of pressure in the 'cavern'
	http://www.gl-nobledenton.com/assets/downloads/3.Hazard_Operability_Studies_external.pdf
North Sea	'Ninian Third Party Entrants Project' The tying back of three new offshore oil fields – Lyell Staffa and Strathspey – to two platforms at an exisiting mature field
	http://www.onepetro.org/mslib/servlet/onepetropreview?id=00028886
Kuwait	Design and siting of a new aviation fuel depot
	http://bharatpetroleum.com/EnergisingBusiness/FS_Overview.aspx?id=1

The FK model expresses the critical condition in terms of the parameter δ:

$$\delta = \frac{r_o^2 QE\sigma A \exp(-E/RT_o)}{kRT_o^2}$$

where T_o = ambient temperature (K)
r_o = reactant dimension (m)
Q = heat of combustion (J kg^{-1})
σ = bulk density (kg m^{-3})
A = pre-exponential factor (s^{-1})
E = activation energy (J mol^{-1})
k = thermal conductivity (W m^{-1}K^{-1})
R = gas constant (8.314 JK^{-1}mol^{-1})

The critical condition is that for an ignition:

$$\delta > \delta_{crit}$$

Because of the space co-ordinates in the conduction part of the formulation, the value of δ_{crit} depends upon the reactant shape. For a cylindrical assembly long enough for radial conduction to dominate to the exclusion of longitudinal it is 2.00, with r_o the radius. The boundary condition is that the outside surface of the sample shall remain at T_o indefinitely for $\delta < \delta_{crit}$ and until after ignition is established for $\delta > \delta_{crit}$.

4.2.7.2 Application to a cetane enhancer

The counterpart for a compression ignition engine of an octane enhancer for a spark ignition engine is called a cetane enhancer. Oxley *et al.* [7] apply the FK model to a number of cetane enhancers including isopropyl nitrate (IPN), structural formula $(CH_3)_2CH—O—NO_2$. They use the following values for the quantities required for application of FK theory:

$$Q = 3.1 \, MJ \, kg^{-1}$$
$$A = 1.06 \times 10^{17} \, s^{-1}$$
$$E = 177 \, kJ \, mol^{-1}$$
$$\sigma = 1050 \, kg \, m^{-3}$$
$$k = 0.31 \, W \, m^{-1} K^{-1}$$

all quantities being for neat (not diluted) IPN. Returning to the FK model:

$$\delta = \frac{r_o^2 Q E \sigma A \exp(-E/RT_o)}{k R T_o^2}$$

and it is convenient to work with spherical geometry, for which the critical value of δ is 3.32 [6], rather than cylindrical. Once results are calculated on the basis of one geometry it is simple to adjust them for another. If the precise size and shape of any vessel proposed for transporting or storing IPN are available, fine-tuning of calculations for spherical geometry is fairly trivial. The effect of container geometry is significant without being huge. Imagine that a tank of IPN is stored in the open air in a location where conditions are hot and ambient temperature can be as high as 45°C (318 K). How big a sphere of IPN would be critical at that temperature?

Inserting the values for A, E, k etc. and $T_o = 318$ K gives:

$$r_o = 41 \, m$$

The quantity of IPN which could be stored in a sphere of this radius is about 350 000 tonne. Such substances are often stored in quantities of a few tens of tonne in which case it is evident, even without fine-tuning for container geometry, that simple storage at atmospheric temperatures will not lead to thermal instability and ignition of IPN.

FK theory can be extended to situations in which a warm region – a 'hot spot' – has developed in an initially stable assembly of an ignitable substance. This first requires introduction of the dimensionless temperature θ_o:

$$\theta_o = E/T_i^2\{T_i - T_A\}$$

where T_i is the hot spot temperature and T_A the temperature of the bulk of the material in isolation from the hot spot. The form of δ which applies here is:

$$\delta = \frac{QA\sigma Er^2 \exp(-E/RT_i)}{RT_i^2 k}$$

with r the hot spot radius. With T_i the minimum hot spot temperature for ignition, δ becomes δ_{crit}. The simplest form of the expression for δ_{crit} [8] is:

$$\delta_{crit} = 25 \text{ for all } \theta_o$$

Imagine that in the pumping of IPN, bulk temperature $T_A = 313$ K an element of the fluid had become heated to 200°C (473 K). How large would this element have to be in order to cause the IPN to explode? Putting in our values of Q, A etc., with $\delta = 25$ and solving for r gives:

$$\delta = \frac{QA\sigma Er^2 \exp(-E/RT_i)}{RT_i^2 k} = 25 \implies r = 3 \text{ mm}$$

so a hot spot of 3 mm radius at 473 K would cause the IPN to explode. Note that in the above calculations no use had to be made of the fact that the surrounding temperature was at 313 K; the treatment used simply requires that $T_i \gg T_A$. There are however other forms of the expression for δ_{crit} which generalise to:

$$\delta_{crit} = f(\theta_o)$$

and which therefore do require a value for T_A. The reader can find these in [8], but should note that they seldom if ever give values of the critical radius very different from those given by the critical condition used herein. In any case there might be no grounds for ascribing greater reliability to values obtained from the more advanced critical conditions, at least not without knowledge of the physical mechanism of hot spot generation; these mechanisms would include viscous heating and heat transfer from an overheating part of the pump[14].

This calculation has predicted that a hot spot of 200°C would need to be of the order of half a centimetre in diameter to ignite IPN. However, hot spots can be much smaller than this, even micron size. We now perform the above calculation in reverse and predict what temperature a hot spot of one micron radius would need to attain in order to ignite the IPN. Returning to:

$$\delta = \frac{QA\sigma Er^2 \exp(-E/RT_i)}{RT_i^2 k}$$

and inserting the quantities including $r = 10^{-6}$m gives:

$$T_i^2 \exp(21289/T_i) = 9.48 \times 10^{17}$$

a transcendental equation which has to be solved by trial and error to give:

$$T_i = 756.5\,\text{K}\,(483.5°\text{C})$$

which is how hot a one micron radius hot spot would have to be in order to cause the IPN to explode.

In both of the above examples the hot spot has been a reactive one, an element of the unstable material itself having become hotter than the bulk. There is, of course, such a thing as an inert hot spot, e.g., a small piece of metal from an overheating part of a pump which finds its way into the unstable material being pumped. A calculation apropos of this is in the appended numerical examples.

4.2.7.3 Signage for stored flammable materials

Figure 4.3 shows three signs issued by Occupational Health and Safety Administration (OSHA) in the US to warn of stored flammable substances. American National Standards Institute (ANSI) standards are followed in the production of such signs and they include the absence of lamination over the sign for elimination of reflective effects. The third of the signs refers to dangers from static electricity, a point mentioned earlier in the chapter.

4.3 Refining

4.3.1 Introduction

At a refinery the primary process is separation of the various constituents of crude oil in a distillation column. Crude oil, already having received some heat by exchange with residue from the column, is admitted part way up the vertical column. Gas, convertible to LPG, comes over first. The lighter liquid fractions – gasoline, naphtha – rise and are tapped off in the upper half at temperatures up to about 150°C. Heavier fractions – kerosine and diesel – are tapped off lower down. If they rise initially they will subsequently descend by condensing on to the plates and spilling over. Residue is taken off at the base,

Figure 4.3 *OSHA signs for stored flammable substances. Note in the second one the continued use in the USA of non-SI units.*

and its heat exchanged with incoming crude. The net effect is that the distillation column develops a temperature profile ascending from about 150°C at the top to about 300°C at the base and fractions are taken off accordingly.

As we saw in Chapter 1, crudes are classified as paraffinic, naphthenic, aromatic or mixed base according to which hydrocarbon type is in preponderance. A particular refinery might be receiving crudes from diverse sources. In order to keep the various fractions within specification there will need to be some flexibility in processing in response to fluctuations in the crude oil composition. For example, a crude with a relatively high proportion of light aromatics and/or light branched paraffins would be expected to give a straight-run gasoline of fairly good octane rating. If such a crude is followed by one poorer in these classes of component there will be a need to supplement the gasoline fraction with octane enhancers, possibly obtained by reforming naphtha.

The scale of refining and the consequent need to obtain crudes from widely different sources, some relatively close to the refinery and some very distant from it, become very clear to the visitor to the Dutch coast near Rotterdam. There is a 50 km stretch of coastland, beginning at the natural coast with the North Sea and then diverting east into an estuary created by land reclaimed from the sea, which is fully occupied by refineries and associated plant. Very many major petroleum companies are represented and some of the facilities for handling the refined material are shared by more than one such company. Incidentally, the Dutch sector of the North Sea is more productive of gas than of oil.

The area of the Netherlands briefly described above is known as the Rijnmond, and there has been a Rijnmond Report comparable to the Canvey ones (see Appendix).

At any refinery there will handling operations including pumping of crude oil from wherever it is received to the distillation column and conveyance and temporary storage of the fractionated material. Each of these is fraught with fire and explosion hazards. Over the years a surprising number of refinery accidents have occurred in the process of pumping the crude. There is a further hazard associated with this: heavy current is required, and in troubleshooting a pumping process there is a risk of electrocution. Once hydrocarbons are produced by refining, they have to be conveyed to wherever they are required either as liquid fuels or as chemical feedstock. As the information in Table 4.1 shows, hazards associated with transportation of hydrocarbons are considerable.

Distillation columns contain large quantities of boiling and refluxing hydrocarbon inventory. Associated hazards include excess pressure if cooling at the condenser drops, which necessitates pressure relief valves as safety features. On the other hand, the pressure might become too low if the evaporation rate of influx hydrocarbon drops; injection of an inert gas can be used to avert a serious accident in those circumstances. Corrosion of plant by salt in the crude oil is also sometimes a difficulty. The solution is to wash the crude with fresh water at ≈120°C.

Plain carbon steel is the most common choice of material for pipe work within refineries. We saw in Chapter 3 that many ASTM standards apply to hydrocarbon safety. In fact when ASTM was first founded in 1898 its primary concern was steel for railroads, and this strand of its activity continues. ASTM A106 / A106M – 11: Standard Specification for Seamless Carbon Steel Pipe for High-Temperature Service is widely applied to refineries. It is concerned with effects on the properties of steel of such operations as welding, flanging and bending.

4.3.2 Accidents at refineries

Table 4.4 gives summaries of case studies of selected refinery accidents, and three more recent ones are discussed after the table. Important questions arising from the first of these are addressed.

Isomerisation is amongst the numerous processes at modern refineries, and it was at an isomerisation unit that the 2005 accident at the BP Texas City

Table 4.4 *Details of refinery accidents*

Location and date	Details	Cause/lessons learnt
Texas, USA, 1980	Explosion at a refinery caused by rupture of a hydrocarbon-bearing vessel. Extensive damage, shutdown of the entire refinery, 41 injuries	The safety valve on the vessel had been rendered ineffective by ice
Wales, 1983	Fire in a refinery storage tank containing light crude oil. Extinguishment only after two days of firefighting. Firemen treated for skin burns	Leakage of oil vapour through the tank roof and ignition by hot particles from the nearby refinery flare
Corton refinery, Essex, England 2007	Fire commencing at a fractionation column. No deaths or injuries	Refinery operating at 10 million tonne per year (180 000 barrels per day) at the time of the fire

refinery occurred. There were fifteen deaths. Venezuela is a major producer of oil and one of the founding members of OPEC in 1960. Accordingly downstream operations are on a large scale, and its Amuay refinery has a nameplate capacity of 0.65 million barrels per day, well into the big league. There was a fire at this refinery in August 2012 in which 41 people died and 80 were injured. There were major overpressures which caused damage to nearby buildings. Little has been reported on the cause beyond that the initial explosion involved natural gas.

4.3.3 *The Marcus Hook and Richmond CA refinery accidents*

The 2009 explosion at the Marcus Hook refinery will be considered first. This refinery, on the Delaware River, opened in 1902 and its capacity rose over the decades to >175 000 barrels per day. There had been a major incident there in 1946, resulting in seven deaths. This was at a facility within the refinery for making fuels for the spark ignition engines used by the aircraft of that period. In 2008 ownership of the refinery passed to Sonoco. About three years earlier $285 million had been spent improving the refinery in terms of its emissions of toxic substances including benzene, and it was purely fortuitous that a fire there in 2006 caused no injury or major damage. These difficulties were attributed to the age of the refinery, and this leads to a point of some general importance. There has been a strong tendency in the US to extend existing refineries instead of building new ones, and the first totally new refinery in the US since 1976 came into operation in the early 2000s. In 1902 when

Marcus Hook began oil processing the American Civil War was still within living memory, and it was to be another year or so before Ford and General Motors began producing automobiles. A refinery whose basic configuration belonged to that period had been extended for continued use until a serious fire there in May 2009 led to a serious review of its future. The fire referred to was in the ethylene plant and was traced to pipe corrosion; such corrosion had occurred previously. The refinery closed after the accident, but in September 2012 Sunoco announced that it would reopen as a facility to process products including condensate.

Over the lifetime of a chemical plant difficulties are expected whilst the plant is being 'shaken down' after commissioning. This is followed by a much longer period during which failures are few, which in turn can be followed by an increase in failures through decline in plant condition. Factors leading to such decline include corrosion and contamination. Marcus Hook was a clear example of such decline, which should be should be distinguished from obsolescence, where the plant remains fit for purpose even though more recent designs have become available. Adoption of such designs will usually have a financial benefit after the payback period.

There was a refinery fire at Richmond CA in 2012, when a finger was pointed at a particular alloy substance that had been used to convey inventory. The alloy is known as 'chromium 9', and was seen as an improvement over carbon steel in terms of resistance to corrosion through sulphur, a view said to have been endorsed by the American Petroleum Institute (API). At least as good as 'chromium 9' at resisting corrosion is stainless steel of certain specification, but with that embrittlement at fractionation temperatures can be a difficulty. In all 15 000 residents of the area reported to local hospitals for breathing difficulties and headaches as a result of this incident.

4.3.4 Possible process integration in refining

In hydrocarbon processing, several operations can be integrated not only in mass flow terms but also in energy balance terms. At a refinery, there will be pumping, fractionating, heat exchange and possibly reforming and hydrotreating. Process integration in energy terms, in particular heat exchanger networks (see following section), has been one of the most important areas of chemical engineering R&D in the last 15–20 years [9]. A heat exchanger is a device by means of which heat is transferred between fluids. This is of course important in energy efficiency. If a fluid, at its production temperature, is hot simply to let it cool is wasteful. Its heat can be transferred to something requiring heat by means of a heat exchanger.

Heat exchanger design is an advanced and specialised topic [9, 10] and is touched on further in Chapter 6. The mode of action of heat exchangers of standard design is discussed later in this chapter.

4.4 Stirring and mixing

We return to our HAZOP example above as a background for some of the principles of stirring. There, a liquid was transferred from a batch reactor to a tank. Imagine that the liquid once transferred to the tank is stirred, perhaps to prevent settlement of tiny amounts of suspended spent catalyst. Classical chemical engineering texts such as those cited previously give power requirements for stirring according to liquid viscosity, and the stirring process lends itself to very direct analysis, in energy terms, by the First Law of Thermodynamics. The working below illustrates this.

Effective stirring of a hydrocarbon liquid requires of the order of 5 horse power per 1000 US gallons of liquid which, in SI, translates approximately to 1 kW per m³ of liquid. The power is that delivered at the shaft of the stirrer, measurable in principle by applying, by means of a brake, just enough force to prevent rotation. The First Law states that, for a closed system (which our example is, being a single quantity of hydrocarbon which can exchange energy but not mass with its surroundings):

$$\Delta U = Q + W \qquad \text{Eq. 4.1}$$

where ΔU = rise in internal energy (U) of the system
$\quad Q$ = heat transferred to or from the surroundings
$\quad W$ = work done on or by the system

The quantity W is calculated simply from the power rating of the stirrer and the time for which the stirrer is applied, but we have no handle on Q except that in stirring applications it is often taken to be nil. This is on the basis of experience from temperature measurement during stirring that most of the energy effect of the stirring is retained by the system and that to assume this gives a result which errs on the safe side. The calculation below continues application of the First Law to our HAZOP example.

Imagine that the liquid in our HAZOP study has a specific heat[*] (C) of 1650 J kg^{-1}K^{-1} and a density of 880 kg m^{-3}. If it is stirred for an hour with power 1 kW per m³ of liquid, the work done is:

$$1000 \text{ W} \times 3600 \text{ s} = 3600000 \text{ J per m}^3 \text{ of liquid}$$

$$\text{work done per kg of liquid} = 4090 \text{ J}$$

applying the First Law and letting $Q = 0$,

$$\Delta U = W = C\Delta T$$

where ΔT is the temperature rise

[*] Strictly, since internal energy is being considered, specific heat at constant volume.

$$\Downarrow$$
$$\Delta T = 2.5\,\text{K}$$

that is, the temperature will rise by up to 2.5 K as a result of the stirring. In some stirring and mixing applications the temperature effect is greater than this and cooling is necessitated. This is further developed in one of the appended numerical examples.

4.5 Heat exchange[15]

4.5.1 Introduction

Clearly if either a waste stream or a product stream in a process is at high temperature, simply to let it cool naturally is wasteful in energy terms and it is preferable to use a heat exchanger to transfer the heat to something that needs it. A simple and very common application is that a hot organic stream – liquid or vapour – will exchange its heat with cold water, the hot water so yielded being subsequently put to some use. The fluid losing heat and that receiving it are of course kept apart, and obviously one fluid drops in temperature as a result of the heat exchange and the other rises in temperature. Sometimes there is a phase change in one or other fluid. Heat exchangers come in different sizes and configurations, and undergraduate texts on heat transfer can be consulted for descriptions of the basic types: counterflow double pipe, parallel-flow double pipe, shell-and-tube and so on. Figure 4.4 shows a shell-and-tube heat exchanger.

Figure 4.4 *Shell-and-tube heat exchanger, showing the array of tubes within the shell. Reproduced courtesy of A. Forsyth and Son, Rothes, UK.*

Table 4.5 *Summary of possible difficulties with heat exchangers*

Nature of the malfunction	Possible causes	Preventative or remedial measures
Blockage	'Scale' from inorganics (e.g., calcium ions) in aqueous fluids. Polymerisation reactions in organic streams and deposition on to heat exchanger surfaces. Carryover of debris from upstream of the exchanger	Cleaning, involving partial dismantling (see the final row of the table) and where possible use of chemical (aqueous and non-aqueous) cleaning agents. Mechanical means, e.g. scraping, best used sparingly*
Vibration during operation, leading to damage	A heat exchanger tube approximates to a metal beam clamped at each end, and as such can resonate. Also, possible transmission of vibrations from a pump or compressor to the heat exchanger	Vibration monitoring during operation possible. Flow conditions for a gas inside the exchanger which will promote vibration can possibly be predicted and avoided
Structural embrittlement through temperature swings	Stoppage of fluid flow to a heat exchanger as a result of delays elsewhere in the plant, causing the exchanger to drop from operating temperature to ambient temperature in a short time	Steam supply to the heat exchanger during plant interruptions
Corrosion	Materials used to fabricate exchangers – chiefly carbon steel and stainless steels – all oxidisable	Regular cleaning. Avoidance of outdoor storage of newly manufactured heat exchange equipment Drainage of an out-of-service heat exchanger Avoidance of junctions of different metals in the construction of an exchanger, as these can lead to electrochemical effects which will accelerate corrosion
Tube leak or rupture	'Manhandling' during removal in maintenance	Adherence to approved removal procedures

* This advice applies when cleaning an out-of-service heat exchanger. Some are built to be continually scraped during use. These are most common in the food industry.

4.5.2   Hazards with heat exchangers

Where a heat exchanger is applied to a hydrocarbon process, loss of containment from a heat exchanger can lead to fire and explosion hazards, as of course can air ingress. The latter was the origin of a fire at a plant in the USA in 1976 (Lees, *op. cit.*). Table 4.5, which has also drawn on Lees' coverage amongst other sources, outlines some of the difficulties which can jeopardise heat exchanger safety.

4.6   Refrigeration

4.6.1   Introduction

Refrigeration is applied to storage vessels in order to maintain the contents at a required sub-ambient temperature, or may be used to provide a coolant which is subsequently reticulated around a plant. Whereas the fluid in the refrigeration cycle itself is, by definition, the refrigerant, the fluid by means of which the cooling extended to plant is called a brine. These are often aqueous solutions of inorganic salts although organic substances including ethylene glycol are sometimes used for this purpose.

4.6.2   The provision of cooling water for plant

The principles of a simple refrigeration cycle can be described within the framework of a temperature–entropy diagram, and the reader is referred elsewhere (e.g., [4]) for a straightforward coverage of this. This theoretical treatment of refrigeration is known as the ideal reversed Carnot cycle, and is a benchmark against which the performances of other refrigeration cycles are measured. In such a cycle, the refrigerant fluid is first compressed, then condensed, then allowed to expand and evaporate. The coefficient of performance (c.o.p.) of the cycle is:

$$\text{c.o.p.} = T_c/(T_h - T_c) \qquad\qquad \text{Eq. 4.2}$$

where T_c is the temperature of the refrigerated space and T_h the temperature of the surroundings (both in Kelvin). When the refrigeration is, so to speak, passed along to water or brine, there can be a difference of up to 5°C between T_c and the temperature which the brine attains by heat transfer with the refrigerant. These ideas will be reinforced by a related calculation below.

Water, for subsequent use in process temperature control, is initially at 20°C and is to be cooled to 7°C by means of a refrigerator which uses refrigerant R-12 (dichlorodifluoromethane). 150 kg per minute of the water so cooled are required. Neglecting the temperature step between refrigerant and water, determine the pressures of refrigerant in the compressor and in the evaporator. Determine also the circulation rate the refrigerant necessary.

Solution
From tables, for example in Hewitt *et al.* [9] the saturation vapour pressure of R-12 at the respective temperatures has the following values[4]:

$$20°C \ (293 \ K), \ P_{sat} = 575 \ kPa \Rightarrow \text{compressor pressure}$$
$$7°C \ (280 \ K), \ P_{sat} = 393 \ kPa \Rightarrow \text{evaporator pressure}$$

The quantity of heat Q removed per kg of refrigerant fluid, known as the 'refrigerating effect', is:

$$T_c(s_g - s_f)$$

where s_g is the specific entropy of the refrigerant vapour at T_h and s_f is the specific entropy of the refrigerant liquid at T_h. From tables:

$$s_g = 1552 \ J \ kg^{-1} K^{-1}$$
$$s_f = 1067 \ J \ kg^{-1} K^{-1}$$

from which:
$$Q = 280(1552 - 1067) \ J \ kg^{-1} = 135\,800 \ J \ kg^{-1}$$

Required rate of heat removal from the water
$$= (150/60) \ kg \, s^{-1} \times 4180 \ J \ kg^{-1} K^{-1} \times 13 \ K = 135\,850 \ W$$
$$\text{Rate of circulation of refrigerant} = (135\,800/135\,240) = 1.0 \ kg \, s^{-1}$$

4.6.3 Accidents due to refrigeration failure

It must not be forgotten that some refrigerants are simple hydrocarbons such as propane, n-butane and isobutane and are therefore highly flammable. About 25 years ago a fire originating in a refrigerator caused extensive damage to an olefins plant [1].

4.7 Site layout

All of the processes above, as well as those in the following chapter, come within the scope of the important topic of site layout. Standards will again be applied. These are BN (base number) design standards (DS) and the particular example here will be:

BN-DS-C69 Minimum spacing for refineries and petrochemical plants.

Figure 4.5 appertains to this. The numerals on the slanting side of the chart have the meanings given in the boxed area below the chart. One or two of these require clarification. 'HC' of course means hydrocarbon. 'Above ignition temperature' means containing a flammable liquid above its flash point and 'below ignition temperature' means carrying a liquid below its flash point. A KO pot ('port', in the boxed area, is probably a misprint in the original) is a vapour-liquid separator. A simple application of the chart follows in the shaded area below.

How far apart would a heat exchanger below ignition temperature and pipe ways need to be according to BN-DS-C69?

Heat exchanger below ignition temperature: 25

Pipe ways: 31

From intersection of the column for 25 and the row for 31 on the chart, these need to be a minimum of 5 m apart.

In the chart 'na', means not applicable: that the situation is outside the scope of the standard and that a more suitable standard must be consulted. 'nm' means no spacing requirements: the standard imposes no minimum spacing beyond that recommended by the engineer responsible. Note that for some pairs, e.g. a high pressure storage tank (descriptor 35) and a property boundary (descriptor 12) a user of the chart is referred to another standard, actually NFPA (National Fire Protection Association, a US body) 30. The full title of this is

NFPA 30:   Flammable and combustible liquid code 2012

a revised version of which is expected in 2015. Two of the appended numerical examples are concerned with BN-DS-C69.

This section draws *inter alia* on [12], within the scope of which is orientation of units and components with respect to each other and minimum separations at refineries and other utilities containing hydrocarbons in large quantities e.g., petrochemical plants. Fire safety is the theme of the code and its *raison d'etre*. It is concerned with access ways, and freedom of restriction of them due to piping. In fact it specifies a minimum 1.05 m between piping and an item of equipment. Any part of a site which might be occupied by a moving vehicle must allow for 3 m between the vehicle and any installation. An interesting and important point is that where there are two parallel processing trains a mirror image approach is to be preferred as this provides access of fire trucks and the like to either via a single central aisle.

Reference [12] has a good deal to say about racks for piping. A generic illustration of a pipe rack is given below as Figure 4.6. The following is a

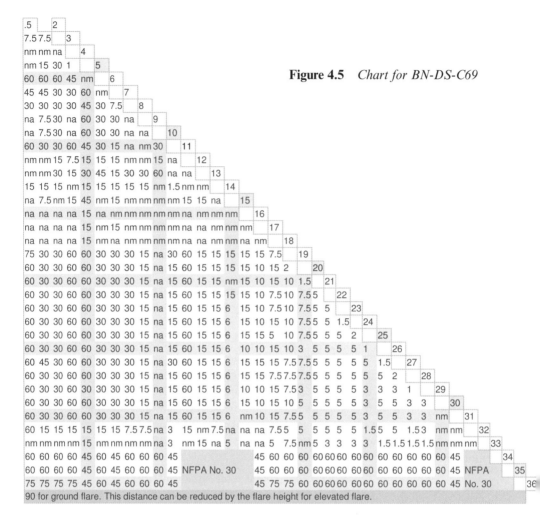

Figure 4.5 *Chart for BN-DS-C69*

90 for ground flare. This distance can be reduced by the flare height for elevated flare.

Distances are in metres.   na = not applicable, no distance can be determined   nm = no minimum spacing requirements, use engineering judgement for spacing

1. Administration and office building 2. Shops, warehouses, shipping building, laboratories, etc. 3. Main plant substation 4. Fire pumps and fire station 5. Tank truck and rail car loading racks, piers 6. Boilers and power generation, instrument air compressors 7. Cooling towers 8. Main process control house 9. Process unit control house 10. Process unit battery limits 11. Main plant roads 12. Property boundary and public roads 13. Hydrants and fire monitors 14. Process unit substation 15. Explosion proof electrical switch racks 16. Water spray control valves and emergency shutdowns 17. Unit isolation, depressurising, control and snuffing steam valves 18.Fired process heaters and open flame equipment 19. HC compressors and expanders 20. Desalters 21. Internally insulated reactors (above ignition temperature) 22. Externally insulated reactors (above ignition temperature) 23. Reactors (below ignition temperature) 24. Heat exchangers (above ignition temperature) 25. Heat exchangers (below ignition temperature) 26. Process pumps handling LPG products 27. Process pumps handling HC above autoignition temperature 28. Process pumps handling HC below autognition temperature 29. Fractionation towers, absorbers, drums, accumulators, KO ports, etc. 30. Air coolers 31. Pipe ways 32. Equipment handling non-flammables 33. Atmospheric storage tanks 34. Low pressure storage tanks (not over 15 PSIG) 35. High pressure storage tanks (spheres and bullets) 36.Flare

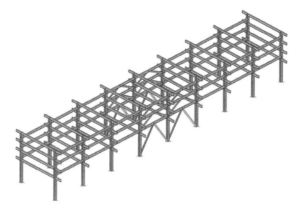

Figure 4.6 *Refinery pipe rack showing transverse frames connected by longitudinal struts.*

quotation from [12]: *Allow a continuous clear area of 4 meters high by 4 meters wide below main [pipe] racks in process units for maintenance access ways.*

In the boxed area below are given some of the other recommendations. Of course [12] is not the sole source of guidance on such matters and interested readers can find others and for comparison.

Minimum widths of access way shall be as follows:

Vehicular access ways within units: 4.0 m

Pedestrian access ways and elevated walkway: 1.2 m

Stairways and platforms: 0.8 m

Footpaths in tanks areas: 0.6 m

Maintenance access around equipment: 1m

Fire truck access way: 6m

Minimum headroom clearance for access ways shall be as follows:

Over railways or main road: 6.8 m

Over access roads for heavy trucks: 6 m

For passage of truck: 4 m

For passage of personnel: 2.1 m

Over fork-lift truck access: 2.7 m

4.8 Concluding remarks

Many of the themes of this chapter, including heat exchange, are resumed in later chapters including that concerned with design principles. The next chapter outlines some basic chemical processes to which hydrocarbons are subjected. Physical processes (this chapter) and chemical processes (the following one), fundamentally distinct though they are, come together in process safety. A substance to be chemically processed is previously stored and conveyed to a reactor, having quite possibly been heat exchanged along the way. It is helpful to consider physical and chemical operations separately but there needs to be an awareness of their interdependence in terms of hazards.

References

[1] Lees F.P. *Loss Prevention in the Process Industries*, Second Edition. Butterworth-Heinemann, Oxford (1996)

[2] Moore W.J. *Physical Chemistry*, Fourth Edition. Longman, London (1970)

[3] *Handbook of Physics and Chemistry*, Eightieth Edition. CRC Press, New York (1999)

[4] Jones J.C. 'Dictionary of Fire Protection Engineering' 297 pp. Whittles Publishing, Caithness (2010)

[5] Jones J.C. *The Principles of Thermal Sciences and their Application to Engineering*. Whittles Publishing, Caithness, UK and CRC Press, Boca Raton, FL (2000)

[6] Jones J.C. *Combustion Science: Principles and Practice*. Millennium Books, Sydney (1993) Chapter 1

[7] Oxley C.J., Smith J.L., Rogers E., Ye W. Heat-release behaviour of fuel additives *Energy and Fuels* **15** 1194–1199 (2001)

[8] Thomas P.H. A comparison of some hot spot theories *Combustion and Flame* **9** 369–371 (1965)

[9] Linnhoff B. *et al.* 'User Guide on Process Integration for the Efficient Use of Energy' Institution of Chemical Engineers, Rugby, UK (1994)

[10] Hewitt G.F., Shires G.L., Bott T.R. *Process Heat Transfer*. CRC Press, London (1994)

[11] Gol'dshleger U.I., Pribytkov K.V., Barzykin V.V. Ignition of a condensed explosive by a hot object of finite dimensions *Combustion, Explosion and Shock Waves* **9** 99–102 (1973)

[12] http://kolmetz.com/pdf/ess/PROJECT_STANDARDS_AND_SPECIFICATIONS _layout_and_spacing_Rev1.0.pdf

Numerical examples

1. Diethyl ether has a normal boiling point 35°C (308 K) and a heat of vaporisation of $27\,kJ\,mol^{-1}$. A quantity of this substance is stored in a vessel fitted with a safety valve. The safety valve needs to be set to open when the vapour pressure of the diethyl ether reaches 4 bar ($4 \times 10^5\,N\,m^{-2}$). The valve opening is activated by the e.m.f. from a thermocouple circuit which comprises a type K thermocouple immersed in the liquid diethyl ether connected to the instrumentational equivalent of a cold junction at 0°C. At what e.m.f. must the valve be programmed to open? Use thermocouple tables. *From MSc (Safety and Reliability) Examination, University of Aberdeen, 2000.*

2. A vessel contains benzene at ordinary outdoor storage temperatures. There is an accidental fire nearby, and consequent warming of the vessel and contents. At what temperature of the contents will the vessel start to release benzene vapour if the safety valve is set to open at 7 bar. Use the data given below.

normal boiling point of benzene = 78°C

heat of vaporisation = 34 kJ mol⁻¹

From MSc (Safety and Reliability) Examination, University of Aberdeen, 1999.

3. Vinyl chloride, CH_2CHCl, is the monomer of PVC. Consider the storage of vinyl chloride as a liquid under its own vapour pressure. Its normal boiling point is – 13°C and its heat of vaporisation 20.8 kJ mol⁻¹. Calculate:

(a) The equilibrium vapour pressure at 25°C.

(b) The temperature at which, in an accidental fire, a valve fitted to the container will start to release vapour if it is set to do so at a pressure of 8 bar.

4. Hydrocarbon vapour at a refinery explodes, and in the follow-up it is reported that 'the TNT equivalence was 30 tonne'. Using a value of 4.2 MJ kg⁻¹ for the blast energy of TNT and making reasonable estimates of your own for other quantities required, calculate how much hydrocarbon must have reacted in the explosion. *From MEng/BEng Level 4 Examination, University of Aberdeen, 2001.*

5. In a continuous process, a spherical tank of diameter 0.5 m standing out of doors is used to hold a hydrocarbon liquid without refrigeration. As hydrocarbon is drawn from the tank for processing, it is replaced by a fresh supply added at the base of the tank. The tank is, at all times, close to being full. The residence time of liquid[16] in the tank is 1 minute. The tank is open to the atmosphere, so any vapour entering the space above the liquid surface is assumed to become an 'evaporative loss'.

The tank is made of stainless steel and its outside surface has, over time, become very tarnished so that it has an absorptivity of 0.8. It is proposed that, next time the tank is emptied for maintenance, the steel outside surface shall be professionally polished, in which case its absorptivity could be reduced by an order of magnitude to 0.08. Were this modification to be effected, what would be the benefit in terms of reduced atmospheric contamination in an 8-hour day time shift? Use a value of 500 kJ kg^{-1} for the heat of vaporisation. Also use the value of the solar flux given in the main text.

6. A spherical tank of m-Xylene of radius 0.5 m is initially uniformly at 15°C and is vented to the atmosphere. It is two-thirds full of the hydrocarbon liquid. The tank, which has a surface which approximates to being 'black' in the sense of that term in thermal radiation, is then placed at a site such that it receives solar flux of 700 W m^{-2}. For how long will the tank need to remain exposed to the flux in order for there to be a flash hazard if a flame enters the vapour in the tank? The boiling point of m-Xylene is 139°C, the heat of vaporisation 43 kJ mol^{-1}, the liquid density 860 kg m^{-3} and the specific heat of the liquid is 180 J mol^{-1}K^{-1}.

7. A hydrocarbon vessel stands in the open air, and is being stirred to prevent settlement of sludge. The vessel contains 4 tonne of hydrocarbon inventory of density 900 kg m^{-3}, and the outside surface area of the vessel is 15 m^2, the surface being highly emissive. The surrounding air is at 30°C, and the outside surface of the tank transfers heat to it with convection coefficient 10 W m^{-2}K^{-1}. The sludge is believed to contain some unstable chemical species such as peroxides, therefore temperature regulation is important. The stirrer is rated at 0.5 kW per m^3 of inventory. Taking the surroundings to be at a temperature of 20°C and neglecting the thermal resistance of the tank wall, what will be the steady temperature of the liquid in continuous stirring?

8. Return to the calculation in section 4.6.2. Retaining R-12 as the refrigerant but using instead of water a brine comprising a 30% aqueous solution of calcium chloride, calculate the circulation rate of refrigerant required for the same initial and final coolant temperatures and the same quantity of coolant per minute. In the temperature range of interest the brine has a specific heat of 2730 J kg^{-1}K^{-1}.

9. Nitrobenzene is made in a continuous reactor which has an influx of 10 kg of benzene per minute plus the necessary amounts of nitric acid and sulphuric acid. The reactor is required to operate isothermally, and to this end brine of heat capacity 3200 J kg^{-1}K^{-1} is supplied as a coolant. The coolant is, by refrigeration, supplied to the reactor at −3°C and must return at + 27°C the refrigerator for cooling back to −3°C. The refrigerant is R-11, trichlorofluoromethane. For safe operation, i.e., isothermal conditions as intended, determine:

the necessary reticulation rate of the brine
the evaporator and condenser pressures at the refrigerator
the necessary circulation rate of refrigerant

The heat released per kg benzene reacted is 1.8 MJ. Assume that the refrigerator acts as an ideal reversed Carnot cycle, and that there is a temperature step of 3°C between the refrigerant and the brine at the stage where they exchange heat.

10. A petroleum distillate is stored in a spherical container of 1 m diameter. Its outside surface has become quite tarnished through formation, over time, of a thin layer of metal oxide. A type K thermocouple is immersed in the liquid which the container holds and at daybreak this records a temperature of 15°C. It subsequently receives solar flux of 1000 W m^{-2} for 8 hours in the day. For the contents to reach 80°C at any time of day would cause incipient boiling which has to be avoided, consequently if the temperature of the bulk liquid rises to this value the thermocouple output activates an alarm. Determine:

(i) How long it will take for the contents of a vessel which is four fifths full to reach 80°C[17]. State any assumption made in your calculation. Use a value for the specific heat of the liquid of 2000 J kg^{-1}K^{-1}. Make an estimate of your own for the liquid density.

(ii) To what value the absorptivity of the vessel surface must be reduced, by painting white or by polishing, to ensure that the contents do not reach 80°C during daylight hours.

(iii) The minimum voltage in the thermocouple measurement circuit to which the alarm must be set to respond. Take it that the circuit contains an 'instrumentational cold junction' at 0°C and use thermocouple tables.

11. A proprietary cetane enhancer has the following properties (symbols as in the main text):

$$A = 2.12 \times 10^{11} \text{s}^{-1}$$
$$E = 198 \text{ kJ mol}^{-1}$$
$$Q = 2.1 \text{ MJ kg}^{-1}$$
$$k = 0.3 \text{ W m}^{-1}\text{K}^{-1}$$
$$\sigma = 1380 \text{ kg m}^{-3}$$

If the material is to be stored in a full spherical container, calculate the radius of the container which would be required to hold a quantity which would be thermally critical at 323 K.

12. For the cetane enhancer details of which are given in question 11, at what temperature would a full spherical container containing 100 tonne become thermally critical?

13. The formulation for criticality due to an inert hot spot is given in the box below having been taken from [10]:
A spontaneously unstable organic substance has the following properties:

$$E = 125\,kJ\,mol^{-1} \quad k = 0.027\,W\,m^{-1}K^{-1}$$
$$A = 2 \times 10^{11}s^{-1} \quad Q = 2.2\,MJ\,kg^{-1}$$
$$\text{specific heat } (c)\; 1860\,J\,kg^{-1}K^{-1} \quad \sigma = 900\,kg\,m^{-3}$$

This substance is being stored at 273 K. An electrical fire starts nearby and causes a blob of alloy at its melting point of 1200°C (1473 K) to find its way into the substance. At that temperature the density of the alloy is 7500 kg m⁻³, its specific heat 600 J kg⁻¹K⁻¹ and its thermal conductivity 25 W m⁻¹K⁻¹. Assuming spherical geometry, how large will the blob need to be to cause the organic substance to explode? Having calculated the size of the inert hot spot at 1473 K required to cause explosion, calculate the temperature which a reactive hot spot of the same size would need to be in order to cause explosion.

14. (a) A pump transfers 100 m3 per minute of gasoline from a refinery exit to a terminal. What power of pump will be required? The pressure is 2 bar. Assign a reasonable value of your own to the density of the gasoline.
(b) If the pipe which conveys the gasoline in part (a) is 0.5 m in diameter what is the Reynolds number?
(c) The gasoline enters the pipe at 60°C, the surrounding temperature being 10°C. If the gasoline is required to cool to 50°C as it flows along the pipe, what will be the required pipe length? For convection heat transfer from the pipe use the correlation:

$$Nu = 0.023Re^{0.8}Pr^{0.3}$$

where Nu = Nusselt number = hd/k, where h is the convection coefficient (W m⁻²K⁻¹), d the diameter and k the thermal conductivity of the fluid, Re is the Reynolds number and Pr the Prandtl number. Values for k and Pr are 0.14 W m⁻¹K⁻¹ and 10 respectively. Use a value of 2000 J kg⁻¹K⁻¹ for the heat capacity of the gasoline.

15. Water can be used as a refrigerant, and is coded R718. Imagine water as a refrigerant operating between 40°C and 5°C. Determine the evaporator pressure, the compressor pressure, the coefficient of performance and the refrigerating effect. It will be necessary to consult tables of the properties of water.

16. At the Amuay refinery accident described in the main text the amount of hydrocarbon inventory present was about 0.6 million barrels. In the hypothetical limit that this entire amount exploded what would have been the TNT equivalence?

17. The manufacturer of gas cylinders referred to in section 4.2.1 also makes available hydrogen (calorific value 143 MJ kg^{-1}) in cylinders of 146 cm height and 23 cm diameter. If the pressure was 200 bar, what would be the fire load due to such a cylinder?

18. It was reported in 2010 that Samsung had built a tank capable of holding 40 000 tonnes of LPG. How far apart would two such tanks need to be in order to comply with the IP code given in the main text? Take the tanks to be spherical and 90% full when holding the nameplate quantity of LPG. Use a value of 550 kg m^{-3} for the density.

19. A Land Rover, width 1.8 m, can be adapted into a small fire fighting vehicle. What is the maximum amount of LPG which can be held in adjacent spherical tanks of the same diameter for a separation of 2 m, which would permit passage of the Land Rover between the tanks with a 10 cm margin of safety at each side.

20. Determine from the chart above the minimum distance between each of the following pairs for conformity with standard BN-DS-C69.

A fractionation tower and a low-pressure storage tank
A hydrocarbon compressor and a heat exchanger 'above ignition temperature'.
A pipe way and a reactor 'below ignition temperature'.

21. According to BN-DS-C69:
How far would a flare need to be from a fractionation tower?
How far would a fire hydrant need to be from a reactor below ignition temperature?
How far would a process pump handling LPG need to be from a flare?
How far would a hydrocarbon compressor need to be from the main plant roads?
How far would equipment handling non-flammables need to be from open flame equipment?

Go on to apply the chart to open flame equipment and property boundary.

Formulation for an inert hot spot in an ignitable medium [11]

$$\sqrt{\delta^*_{crit}} = 0.4\sqrt{b^2 + 0.25j(j+1)(b+0.1b^3)}\ [\theta_i + 2.25(j-1)]^2[1+0.5\varepsilon\theta_i]$$

$$\sqrt{\delta_{crit}} = \sqrt{\delta^*_{crit}} \times \left[1 + \frac{(\theta_i - 3)^2 b(j+1)}{30k_r^{2/3}\,(1+3b^{2/3})}\right]$$

$$b = \frac{(\text{heat capacity} \times \text{density})_{\text{reacting medium}}}{(\text{heat capacity} \times \text{density})_{\text{hot spot}}}$$

$\varepsilon = RT_i/E$ where $T_i =$ hot spot temperature

$$k_r = \frac{\text{thermal conductivity of the hot spot material}}{\text{thermal conductivity of the reacting material}}$$

$j =$ shape factor $= 2.0$ for a spherical hot spot and, as for a reactive hot spot:

$$\delta = \frac{QA\sigma Er^2\,\exp(-E/RT_i)}{RT_i^2 k}$$

$$\theta_i = (E/T_i^2)\{T_i - T_A\}$$

Endnotes

[14] Note that the hazard of hot spots in spontaneously unstable substances is recognised in approved pumping procedures for such substances. Pumping methods, possibly requiring particular designs of pump, are prescribed which will ensure that conditions remain sufficiently cool to prevent hot spot formation.

[15] Aspects of this are dealt with in later chapters, in particular when recuperative burning of waste hydrocarbon is covered in Chapter 11.

[16] Residence time defined as: {(volume of tank × density of liquid)/mass flow rate in and out of tank}

[17] Note that a large rather than small quantity of hydrocarbon, rather counterintuitively, provides for safety in that it raises the time for hazardous temperatures to be reached. The greater risk with partly filled tanks is in fact recognised in the Rijnmond Report (see Appendix). In this report, for propylene, ammonia (both flammable) and chlorine the frequency of catastrophic failure of a partly filled vessel is estimated in each case as being about an order of magnitude higher than the frequency of catastrophic failure of a full vessel.

5

CHEMICAL OPERATIONS ON HYDROCARBONS AND HYDROCARBON DERIVATIVES

5.1 Introduction

Starting with cracking, this chapter will consider important chemical operations on hydrocarbons and hazards associated with them. The text by List [1] has been drawn on considerably.

5.2 Cracking and hydrocracking

Cracking is the way in which molecular weight adjustment of the hydrocarbons in a petroleum fraction is achieved. It may be thermal only, or it may involve a catalyst. Ethylene ($H_2C=CH_2$) is a major cracking product. Further products can be made from ethylene and other alkenes by addition reactions. Precautions taken at an ethylene plant include elimination of ignition sources in places where the risk of leak is greatest (such sources include electrical switching and 'hot work'). There will also be the facility to isolate hydrocarbon inventory in the event of a fire elsewhere in the plant.

The numerical example in the shaded area below appertains to cracking.

In a cracking plant hydrocarbon liquid exits the process at 250°C and enters a cylindrical container of height 10 m and radius 1.5 m which it fills. Inside the cylinder the contents are stirred, and they must cool to below 90°C before being passed along for subsequent chemical processing. The heat capacity of the oil is 2000 J kg^{-1}K^{-1} and convection with coefficient 150 W m^{-2}K^{-1} takes place at the curved and flat outside surfaces of the cylinder by reason of light sprinkling with coolant. The surrounding temperature is 298 K. The cylinder is made of metal and its thermal resistance is negligible in comparison with that of the oil. How long should be allowed for the cooling? Use a value of 850 kg m^{-3} for the density of the oil.

Solution

The significance of the stirring is that the oil has effectively infinite thermal conductivity therefore, at any one time, its temperature profile is flat. Heat balance on the cylinder:

heat transferred from the surface = heat lost by the contents by convection

$$hA(T_o - T(t)) = c\rho V dT(t)/dt$$

where h = convection coefficient ($W\,m^{-2}K^{-1}$), A = area (m^2), T_o = surrounding temperature (K), $T(t)$ = temperature of the oil at time t, c = heat capacity of the oil ($J\,kg^{-1}K^{-1}$), ρ = density of the oil ($kg\,m^{-3}$), V = volume of the oil (m^3).

Rearranging:

$$\frac{dT(t)}{dt} = [hA/c\rho V]\,(T_o - T(t))$$

$$\int_{T_i}^{T} dT(t)/(T_o - T(t)) = [hA/c\rho V]\int_0^t dt$$

where T_i is the initial temperature of the oil. Integrating:

$$-\ln\frac{(T_o - T(t))}{(T_o - T_i)} = [hA/c\rho V]t$$

Now for the cylinder $V = \pi r^2 L$ where L is the height in m, and

$$A = 2\pi r(L + r)$$

$$\Downarrow$$

$$A/V = 2(L + r)/rL = 1.53\,m^{-1}$$

$$[hA/c\rho V] = 1.35 \times 10^{-4}\,s^{-1}$$

put $T(t) = 363\,K$, $T_o = 298\,K$, $T_i = 523\,K$

$$\Downarrow$$

$$t = 9197\,s \text{ (about 2.5 hours)}$$

In the context of the above example some principles of heat transfer have been introduced. It is worth reiterating the point made above that the stirring enables the assumption to be made that at all times the vessel contents have a flat temperature profile. Had this not been so, there would have been a need to consider both conduction and natural convection and a temperature gradient between the cylinder walls and centre.

Amongst recorded serious accidents during cracking processes is one in the Netherlands in 1975. Escape of hydrocarbon vapour led to a v.c.e., and consequent fires in storage areas and pipelines. There were 14 fatalities and 107 injuries.

Ethylene and propylene, obtained either by cracking of naphtha or by dehydrogenation (of ethane and propane to produce ethylene and propylene respectively), are the petrochemical industry's primary building blocks, and production of these chemicals is massive. In North America, ethylene and propylene facilities tend to be concentrated in two regions: the Gulf Coast and Alberta. A single ethylene plant in Alberta, which began production in 2000, produces 1.3 million tonne of ethylene per year. The current situation is that dehydrogenation accounts for more than half of the North American ethylene production whereas in Europe naphtha cracking is prevalent.

In circa 1915 a US patent for cracking heavier petroleum material to bring it within the boiling range of gasoline was granted. Ten to fifteen years later cracking processes to make alkenes for petrochemical manufacture came into being and involved catalysts. Cracking continues to advance, and a major cracking process of recent introduction is fluid catalytic cracking – FCC – of higher boiling material to make it suitable for blending with gasoline, an improvement on the 'reforming' processes which had previously been used to this end. Most current R&D into cracking is catalyst focused.

The formula of a heavy petroleum product – higher boiling distillate or residue – approximates to C_nH_{2n} and this is often taken to be true of crude oil itself. An alkane has formula C_nH_{2n+2}. Cracking to produce say ethane from a heavy feedstock requires incorporation of some hydrogen gas in which case the process is called hydrocracking. The discussion continues in the shaded area below.

The chemical equation for the process summarised in the paragraph above is:

$$C_nH_{2n} + (n/2)H_2 \rightarrow (n/2)C_2H_6$$

per kilogram of the material cracked:

$$(12/14) \text{ kg of carbon} = 857 \text{ g} = 71 \text{ mol}$$
$$(2/14) \text{ kg of hydrogen} = 143 \text{ g } 71 \text{ mol (expressed as } H_2)$$

Hydrogen requirement 71 mol per kg of material cracked, $(71/40)$ m^3 at 15°C, 1 bar = 1.8 m^3

Considering now hydrocracking to make propylene (strictly propene according to IUPAC nomenclature).

$$3C_nH_{2n} + nH_2 \rightarrow nC_3H_8$$

As before, per kilogram of the material cracked:

(12/14) kg of C = 857 g = 71 mol requiring 71/3 mol of hydrogen = 24 mol hydrogen 0.6 m^3

When crude oil is hydrocracked, products include diesel as well as the simple organics discussed. At the present time there is significant activity in hydrocracking crude oil to make diesel and other fractions as an alternative to refining. One example amongst very many is the hydrocracker at the Repsol refinery in Tarragona, Spain. It converts 25 000 barrels of crude oil per day to the equivalent of distillates.
Returning to the process:

$$C_nH_{2n} + (n/2)H_2 \rightarrow (n/2)C_2H_6$$

The benchmark compound for diesel performance is cetane $C_{16}H_{34}$. A simplified equation for the conversion from crude oil to diesel would be:

$$C_nH_{2n} + (n/16)H_2 \rightarrow (n/16)C_{16}H_{34}$$

Hydrogen generation at a hydrocracker is by steam reforming of hydrocarbon. The presence of hydrogen at a hydrocracker makes for hazards because of the extremely high reactivity towards combustion of hydrogen. There was an explosion at a hydrocracker in 1997 which caused one death and 46 injuries [2].

5.3 Hydrodesulphurisation and hydrodenitrogenation

When crude oil is refined, sulphur is not distributed equally between the products but is concentrated in the heavier fractions. There is therefore a tendency, when the sulphur content of the crude is at all high, for the residual material to have a sulphur content so high as to preclude its use as a fuel oil unless it is desulphurised.

Such a fuel oil is 2% sulphur, and is to be burned in an area where the local emission standard is 1 kg of sulphur dioxide per 2 GJ of heat. To what degree will the fuel oil need to be desulphurised? What quantity of hydrogen will be needed per tonne of the fuel oil? Use a value of 42 MJ kg^{-1} for the calorific value and express the answer in cubic metres of hydrogen measured at 1 bar pressure and 15°C

$$2GJ = 2 \times 10^9 \text{ J obtainable from } (2 \times 10^9)/(42 \times 10^6) \text{ kg of the fuel oil}$$
$$= 48 \text{ kg of the fuel oil containing 1 kg of sulphur} \rightarrow 2 \text{ kg SO}_2$$
and the following points should be noted.

• The sulphur in any fuel is converted on combustion in quantitative yield to sulphur dioxide.
• The molar masses of elemental sulphur and of sulphur dioxide are respectively 32 and 64, and that is the origin of the factor of two in the working above.

Clearly, to meet the emission standard the sulphur content will have to be halved. This involves:

$$\Phi\text{-SH} + H_2 \rightarrow \Phi H + H_2S$$

where Φ denotes the organic structure to which the sulphur is bonded, analogously to a simple mercaptan.

Per kg of fuel there are 20 g sulphur 10 g of which need removal or 0.31 mol requiring 0.31 mol of hydrogen. Per tonne fuel 310 mol of hydrogen are needed, roughly 8 m^3.

Hydrodesulphurisation is continued in the appended numerical examples. Sulphur in natural gas is present as hydrogen sulphide, making the gas 'sour', and can be removed by 'hydrogen sulphide scavenging' with an aqueous reagent or by a catalyst. Amines are commonly used, and the process is an acid-base reaction:

$$2R_3N + H_2S \rightarrow [(R_3NH)^+]2S^{2-}$$

where R denotes alkyl groups which, contrary to usual conventions in organic formulae, are not necessarily all the same and where R might denote a substituted alkyl (see numerical examples). Hydrogen sulphide can also be removed from natural gas by an adsorbent carbon, and here again quantitative treatment is deferred until the numerical examples.

Hydrodenitrogenation is analogous to hydrodesulphurisation:

$$\phi N + 2H_2 \rightarrow \phi H + NH_3$$

Removal of nitrogen by this means is usually from refined fractions rather than from crude oil and is necessitated not only by the need to control NO_x but also by the tendency of nitrogen-containing compounds in such a fraction to contaminate catalysts in subsequent processing. For this reason it is sometimes necessary to take the nitrogen content down to as low as 1 p.p.m. There is always a catalyst involved. A related numerical example follows.

A petroleum fraction of 0.5% nitrogen and density $800\,kg\,m^{-3}$ needs to be reduced in nitrogen content to 5 p.p.m. Per barrel of the fraction how much hydrogen is needed for this process?

$$1 \text{ barrel } = 0.159\,m^3 \text{ equivalent to } (0.159 \times 800)\,kg = 127\,kg$$

$$\text{Weight of nitrogen} = (127 \times 0.5/100)\,kg = 0.636\,kg$$

Weight of nitrogen in the denitrogenated product = $(127 \times 5 \times 10^{-6})\,kg = 0.0006\,kg$, negligible in comparison with the initial amount

Mol nitrogen needing removal per barrel = $(0.636/0.014) = 45$ expressed as atomic N
$$\text{moles hydrogen required} = 90 \text{ or } 2.3\,m^3$$

The importance of the two processes covered in this section to safety is that they both involve hydrogen gas. The appalling explosion at the Amuay in Venezuela in 2012 was reported in the Chapter 4. Only just over a year earlier there had been a fire at the hydrodesulphurisation unit there.

5.4 Partial oxidation

Examples of partial oxidation of hydrocarbons of importance in the manu-facturing industries include:

butane → acetic acid etc.
↓
e.g., vinyl acetate
used in adhesives and paints

toluene → benzoic acid
↓
phenol
used in synthesis of resins, adhesives and fibres

> xylene → terephthalic acid
> *raw material in the manufacture of polyesters*

> cyclohexane → cyclohexanone → adipic acid
> *Flixborough!*

Such oxidations may be vapour phase or liquid phase. Catalysts are often involved. Vapour phase reactors have the advantage that, because of the lower density of vapours, they contain a smaller amount of hydrocarbon inventory and the fire hazard in the event of reactor failure is therefore less.

Partial oxidations are exothermic, often powerfully so, e.g., for the process:

> naphthalene $\overset{\text{air}}{\rightarrow}$ phthalic anhydride
> *400°C, vanadium catalyst*

12 MJ of heat are released per kg naphthalene reacted. There is therefore an ignition hazard within the reactor in vapour-phase oxidations unless the mixture is kept outside the flammable range and the temperature is kept relatively low. Low temperatures may in any case be necessary to minimise side reactions.

Sometimes the extent of conversion per reactor pass is small, and there have to be several passes to obtain a satisfactory yield of product. This limits the rate of heat release and prevents thermal runaway. Also, for vapour-phase reactions fluidised beds can be used and the fluidising solid takes up some of the heat which might otherwise have caused thermal runaway. In liquid-phase reactions, the space above the liquid becomes occupied by vapour, and ignition is possible if oxygen enters this. There are also such risks in multiple pass procedures. Ingress of liquid into air pipes can result in escape of the contents and ignition. In liquid or vapour-phase oxidation, highly reactive species such as peroxides are frequently present as by-products, and these are a severe explosion hazard.

The following worked example relates to oxidation in the vapour phase, and is also used as a context within which to illustrate what is an important issue in process safety and indeed of other disciplines including fuel technology: the possible unreliability of temperature readings because of radiation effects.

In a vapour-phase hydrocarbon oxidation, with air as the oxidant, a thermocouple is installed within the reacting mass, which is in an enclosed metal container. The thermocouple is wired into circuitry to initiate emergency responses when necessary. These are set to operate when the 'reacting temperature' is 280°C. During one particular operation the thermocouple in the vapour reads 270°C. A thermocouple on the inside wall reads 265°C. In the vapour, heat is transferred to the thermocouple tip by convection with a coefficient of $10\,W\,m^{-2}\,K^{-1}$. What is the true temperature of the vapour? Make the assumption that the hydrocarbon vapour is sufficiently dilute to be transparent to thermal radiation and use a value of 0.95 for the emissivity of the thermocouple tip. Comment on your answer.

Solution
Heat balance at the t.c. tip:

$$\text{heat transferred to the tip} = \text{heat transferred from the}$$
$$\text{by convection} \qquad \text{tip by radiation}$$

$$hA\,(T_g - T_t) = \varepsilon\sigma A(T_t^4 - T_w^4)$$

h = convection coefficient $(W\,m^{-2}K^{-1})$, A = tip area (m^2), T_t = tip temperature (K), T_g = gas temperature (K), T_w = wall temperature (K), ε = emissivity, σ = Stefan's constant = $5.7 \times 10^{-8}\,W\,m^{-2}K^{-4}$

Rearranging:
$$T_g = (\varepsilon\sigma/h)\{T_t^4 - T_w^4\} + T_t$$
giving:
$$T_g = 560\,K\ (287°C)$$

i.e., the temperature is in excess of the safety threshold but the t.c. does not detect this!

In the above calculation, the heat transfer to the tip of the thermocouple from the reacting vapour is, in steady or 'quasi-steady' operation of the reactor, balanced by heat transfer by radiation from thermocouple tip to the cooler enclosure wall. The resultant temperature, the 'thermocouple reading', is significantly below the gas-phase temperature. In this case, the temperature is actually in excess of that at which the valve is set to operate but the thermocouple, the e.m.f. from which controls this, has not recognised this. Temperature readings in the gas phase, even at close to room temperature, always need to be assessed for possible radiation errors.

As recorded earlier in this section, the process occurring at Flixborough when the 1974 accident happened was partial oxidation, actually cyclohexane to cyclohexanone. At Antwerp in 1989, there was an explosion at a plant manufacturing ethylene oxide (C_2H_4O). Though there were no fatalities, five employees suffered minor injuries.

5.5 Chlorination

Chlorination of hydrocarbons can be either liquid- or vapour-phase. Examples of chlorinations of importance are:

$$methane \rightarrow chloromethanes$$

$$ethylene \rightarrow ethylene\ dichloride$$

$$ethylene \rightarrow vinyl\ chloride\ (\rightarrow PVC)$$

Chlorine is an oxidising agent, and chlorinations in some ways resemble partial combustion. They are quite strongly exothermic, e.g., for the process:

$$CH_4 + Cl_2 \rightarrow CH_3Cl + HCl$$

1.5 MJ of heat are released per kg methane reacted. Heat removal and temperature control can therefore be a difficulty. Also, hydrocarbons will burn totally in chlorine if conditions are suitable (see following paragraph).

Precautions followed in vapour-phase chlorinations include admittance of the chlorine in stages, not all at once, and incorporation of a diluent into the reacting mixture to control the temperature. Hydrocarbon/chlorine proportions must be kept outside the flammable range, which are usually similar to the flammability limits of the same hydrocarbon in oxygen. A related numerical example follows.

Dichloromethane (CH_2Cl_2) is made from methanol by the vapour-phase process:

$$CH_3OH + Cl_2 \rightarrow CH_2Cl_2 + H_2O$$

for which the heat of reaction is 127 kJ mol^{-1} reacted (4.0 MJ per kg methanol reacted). There is a safety requirement that the temperature does not, during reaction, rise by more than 700 K. This is ensured by incorporating some nitrogen. Assuming that the reaction goes to completion and neglecting heat losses to the outside (i.e., taking the reactor to operate adiabatically), determine how much nitrogen must be incorporated per kg methanol reacted.

Specific heats/$J\,K^{-1}\,mol^{-1}$:

$$H_2O\ (g) = 34$$

$$CH_2Cl_2\ (g) = 51$$

$$N_2\ (g) = 29$$

Solution

If the reaction goes to completion, a mole (0.032 kg) of methanol reacted releases 127 kJ, which, under adiabatic conditions, is all supplied to a gas mixture of total heat capacity C_{tot}, given by:

$$C_{tot} = C\{CH_2Cl_2\} + C\{H_2O\} + \phi C\{N_2\}$$

where the Cs denote the specific heats of the chemical species present and ϕ is the moles of nitrogen incorporated per mole methanol reacted.

$$\Downarrow$$

$$C_{tot} = (85 + 29\phi)\ J\,K^{-1}$$

Since the maximum allowable temperature rise is 700 K:

$$700\,K = [127000\,J/(85 + 29\phi)\ J\,K^{-1}]$$

$$\Downarrow$$

$$\phi = 3.3$$

Therefore per mole (0.032 kg) of methanol reacted 3.3 mol (0.093 kg) of nitrogen are required, or 2.9 kg of nitrogen per kg methanol.

The calculation has erred on the safe side in the following three ways:

• Its assumption of complete reaction
• Its assumption of adiabatic conditions
• The use of heat capacities at room temperature, which the above values are. The truly correct value over the temperature range will be somewhat higher.

In addition to thermal hazards, chlorine leakage is also a major hazard with chlorination processes. Chlorine leaks can easily result in deaths: 100 p.p.m. for 10 minutes can be fatal. Not all industrial chlorine leaks are associated with the petrochemical industry. The case study below is however a petrochemical example, and is reported more fully in Lees *op. cit.*

> Batch reactor for an organic chlorination process
>
> *Failure of a thermocouple in the control circuitry*
>
> ↓
>
> Reactor shut off, cooling system turned off, explosion from the residual chlorine/hydrocarbon mixture in the reactor during attempted repair
>
> Fatalities from the explosion, and rupture of the chlorine line resulting in its leakage

The toxic hazards of elemental chlorine are discussed more fully in Chapter 10. Other halogens can be reacted with hydrocarbons, for example bromobenzene C_6H_5Br is an important intermediate in the organic chemistry industry, notably in manufacture of the anaesthetic phencyclidine. It is shown in one of the appended numerical examples that the exothermicity of the manufacture of this is small without being negligible in terms of hazards.

Signage for 'active' chemical plants, as opposed to 'passive' storage (Chapter 4) have been designed by the European Chemical Bureau, and the one for toxic, shown in Figure 5.1, would apply where chlorine is in use. Anticipating section 5.8, one or other of the signs for a corrosive chemical hazard would be expected at a nitration plant.

5.6 Gasification

Gasification is conversion of solid or liquid starting product to fuel gas or synthesis gas, as shown in the box below. Gasification of coal and coke, to make either a fuel gas or a carbon monoxide/hydrogen blend for further chemical processing, is a nineteenth-century technology, though by no means obsolete in the twenty-first. Gasification processes often involve very high pressures, with concomitant risk of loss of containment. Where, in reaction with steam, the feedstock is already a gas (e.g., natural gas), the term 'steam reforming' applies. Our concern in this book is with gasification of hydrocarbon liquids with steam. It is because the diesel fraction of crude oil has found particular application in gasification that diesel is often called 'gas oil'. Also, gas made from coal is sometimes supplemented with hydrocarbons made by racking low-value refinery products.

Petroleum distillate was being used in the manufacture of city gas in Sheffield, UK, in 1973, when three people lost their lives and 29 were injured [3]. There was damage over a half-kilometre radius. Flame-cutting operations were taking place on a tank previously believed to be empty, whereas there was some distillate present and it was above its flash point.

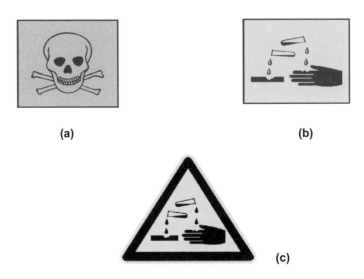

(a) (b)

(c)

Figure 5.1 *(a) sign for toxic chemicals. (b) and (c) signs for corrosive chemicals.*

Gasification provides a means of utilisation of low-value hydrocarbon products including refinery residues and cracking by-products. This is discussed more fully in Chapter 11.

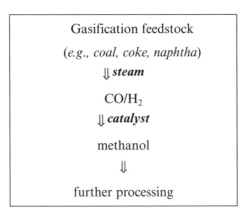

Gasification feedstock

(*e.g., coal, coke, naphtha*)

⇓ *steam*

CO/H_2

⇓ *catalyst*

methanol

⇓

further processing

5.7 Hydrogenation

5.7.1 Introduction

A chemical engineer or chemical technologist on encountering the term 'hydrogenation' is likely to think first of all of processes for the conversion of solid fuels to liquid, e.g., lignite (brown coal) to aviation fuel. A good deal of the aviation fuel used by Germany during World War II was produced this way, and hydrogenation of solid fuels continues to be an important area of research and development in fuels processing. The hydrogen is often supplied not as hydrogen gas but from a hydrogen-donating solvent; tetralin is perhaps the best known. Particular 'donor solvents' need to be matched to particular coals, and a fruitful approach is an adaptation of the thermodynamics of monomer–polymer interactions, the solvent being the monomer and the coal the 'polymer'. All of this is discussed a little more fully in the final chapter of the book, which deals with ways of obtaining hydrocarbons other than from crude oil.

In the context of hydrocarbon technology, hydrogenation means addition of hydrogen, supplied as hydrogen gas, to organic structures, for example, the conversion of an alkene to an alkane.

5.7.2 Process details

Consider the hydrogenation of benzene to cyclohexane:

$$C_6H_6(\text{vapour}) + 3H_2 \rightarrow C_6H_{12}(\text{vapour})$$

The enthalpy change for the reaction as written is $-206\,\text{kJ}\,\text{mol}^{-1}$. Hydrogenations of hydrocarbons are usually exothermic and are classified as such for hazard assessment purposes. Hydrogenation requires high pressures (up to several hundred bar) and a catalyst [1]. There are therefore hazards due to the high pressures, as there are with gasification. There are some obvious similarities between gasification and hydrogenation.

During the course of a hydrogenation reaction the pressure will change. This is illustrated in the calculation in the shaded area below.

Ethylene and hydrogen, comprising an equimolar mixture at 25°C, are fed into a rigid reactor for hydrogenation, and the initial pressure in the reactor is 30 bar. The reaction proceeds to 85% completion, at which stage the temperature is 305°C. By how much does the pressure rise as a result of the reaction?

Solution

Calling the amount of reactant initially admitted z mol, the stoichiometry is:

$$\begin{array}{ccccc} C_2H_4 & + & H_2 & \rightarrow & C_2H_6 \\ 0.5z(1-\alpha) & & 0.5z(1-\alpha) & & 0.5\alpha z \end{array}$$

where α is the fractional extent of reaction (=0.85)

initially	after reacting
$0.5z$ mol C_2H_4	$0.5z(1-\alpha)$ mol C_2H_4
$0.5z$ mol H_2	$0.5z(1-\alpha)$ mol H_2
no C_2H_6	$0.5\alpha z$ mol C_2H_6
total z mol	total $z(1-0.5\alpha)$ mol

Applying the ideal gas equation to the initial and final states (assigned subscripts 1 and 2 respectively):

$$P_1V_1 = n_1RT_1 \quad \text{and} \quad P_2V_2 = n_2RT_2$$

Now since the vessel is 'rigid' $V_1 = V_2$, also:

$$n_1 = z$$

$$n_2 = z(1-0.5\alpha)$$

$$\Downarrow$$

$$P_2 = P_1 \times (1-0.5\alpha)(T_2/T_1) = 33.45\,\text{bar}$$

That is, the pressure rises by 3.45 bar

Accidents during hydrogenation include one in Canada in 1987, where there was loss of containment and a flash fire. Though damage was very extensive, there were no deaths or injuries.

5.8 Nitration

In the manufacture of substances such as TNT, nitroglycerine and nitrocellulose, nitration of the starting material is usually in the liquid phase with nitric acid the nitrating agent, with some sulphuric acid as a dehydrating agent. The reaction is powerfully exothermic and both the intermediate products in the reactor and the end products are unstable and capable of very powerful explosion. Moreover, nitric acid is extremely corrosive and dangerous if leaked. On the basis of accident records, nitration can be viewed as the most hazardous process in the entire chemical industry (Lees, *op. cit.*). Another common nitration process is the production of nitrobenzene (boiling point 210°C), an important starting material for dye manufacture. A related numerical example follows.

Nitrobenzene ($C_6H_5NO_2$) is manufactured from benzene and nitric acid in the liquid phase according to:

$$C_6H_6 + HNO_3 \rightarrow C_6H_5NO_2 + H_2O$$

for which the heat of reaction is $-144\,kJ\,mol^{-1}$. Some plants for this process use cooling coils, with a suitable brine as the cooling liquid, whilst others operate close to adiabatically by including in the reactant mixture some water to absorb the heat. Good agitation is essential to ensure heat transfer between the aqueous and non-aqueous phases.

Imagine that, in close-to-adiabatic manufacture, the reactor has inventory such that, with 100% yield, one tonne of nitrobenzene will be produced. The nitrating agent is supplied as 35% nitric acid (balance water). Sulphuric acid (10%) is supplied in a quantity equal to that of the nitric acid solution. What will be the temperature rise? Specific heat of nitrobenzene $= 2000\,J\,kg^{-1}\,K^{-1}$.

Solution
The exothermicity of $144\,kJ\,mol^{-1}$ converts to $1.8\,MJ$ per kg benzene nitrated.

Production of one tonne of nitrobenzene requires (78/123) tonne $\equiv$ 634 kg of benzene. The heat released will be $1.14\,GJ$.

Now 634 kg benzene $\equiv$ 8128 mol, requiring 8128 mol of nitric acid, yielding 8128 mol each of nitrobenzene and water.

8128 mol of nitric acid $\equiv$ 512 kg. If it is supplied as a 35% solution, the total weight of solution is 1462 kg, there being 950 kg water. 1462 kg of 10% sulphuric acid are therefore also supplied, there being 146 kg of the acid and 1316 kg water.

Hence the final composition of the reacting mixture is:

nitrobenzene 1000 kg

> product water 146 kg
> diluent water from HNO_3 950 kg
> sulphuric acid 146 kg
> diluent water from H_2SO_4 1316 kg

$\Downarrow$

A 6% solution of sulphuric acid, total quantity 2558 kg, specific heat (Perry, *op. cit.*) $3950\,J\,kg^{-1}°C^{-1}$

Heat capacity $= [(1000 \times 2000) + (2558 \times 3950)] = 1.2 \times 10^7\,J\,K^{-1}$

Temperature rise $= (1.14 \times 10^9\,J/1.2 \times 10^7\,J\,K^{-1}) = 94°C$

5.9 Polymerisation

Hazards due to the polymerisation process itself have to be distinguished from those at polymerisation plants which are related simply to the presence of large amounts of hydrocarbon and possible loss of containment. A fatal accident at a polymerisation plant in Houston in 1989 occurred during maintenance, not during operation. Figure 5.2 shows a polymerisation plant in Geelong, Australia. It is used to manufacture polypropylene.

Polymerisation is significantly exothermic, necessitating cooling during operation of a polymerisation process. For example, the polymerisation of vinyl chloride[18] is exothermic to the extent of 1.5 to 1.6 MJ kg^{-1} [4]. PVC is often manufactured from a suspension of vinyl chloride monomer (VCM) in

Figure 5.2 *Polypropylene plant. Reproduced courtesy of Shell International.*

water, typically 1 part by weight VCM to 1.4 parts by weight water, with other minor constituents (buffer, inhibitor, 'protective colloid') in very small amounts. The reagents are admitted to an autoclave and pressurised to several bar. Cooling is by a jacket of (possibly refrigerated) water around the autoclave.

An alternative to using a jacket of cooling water in suspension polymerisation of vinyl chloride is for the unreacted VCM, during the manufacture, to be diverted several times to a heat exchanger (Burgess, *op. cit.*). Unreacted VCM, having absorbed the heat of polymerisation liberated, will thereby evaporate and can be made to condense in the heat exchanger.

The numerical example below relates to PVC production in an autoclave.

In an autoclave operating at 80°C and initially containing 3 tonne of VCM there is continuous passage of the unreacted monomer vapour, at 80°C and its saturated vapour pressure at that temperature, through a shell-and-tube heat exchanger. The other fluid in the exchanger is water, which enters at 25°C and exits at 60°C. The autoclave conditions are maintained for 5 hours, and the polymerisation goes to 90% completion.

Determine:
• At what rate the VCM must be recirculated through the heat exchanger.
• At what rate the water must flow.
• The required area of the heat exchanger, if the heat transfer coefficient at the heat exchanger (symbol **U**) is $200 \, W \, m^{-2} \, K^{-1}$.

The polymerisation has a heat of reaction of $1.55 \, MJ \, kg^{-1}$ and the heat of vaporisation of VCM at 80°C is $279 \, kJ \, kg^{-1}$. Use a value of $4180 \, J \, kg^{-1} \, K^{-1}$ for the specific heat of water.

Solution

Quantity of heat to be removed over the five-hour period
$$= 0.9 \times 3000 \times 1.55 \, MJ = 4185 \, MJ$$

rate of heat exchange $= [(4185 \times 10^6)/(5 \times 3600)] \, W$
$$= 232500 \, W \, (232.5 \, kW)$$

Total amount of VCM to be condensed $= (4185000/279) \, kg = 15000 \, kg$ (15 tonne), i.e., equivalent to 5 recirculations of the total initial quantity.

flow rate of water $= [232500/(4180 \times 35)] = 1.6 \, kg \, s^{-1}$

Doing a heat balance on the water and VCM and calling the recirculation rate of the latter $y \, kg \, s^{-1}$:

$$1.6 \times 4180 \times (60 - 25) = y \times 279000 \implies y = 0.84 \, kg \, s^{-1}$$

Check: $0.84\,kg\,s^{-1}$ for 5 hours = $15000\,kg$, equivalent to 5 recirculations of the initial amount as previously stated.

Since the VCM maintains phase equilibrium in the heat exchanger its temperature will remain at 80°C. The log-mean temperature difference[19] (ΔT_m) is therefore:

$$\Delta T_m = \frac{(80-25)-(80-60)}{\ln\{(80-25)/(80-60)\}} = 34.6°C$$

The rate of removal of heat $q = (232.5\,kW)$ is then:

$$q = UA\,\Delta T_m = 200 \times A \times 34.6\,W = 232500\,W \Rightarrow A = 33.5\,m^2$$

The polymerisation of ethylene to form polyethylene is more exothermic still, about $4\,MJ$ per kg polymer formed.

5.10 Alkylation

Alkylation in the hydrocarbon industry is the process whereby compounds are made more suitable for inclusion in gasoline, in terms of resistance to engine knock. An example is the alkylation of 1-butene with isobutane:

$$H_2C=CH\text{-}C_2H_5 \quad + CH(CH_3)_3 \quad \xrightarrow[\text{acid catalyst}]{} \quad \begin{matrix} C(CH_3)_3 \\ | \\ CH_3CH\text{-}C_2H_5 \end{matrix}$$

and compounds such as those on the left of the above chemical equation are very effective octane enhancers. Either hydrofluoric or sulphuric acid will be the catalyst. The corrosiveness of each of these is of course an extra hazard. In March 2012 there was a leak of hydrofluoric acid at the BP Texas City refinery where, seven years earlier, there had been a fatal explosion at an isomerisation unit. Water was applied to the leaked hydrofluoric acid in the 2012 incident and there were no consequences.

An accident at a refinery in Delaware City in 2001 originated in flammable material present in spent sulphuric acid alkylation catalyst. 'Hot work' was taking place on a catwalk close to tanks of sulphuric acid one of which, containing spent acid as noted, exploded. An inerting system for the space above the spent acid, using carbon dioxide, was present but ineffective. One of the workmen on the catwalk died. As a result of tank breakage a quantity of 264000 gallons of spent sulphuric acid was released, over a third of which entered the environment. Compounds as high in carbon number as C_{16} are formed in small amounts as side products in acid catalysed alkylation, and these were present in the spent sulphuric acid additionally to unreacted isobutane.

5.11 Safety issues relating to catalysis

Catalysis is an important branch of chemical technology, the subject of ongoing research programmes around the world. Catalysts feature several times in this text and, of course, there are safety issues relevant to catalysis. The final section of the chapter will introduce these. Hazards associated with catalysts include the following [5].

(i) Toxicity hazards. Such may arise from catalyst constituents such as chromium, inhalable as dusts if suitable protection is not used. There is also the possibility of reaction during the cooling down of a catalyst, preceding regeneration or disposal, from reacting temperature to room temperature. For example, nickel-containing catalysts react with carbon monoxide, at temperatures below 150°C, to form the highly toxic (and odourless) nickel carbonyl.

(ii) Asphyxia hazards. A catalyst might be stored before use in an enclosure filled with inert gas. A person entering such an enclosure would lose consciousness almost immediately and die within minutes. Breathing apparatus must be worn by persons entering with good reason, and precautions must be taken against inadvertent and unauthorised entry.

(iii) Fire hazards. A catalyst awaiting disposal can be a self-heating hazard, particularly in the presence of water [6].

(iv) Explosion hazards. A catalyst might be supplied for use in the form of charcoal impregnated with the active ingredient. If this becomes airborne and ignites the resulting deflagration will be very powerful.

5.12 Concluding remarks

Cracking, partial oxidation, chlorination, gasification, hydrogenation and nitration have, amongst other processes, received attention in this chapter. This is a broad overview of reaction types, all of importance in the present-day petrochemical industry.

References

[1] List H.H. *Petrochemical Technology: An Overview for Decision Makers in the International Petrochemical Industry*, First Edition. Prentice-Hall, Englewood Cliffs, NJ (1986)

[2] http://www.h2incidents.org/incident.asp?inc=262

[3] Marshall V.C. *Major Chemical Hazards*, First Edition. Ellis Horwood, Chichester (1987)

[4] Burgess A.R. (Ed.) *Manufacture and Processing of PVC*, First Edition. Applied Science Publishers, London (1982)

[5] Twigg M.V. (Ed.) *Catalyst Handbook*, First Edition. Wolfe Scientific Books, London (1970)

[6] *Recommendations for the Transportation of Dangerous Goods: Manual of Tests and Criteria*, Second Edition, Section 33.3.1.6. Test method for self-heating of substances. United Nations, New York and Geneva (1995)

Numerical examples

1. Consider the conversion of cyclohexane to cyclohexanone. Commonly, the temperature is 155°C, and the pressure 8 bar in a vessel holding 35 tonne of cyclohexane. Per kg cyclohexane converted to product, 4 MJ of heat are released.

(a) Calculate the equilibrium vapour pressure of cyclohexane at this temperature, given that its normal boiling point is 81°C, its molecular weight 84 g mol^{-1} and its heat of vaporisation 33 kJ mol^{-1}.

(b) Calculate the pressure of air in the reactor.

(c) If through an operational failure there is cessation of cooling for a period during which 0.001% of the cyclohexane reacts how much unreacted cyclohexane will be converted to vapour as a result of this heat release?

(d) By what amount will this quantity raise the pressure if the ullage space in the reactor is 10 m^3?

2. In a hydrocarbon processing plant the liquid leaves one particular stage of the process at 200°C and enters a spherical container of radius 2 m which it fills. The sphere contents are stirred, and they must cool to below the flash point of 85°C before being passed along to the next stage of the processing. The cooling is brought about by vigorous flow of air at 298 K, affording a convection coefficient of 150 W m^{-2}K^{-1}. The sphere is made of metal and its thermal resistance is negligible in comparison with that of the oil. Using the information below, estimate how much time should be allowed for the cooling.

$$\text{heat capacity of the oil} = 2000 \, \text{J kg}^{-1}\text{K}^{-1}$$
$$\text{density of the oil} = 865 \, \text{kg m}^{-3}$$

From MSc (Safety and Reliability) Examination, University of Aberdeen, 2000.

3. A pipe bearing chlorine develops a small hole and chlorine leaks into a room of floor area $700\,m^2$ and height $5\,m$ at a rate of $5\,cm^3\,s^{-1}$ (measured at $288\,K$, 1 atm., also the conditions in the room). In the limit of there being no ventilation in the room, how long will it take for the chlorine level to reach $1\,p.p.m.$?

4. Consider the gas-phase chlorination of ethane to chloroethane:

$$C_2H_6 + Cl_2 \rightarrow C_2H_5Cl + HCl$$

for which the heat of reaction is $-119\,kJ\,mol^{-1}$ of ethane.

This is accomplished in a continuous reactor with equimolar influx of the two reactants. The reactor contents, under steady conditions, signify an extent of reaction of 65%. There is a requirement that the steady temperature is not more than $500\,K$ above the admittance temperature and this is accomplished by means of a water jacket, and the water is supplied at $25°C$ and must not rise above $80°C$ by exit. Per kg of ethane admitted to the reactor, how much cooling water is required?

Specific heats/$J\,K^{-1}\,mol^{-1}$:

$$HCl\,(g) = 29$$
$$C_2H_5Cl\,(g) = 63$$
$$Cl_2\,(g) = 34$$
$$C_2H_6\,(g) = 53$$
$$H_2O\,(liq.) = 4180\,J\,kg^{-1}\,K^{-1}$$

5. Return to question 4, and consider removal of the HCl from the product stream. Per kg of ethane reacted how much HCl will require neutralisation? What quantity of lime ($CaCO_3$) would be required for this, allowing for a 20% excess?

6. Consider the hydrogenation of naphthalene to tetralin:

$$C_{10}H_8 + 2H_2 \rightarrow C_{10}H_{12}$$

which is carried out in the presence of a palladium catalyst.

Naphthalene vapour and hydrogen, in proportions 1:2, are admitted to a reactor at $225°C$ and 10 bar pressure. The reaction proceeds to 70% completion at which stage the temperature is $600°C$. What is the final pressure?

7. Nitrobenzene, once produced by the nitration of benzene as discussed in the main text, can be hydrogenated to aniline ($C_6H_5NH_2$) by:

$$C_6H_5NO_2\,(liq.) + 3H_2\,(g) \rightarrow C_6H_5NH_2\,(liq.) + 2H_2O\,(liq.)$$

for which the heat of reaction is $-540\,kJ\,mol^{-1}$. Calculate the heat produced in the manufacture of one tonne of aniline.

8. In PVC production in an autoclave, 5 tonne of the monomer is reacted and the conversion rate is 85%. The process takes 6 hours. Cooling is by chilled water which enters a jacket surrounding the autoclave at 1°C and is required not to be in excess of 55°C on exit. Calculate the flow rate of water required.

9. Return to the problem in section 5.2 of the main text, where hydrocarbon liquid after cracking is required to cool from 250°C to 90°C. Instead of by the means considered previously, it is decided to effect this cooling with a simple counterflow heat exchanger, using as the other fluid water, which enters the exchanger at 20°C. The flow rate of hydrocarbon liquid is $7.5\,kg\,s^{-1}$. Heat exchanger plant suitable for the purpose can be obtained which has a heat transfer surface area per metre of axial length of $70\,m^2$ and a heat transfer coefficient of $65\,W\,m^{-2}\,K^{-1}$. Under such operating conditions, what is the minimum length of heat exchanger required if the water is to exit at a temperature not exceeding 75°C? Use a value of $2000\,J\,kg^{-1}\,K^{-1}$ for the specific heat of the hydrocarbon liquid.

10. Hydrogen is very reactive and amounts of it at refineries for processes such as hydrocracking need to be controlled and factored into safety. Consider the simplified chemical equation for the conversion of crude oil to lighter material by hydrocracking:
$$CH_2 + 0.0625H_2 \rightarrow CH_{2.125}$$

At the Repsol refinery in Spain 25000 barrels per day of crude oil are so processed. What would be the daily hydrogen requirement? Express your answer in cubic metres measured at 1 bar 15°C.

11. Imagine that a refinery produces 100000 barrels per day of lighter distillate of sulphur content 0.5% and that it is required to hydrodesulphurise that to 0.01%. What would be the daily hydrogen requirement?

12. When ethylene is polymerised the heat of reaction is about 4 MJ per kg of polymer formed. In manufacture it is proposed to maintain isothermal conditions by recirculating the unreacted monomer vapour through a heat exchanger. If the heat of vaporisation is $480\,kJ\,kg^{-1}$ how many recirculations of the initial quantity are required during the reaction of a tonne of the monomer?

13. The polymerisation of styrene to make polystyrene is exothermic, but less so on a weight basis than the polymerisation of VCM or of ethylene. In fact the heat of polymerisation of styrene is about $0.5\,MJ\,kg^{-1}$. The heat of vaporisation of styrene monomer is $44\,kJ\,mol^{-1}$ and the molar mass $0.104\,kg$. How many recirculations of the initial quantity with water at room temperature will be needed to maintain isothermal conditions when a tonne of the monomer is polymerised.

14. Under isothermal conditions in a rigid reactor, a hydrocarbon of C_{20} is hydrocracked to make propylene. By what factor will the pressure in the reactor change?

15. In places where natural gas occurs some distance from infrastructure to transport it, conversion to methanol is often carried out and the usual way is by steam reforming. There is however some R&D into direct conversion of methane to methanol by partial oxidation according to:

$$CH_4 + 0.5O_2 \rightarrow CH_3OH$$

Heat of formation of methane and methanol are respectively -75 and $-239\,kJ$ mol^{-1}. What quantity of heat would accompany the formation of $1\,kg$ of methanol by this process?

16. Consider the reaction:

$$C_6H_6 + Br_2 \rightarrow C_6H_5Br + HBr$$

The heats of formation in $kJ\,mol^{-1}$ are C_6H_6 $+49$, C_6H_5Br $+59$ and HBr -36. What is the heat in MJ accompanying formation of a kg of the bromide?

17. The 2001 explosion at the Delaware City refinery was due to unreacted isobutane in the spent acid catalyst, as noted in the main text. On the basis of the half-stoichiometric rule presented in Chapter 3 calculate what the pressure of isobutane in the air above the meniscus of the acid must have been if the total pressure was 1 bar.

18. Methyl diethanolamine (MDEA) is a common choice of reagent for hydrogen sulphide scavenging in natural gas whereupon, according to the generalised scheme given in section 5.3:

$R_1 = CH_3$ $R_2 =$ $R_3 = CH_2OH$

molar mass 0.091 kg

The usual performance criterion for a hydrogen sulphide scavenger is p.p.m. of scavenger required to reduce by 1 p.p.m. weight basis the hydrogen sulphide content of a gas. Calculate this for MDEA.

19. What quantity of activated carbon of internal surface area of $1000 \, m^2 g^{-1}$ would be required to remove totally the hydrogen sulphide from a particular natural gas in which it is present at 100 p.p.m.? The area occupied by one hydrogen sulphide molecule is $0.12 \, nm^2$.

20. An alternative to MDEA in the 'sweetening' of natural gas is diisopropylamine:

$$(CH_3)_2CH\text{-}NH\text{-}CH(CH_3)_2$$

Write the chemical equation for H_2S neutralisation by that and calculate how many p.p.m. by weight would be necessary to reduce the sulphur content of a natural gas by 1 p.p.m.

Endnotes

[18] The toxicity of vinyl chloride is discussed in Chapter 10.
[19] Readers requiring background on the log mean temperature difference (LMTD) should consult: Holman J.P. *Heat Transfer*, McGraw-Hill (any available edition) or equivalent. In the above example the arithmetic mean temperature difference is 37.5°C. Though the two do not differ markedly, the LMTD is better for heat exchanger calculations.

6

SOME RELEVANT DESIGN PRINCIPLES

6.1 Background

Frequent reference to vessels, pipes and other plant components has been made in earlier chapters. An important facet of process safety is the design of such components. In this chapter we return to some of the previous material and supplement it with some design principles.

6.2 Design of pressure vessels

6.2.1 LPG storage

As will be discussed more fully in Chapter 9, LPG is stored under its own very high vapour pressure. The vapour pressure at storage temperatures (0–40°C) will be up to about 14.5 bar. This will be used as a basis for illustrating design principles for vessels.

   Imagine that it is desired to store LPG is a seamless (i.e., no welded joint) spherical container. The thickness of the wall of the sphere required safely to contain the pressure can be calculated from:

$$e = \frac{PD}{4f - 1.2P}$$ Eq. 6.1

where e (m) = wall thickness required, D = sphere diameter (m), P is the pressure to be contained (MPa) and f (MPa) the design stress, to be explained more fully below. The reader should first note that the above, though taken from an authoritative source [1], is not 'the' equation for seamless spherical vessels. There are many such, and in different parts of the world different codes and standards are incorporated into legislation. In North America (including Canada) ASME codes often apply; in the UK British Standards apply.

The design stress f is the maximum allowable stress which a particular construction material can be taken to withstand, and it obviously depends on the temperature. It has the same units as pressure, and is available for many materials of interest in sources such as [1]. Imagine that in our example we are concerned only with storage at ordinary temperatures and that we select plain carbon steel as the material; this is, in fact, the most widely used material for LPG containers. From [1] we learn that up to 50°C the design stress of this material is $135\,\text{N}\,\text{mm}^{-2}$ ($\equiv$ MPa). The pressure which has to be withstood is the pressure of the vapour minus atmospheric, that is 13.5 bar, 1.35 MPa. If the sphere is to be 5 m in diameter, substituting in the above equation gives:

$$\text{thickness } (e) = 0.0125\,\text{m} \ (12.5\,\text{mm})$$

Of course, it is possible to build in safety margins, perhaps allowing for possible corrosion. A related numerical example follows.

LPG is to be stored in a horizontal cylindrical container of diameter 2 m. At each end there is a flat plate, bolted to the cylinder. Find the cylinder wall thickness and plate thickness necessary. Use the following expressions for the thicknesses:

Hollow cylinder

$$e = \frac{PD}{2f - P}$$

where D is the internal diameter of the cylinder, P the pressure inside the cylinder and f is the design stress.

Flat plate, bolted with gasket

$$e = 0.4D\sqrt{\{P/f\}}$$

The coefficient 0.4 in the above equation is a 'design constant'.

The material is plain carbon steel, for which the design stress at the temperatures of interest is 135 MPa.

Solution
As previously, taking the pressure to be 14.5 bar, therefore the pressure difference between vessel interior and surroundings to be 13.5 bar, and using a value for f of 135 MPa, for the cylinder wall:

$$e = 10\,\text{mm}$$

and for the flat ends:

$$e = 80\,\text{mm}$$

Safety margins might well be built in by adding 2–3 mm to each, and by making the nominal pressure difference 10% higher than the actual.

Modifications to the expressions for '*e*' might be required to allow for welds, by means of a 'welding factor' '*J*'. However, a weld having the strength of the metal before welding (the virgin plate) is achievable by radiographic examination of a weld and re-welding of any areas shown to have a defect. If such measures are taken, a *J* value of 1.0 applies. For a sphere with a welded seam the equation:

$$e = \frac{PD}{4Jf - 1.2P} \qquad \text{Eq. 6.2}$$

applies. An application of the *J* value to a cylinder is given in the following section.

The reader should be aware that only one part of 'designing the vessel' has been considered in the above numerical problems, namely the capability of the walls to withstand the internal pressure. There are further issues including the ability of a vessel to support its own weight. For example, there are guidelines [1] for what thickness of metal is required for a horizontal cylinder to support its own weight. Of course, support for the vessel's own weight can be provided for by the fitting of suitable load-bearing structures. Vessel support is further discussed in section 6.4.

6.2.2 Extension to other hydrocarbons

One of the calculations in the main text of Chapter 4 can now be revisited (below) in the light of what has been learnt so far in this chapter about design of vessels.

The calculation in Chapter 4 referred to a quantity of n-decane, stored in a closed and evacuated vessel, the ullage space being occupied solely by the compound's own vapour. A safety valve had been set to open when the internal pressure reaches 5 bar, and it was shown in the solution that the temperature at which this would occur would be 263°C. Take the vessel to be spherical with a diameter of 2.5 m and a welded seam such that a '*J*' value of 0.85 applies. Using carbon steel as the material and reference [1] for design stress data, calculate what thickness the curved steel wall will need to be. The ends are to be flat with a 'design constant' of 0.4. Calculate the thickness of the material required for the ends.

Solution
Applying equation 6.2 for a cylinder with a welded seam:

$$e = \frac{PD}{4Jf - 1.2P}$$

symbols as defined previously. Now $P = 0.4\,\text{MPa}$ (i.e., the pressure difference), $D = 2.5\,\text{m}$. From [1] by interpolation, for carbon steel at 263°C, $f = 92\,\text{MPa}$, hence:

$$e = 3.2\,\text{mm}$$

Possibly add 1–2 mm as a safety margin. Such matters are discussed with the manufacturer in advance.

For the flat ends:

$$e = 0.4D\sqrt{P/f} = 0.0659\,\text{m} \ (66\,\text{mm})$$

Pressure containment is also very important in the design of heat exchangers (see Chapter 5). Lees *op. cit.* reports that in a pressure system comprising a pressure vessel and one or more heat exchangers, any failure is more likely to be attributable to one of the heat exchangers than to the pressure vessel itself.

6.3 Pipes

6.3.1 Liquids in pipes

We return in this section to some of the principles discussed in Chapter 2. In chemical engineering practice, pipe selection is according to schedule numbers:

$$\text{schedule number} = \frac{1000P}{S} \qquad \text{Eq. 6.3}$$

where P = internal pressure (MPa) and S (same units) = the design stress at the working temperature. A great deal of piping in chemical plant is schedule 40. The meaning and application of the schedule number will be explained by means of simple calculations below.

In the initial introduction to HAZOP in Chapter 4 the process under consideration was, in effect, transfer of an aqueous liquid from one vessel to another. Such transfer is common in the chemical industries and might simply be by gravity transfer, or otherwise by means of a suitable pump[20]. Let us suppose that such a process utilises a pump and that the discharge pressure is 6 bar, the rate of transfer of the liquid being typically 700 litre min^{-1} (0.7 m^3 min^{-1}). The maximum temperature which the pipe, between pump and vessel, will experience is taken to be the boiling point of water. It is proposed to use schedule 40 pipe work, therefore the stress is calculable from:

$$1000 \times 0.6/40 \, \text{MPa} = 15 \, \text{MPa}$$

Any metal or alloy commonly used in pipe work has a design stress at 100°C much higher than this, e.g., 145 MPa for 304 stainless steel [1]. Schedule 40 is therefore more than adequate. Clearly the delivery rate, here required to be 0.7 m^3 min^{-1} (approximately 180 US gallon min^{-1}), depends upon the speed with which the liquid travels along the pipe, and for aqueous fluids a speed of 2.5–3 m s^{-1} ($\approx$8–10 ft s^{-1}) is common [1]. Tables of pipe performance (e.g., [3]) give, for various schedules and nominal diameters, flow rates at a velocity of 1 ft s^{-1}. Now the requirement here is a delivery rate of 180 US gallon min^{-1}. The closest match (reference [3], or equivalent tables elsewhere) is nominal 3.5-inch outer diameter schedule 40 pipe, which gives a delivery rate at 1 ft s^{-1} of 23.0 US gallons min^{-1} therefore 180 gallons per minute at a speed of about 8 ft s^{-1}. Such a pipe fabricated of stainless steel would have a wall thickness of 0.216 inch (5.49 mm). Alternatively, an 8.625 inch outer diameter schedule 40 pipe of the same material would give 156 US gallon min^{-1} at 1 ft s^{-1}.

Once fundamental design considerations of this sort for pipes are taken care of, there are further considerations to ensure that the most economical choice is made. A classical chemical engineering tome [4] gives the following formula for the most economic diameter of pipe for given conditions:

$$\text{most economic diameter (inch)} = 0.098 \, m'^{0.45}/(\sigma^{0.31})$$

where m' is the mass flow rate through the pipe (lb hour^{-1}) and σ the fluid density (lb ft^{-3}). In our example above, taking the liquid to be pure water $m' = 90\,136$ lb hour^{-1} and $\sigma = 60$ lb ft^{-3}, therefore the most economic diameter is 4.7 inch (11.9 cm), closer to the lower of the two diameters in the illustrative calculations in the previous paragraph. Factors influencing the final decision include availability and flexibility to accommodate possible future modification. Schedule 40 will allow higher discharge pressures to be used, if that becomes necessary, without replacement of the pipes. Pumps in chemical processing sometimes have discharge pressures of 50 bar.

Depending on the viscosity, liquids other than aqueous ones travel along pipes in chemical plant at 1–3 m s^{-1}. Gases travel at speeds about an order of magnitude higher.

6.4 Vessel support

Support of vessels is commonly either by a 'saddle' support or a 'skirt' support. Saddle support is widely used for vessels which are horizontal cylinders whereas the skirt support is favoured for vertical vessels, also for distillation columns. In a skirt support a cylindrical or conical shell is welded to the base of the vessel. The saddle or skirt transmits the load to the foundations, and has to withstand stresses of two sorts: from the weight of the vessel and its contents and from bending moments. The difference between these is:

$$\lambda_{tens} = \lambda_b - \lambda_w \; N \; mm^{-2} \qquad\qquad Eq. \; 6.4$$

where λ_{tens} = tensile stress; λ_w = weight stress; λ_b = bending stress and their sum:

$$\lambda_s = \lambda_b + \lambda_w \; N \; mm^{-2}$$

is the compressive stress λ_s. The bending stress relates to wind and to 'eccentric loading', that is, loss of symmetry and unevenness of load due to internal and external fittings and accessories. Therefore, for a tall structure exposed to the wind and/or having unevenness of loading, both types of stress apply. For a squat symmetrical vessel, consideration of the weight stress only will suffice.

In the introductory part of this chapter a spherical vessel 5 m in diameter with wall thickness 12.5 mm was considered. This vessel was required to hold LPG and was constructed of plain carbon steel (density $\approx 7780 \, kg \, m^{-3}$). Imagine that it is required to support this vessel with a cylindrical skirt of inner diameter 1 m.

The weight stress (λ_w) is clearly:

$$\frac{weight \; of \; vessel + contents}{flat \; area \; of \; cylindrical \; annulus \; contacting \; the \; foundations}$$

Simple mensuration to evaluate the denominator gives an expression:

$$\lambda_w = \frac{W}{\pi e (D_s + e)} \qquad\qquad Eq. \; 6.5$$

where W is the weight of vessel and contents, D_s is the inner diameter of the support and 'e' is as previously defined though now appertaining to the support not the vessel itself. From equally simple mensuration, using the density of carbon steel given in the previous paragraph, the weight of the vessel is 7676 kg (7.7 tonne).

The internal volume is $65\,m^3$, and to regard the vessel as full of water – significantly denser than liquid propane – would give a margin of safety in the calculated width of the cylindrical skirt support. The design stress of plain carbon steel at temperatures up to 50°C is (as we have already seen) $135\,N\,mm^{-2}$, hence:

$$\lambda_w = \frac{(65000 + 7676) \times 9.81}{\pi e(1000 + e)}\,N\,mm^{-2}$$

$$= \frac{2 \times 10^5}{e(1000 + e)}\,N\,mm^{-2}$$

If we make the metal thickness of the skirt the same as that of the vessel, 12.5 mm, the weight stress is $16\,N\,mm^{-2}$, only just over a quarter the design stress and therefore safe. If a 5 mm thickness skirt were to be used, the weight stress would be $40\,N\,mm^{-2}$, still safe.

Once it is confirmed that the support itself can bear the weight, the surface underneath, to which the support transmits the load, has to be considered. It might be necessary to use a wider support skirt than calculations of the sort outlined above would require in order to reduce the stress on the ground surface. Hard rock can only withstand stresses up to about $40\,N\,mm^{-2}$ [3]. Where the ground as it exists is not suitable, it can be modified for the purpose by excavating and replacement with a stronger material, followed by concrete capping. Even where the ground can withstand the weight there will often be a need for a layer of crushed stone, sand or gravel [3]. The calculation in the shaded area below utilises some of these ideas.

We return to one of the calculations in Chapter 2 which, in the context of Bernoulli's equation, considers a large cylindrical tank used to store crude oil. Take the tank to be, radius 2 m, with a flat base. It holds when full $65\,m^3$ of crude oil. The weight of the cylinder itself is 5 tonne. Local building codes specify that the bearing strength of the ground where it is proposed to site the tank is $0.4\,N\,mm^{-2}$. Perform a calculation to determine whether the ground can withstand the load without reinforcement.

Solution
Using a value of $825\,kg\,m^{-3}$ for the density of crude oil, the weight of the oil is $53\,625\,kg$, to which has to be added the $5000\,kg$ due to the tank. The stress at the base is therefore:

$$\frac{58625 \times 9.81}{4 \times 10^6\,\pi} = 0.05\,N\,mm^{-2}$$

This is an order of magnitude lower than the bearing strength, so there is no need for reinforcement. Note, however, that this calculation has considered only the weight stress. If there is significant bending stress due to wind or unsymmetrical loading, this too has to be considered.

6.5 Design features at the scenes of major accidents

Flixborough 1974

As recorded in an earlier chapter, the 1974 Flixborough accident was caused by failure of a pipe which had been temporarily installed to bypass an out-of-service reactor. The following information, taken from Lees *op. cit.*, and from Marshall *op. cit.*, relates to the plant at the time of the accident and illustrates an application to accident follow-up of the design considerations outlined previously in this chapter.

The operating pressure was $8.8\,\mathrm{kg\,cm^{-2}}$ ($0.86\,\mathrm{MPa}$). The equipment, after installation of the bypass pipe, was pressure tested to $9\,\mathrm{kg\,cm^{-2}}$ ($0.88\,\mathrm{MPa}$). The safety valve was set to open at $11\,\mathrm{kg\,cm^{-2}}$. Calculations were performed to show that a straight pipe of the schedule used could withstand the pressure. In routine operation nitrogen was passed into the reactor, as necessary, to control the composition of the atmosphere.

The 'stubs' on the two reactors which the pipe was required to connect were of 28-inch diameter. The largest pipe available on site was 20-inch diameter. The two stubs were at different heights, so a straight piece of tubing could not be used. Instead, three lengths of pipe were welded together to form a 'dog leg'. Connection of the pipe to the stubs, at each end, was by 'bellows', and the way in which these were supported by scaffolding is thought to have led to failure. Calculations were produced at the enquiry which predicted that the probability of such failure would be high at an internal pressure of $10.6\,\mathrm{kg\,cm^{-2}}$ where the reacting temperature was 150°C.

Texas City 2005

In pretrochemical processing products are sometimes removed by solvent extraction, and the liquid remaining after extraction is called raffinate. Raffinate can be isomerised to useful products, first requiring distillation in what is termed a raffinate splitter. The raffinate splitter at the Texas City refinery was, at the time of the accident, being restarted after a shutdown. It is believed that influx raffinate to the splitter was overheated and that the splitter itself was overfilled. As is routine practice, a blowdown drum was provided for the eventuality of loss of containment by the splitter. The blowdown drum at Texas City is reported to have vented received hydrocarbon into the atmosphere, whereas inherently safer designs divert it to a flare.

Venezuela 2012

It is often helpful as a prelude to a case study to step backwards and review the location briefly, and that will be done here for the case of Venezuela. Venezuela was one of the two founding member of OPEC – the other was Iraq – over 50 years ago. One reason for the founding of OPEC was that foreign investors were taking oil at a price and selling it on at a negotiated price, whereas the countries from which the oil was produced wanted some say in the ultimate price of sale. In much more recent years than the foundation of OPEC, Venezuela's near neighbour Trinidad and Tobago has been the scene of major offshore gas discovery. T&T lacks infrastructure for pipelining and so has had to convert the gas to LNG or to methanol. The obvious approach of renting infrastructure from Venezuela has not been realised. This might of course have led to expansion of such infrastructure to the benefit of both countries. These points apropos of Venezuela as a hydrocarbon producing country are enlarged upon in Chapter 9, where quite a serious accident in LNG production in Trinidad is described.

Forty-one people were killed and 80 injured in a blast at the Amuay oil refinery in Venezuela in August 2012. The refinery has a capacity of 0.65 million barrels per day, putting it in the 'big league'. It was the most serious accident ever to have occurred in an OPEC country. Explosive behaviour of leaked gas at the refinery led to fires in storage tanks of liquid. One press account quoted an employee amongst the survivors as saying that the refinery 'isn't living up to its original design'. The present author interprets that as meaning that expansion of the refinery, which began operations in 1947, has overtaken developments in safety. Here again a general point can be made. New refineries are rare: there has been no new refinery in the US since 1976. A new one is being erected, the Yuma refinery Arizona, but is not producing at the time of writing the project having 'stalled' more than once. Yet think how much the refining capacity of the US has increased since 1976! That can only mean refining is taking place at plants which have, so to speak, evolved from smaller ones. The integrating and interfacing of older designs with new must therefore feature in the safety culture of refineries.

6.6 Design data

6.6.1 Introduction

Throughout this book there has been considerable use of physical quantities of substances in safety-related calculations. These quantities include density, viscosity and thermal conductivity. For a pure chemical substance at a specified temperature and pressure, such information is usually readily available in sources such as the JANAF tables. In chemical engineering design matters are

not always this straightforward, and there are established practices for estimating these properties for materials of interest.

6.6.2 Densities

Densities of substances such as crude oil and petroleum fractions are determined routinely in industrial laboratories. There are several simple ways of
measuring liquid densities very accurately, and these have been made into
standards.

When liquids are mixed, the approximation is usually made in chemical
engineering practice that the resulting blend has a density calculable in a
simple way from the densities of the components and the proportions. For
example, ethanol is miscible in all proportions with water. The density of
ethanol at 25°C is $785\,kg\,m^{-3}$ and that of water at the same temperature
$996\,kg\,m^{-3}$. If ethanol and water were mixed in the proportion by weight 4:1,
the density of this blend would be estimated in the very simple way:
In 1 tonne of the blend:

$$800\,kg \text{ of ethanol, occupying } 1.0191\,m^3$$

$$200\,kg \text{ of water, occupying } 0.2008\,m^3$$

$$\text{density of the blend} = (1000/1.2199)\,kg\,m^{-3} = 820\,kg\,m^{-3}$$

A purist physical chemist might raise an objection to this calculation, and
he or she would in fact be justified in doing so. The point of contention would
be that the implicit assumption has been made that the partial molar volumes
of the two components are the same as the pure liquid molar volumes, and
this is not in general true. The molar volume of pure water at this temperature
is:

$$0.018\,kg\,mol^{-1}/996\,kg\,m^{-3} = 1.81 \times 10^{-5}\,m^3\,mol^{-1}$$

and of pure ethanol:

$$0.046\,kg\,mol^{-1}/785\,kg\,m^{-3} = 5.86 \times 10^{-5}\,m^3\,mol^{-1}$$

The weight composition given for the water–ethanol mixture corresponds
to a mole fraction of ethanol of 0.61, therefore of water 0.39. At this composition, the partial molar volume of water is $1.66 \times 10^{-5}\,m^3\,mol^{-1}$, and that of
ethanol $5.76 \times 10^{-5}\,m^3\,mol^{-1}$ [5]. The unit molar amount of the mixture
contains 0.61 mol (0.0281 kg) of ethanol occupying:

$$0.61\,mol \times 5.76 \times 10^{-5}\,m^3\,mol^{-1} = 3.51 \times 10^{-5}\,m^3$$

and 0.39 mol (0.00702 kg) of water, occupying:

$$0.39 \, \text{mol} \times 1.66 \times 10^{-5} \, \text{m}^3 \text{mol}^{-1} = 0.65 \times 10^{-5} \, \text{m}^3$$

The density of the mixture is therefore:

$$\frac{0.0281 + 0.00702 \, \text{kg}}{(3.51 + 0.65) \times 10^{-5} \, \text{m}^3} = 844 \, \text{kg m}^{-3}$$

and this is 3% higher than that calculated previously by a method which, in effect, does not accommodate the small but not totally insignificant volume change due to mixing. An engineer may or may not regard the 3% difference as negligible. If it is taken to be negligible, design calculations of volumes required for containment of particular quantities of the liquid mixture are also subject to errors of this magnitude. If it is not, there are means of estimating the density of a liquid mixture which, without requiring the formal approach above using partial molar volumes, are an improvement on the very simple method followed at the beginning of this section. One such improved method features in one of the appended numerical examples.

For gases, the ideal gas equation is an adequate basis for density calculation for most purposes. Where it is not, the compressibility factor is used and is obtainable from compressibility plots. These invoke the principle of corresponding states and 'reduced' temperatures and pressures [1, 3]. The ideal gas equation also suffices for vapour density calculations for most routine purposes.

6.6.3 Viscosities

Viscosity appears in a number of the calculations in previous chapters, including one in Chapter 2 appertaining to methane flow. Several of the dimensionless groups routinely used in chemical engineering calculations, including the Reynolds number, contain the viscosity, SI units for which are $\text{kg m}^{-1}\text{s}^{-1}$ for dynamic viscosity and m^2s^{-1} for kinematic viscosity. As with densities, reliable values of viscosity for pure compounds – liquids and gases – are in the literature. That being said, some older (pre-1970s) values for the viscosities of gases are suspect by reason of the fact that they were measured in coiled tubes. This had an effect on the results which was not recognised by the experimentalists of the time.

For complex blends such as petroleum fractions, direct measurement of viscosity is often necessary. For such fractions, measurement of the viscosity by the Redwood method is extremely simple. The time taken for a column of the liquid to descend under gravity in a tube of specified height and diameter is measured, and that is often called the 'Redwood viscosity' (units seconds). Conversion to true units of viscosity, if required, is also simple, though it must be remembered that the Redwood and 'absolute' viscosities are linked by a non-linear relationship. The numerical example which follows illustrates the use of a value for viscosity in a design calculation.

A hydrocarbon vapour of dynamic viscosity $1 \times 10^{-5}\,\mathrm{kg\,m^{-1}\,s^{-1}}$ and density $5\,\mathrm{kg\,m^{-3}}$ enters a pipe of inside diameter 5 cm at a temperature of 130°C. It travels along the pipe at a speed of $1\,\mathrm{m\,s^{-1}}$. It is necessary, for the hydrocarbon, whilst remaining in the vapour phase, to have cooled to below 112°C on exit. The pipe outside surface is maintained at 100°C by contact with saturated steam at 1 bar. The following equation applies for the convection coefficient for transfer of heat from the hydrocarbon[21]:

$$h = 0.008\,\mathrm{Re}^{0.8}$$

where h = convection coefficient $(\mathrm{W\,m^{-2}\,K^{-1}})$, Re = Reynolds number
What is the minimum length of piping for the necessary cooling to be accomplished under steady conditions? Use a value of $2000\,\mathrm{J\,kg^{-1}\,K^{-1}}$ for the specific heat of the vapour.

Solution

The Reynolds number Re is therefore given by:

$$\mathrm{Re} = 1 \times 0.05 \times 5/(1 \times 10^{-5}) = 25\,000$$

Hence the convection coefficient is given by:

$$h = 0.008\,(25\,000)^{0.8}\,\mathrm{W\,m^{-2}\,K^{-1}} = 26\,\mathrm{W\,m^{-2}\,K^{-1}}$$

In order to set up the heat balance equation for the vapour we need the mass flow rate (symbol m') which is clearly:

$$1\,\mathrm{m\,s^{-1}} \times \pi\,(0.025)^2\,\mathrm{m^2} \times 5\,\mathrm{kg\,m^{-3}} = 10^{-2}\,\mathrm{kg\,s^{-1}}$$

rate of heat transfer to the surroundings
= rate of heat loss by the vapour

$$h \times 2\pi r L(T_{\text{vapour}} - T_{\text{wall}}) = m'c\Delta T$$

where r = tube radius = 0.025 m
L = tube length, required
T_{vapour} = vapour temperature, taken as the mean of the entry and exit temperatures i.e., 121°C
T_{wall} = wall temperature = 100°C
m' = mass flow rate of vapour = $10^{-2}\,\mathrm{kg\,s^{-1}}$
c = specific heat of the vapour = $1100\,\mathrm{J\,kg^{-1}\,K^{-1}}$
ΔT = temperature drop of the vapour = 18°C

Substituting and solving for L (and rounding up to the nearest half-metre):

$$L = 2.5\,\text{m}$$

A tube of this length or longer will afford the necessary cooling.

6.6.4 Enthalpies

Thermodynamic data such as heats of vaporisation and heats of combustion, all available from sources such as those referenced herein, are examples of enthalpies. One or two facts from basic thermodynamics, as they relate to such quantities in engineering calculations, will take our discussion forward at this point.

The First Law of Thermodynamics introduces the function of state internal energy, usual symbol U (or u if on a per unit weight basis), whereas the less fundamental function enthalpy (H, or h if on a per unit weight basis) is preferred, for good reasons [6], in many applications. When in processing plant material undergoes heating or cooling, this can be expressed in terms of a rise or fall (respectively) in the enthalpy, as the calculation below shows.

We return to the calculation in the section immediately above, which relates to a hydrocarbon vapour of dynamic viscosity $1 \times 10^{-5}\,\text{kg}\,\text{m}^{-1}\text{s}^{-1}$ and density $5\,\text{kg}\,\text{m}^{-3}$ inside a pipe of inside diameter $5\,\text{cm}$ at a temperature of 130°C. The vapour exits the pipe at 112°C. For this follow-up calculation we identify the vapour as toluene; its dynamic viscosity at the temperatures of interest is (to one significant figure) $1 \times 10^{-5}\,\text{kg}\,\text{m}^{-1}\text{s}^{-1}$, and boiling point is 111°C. Hence the cooling realised by travel along 2.5 m of pipe will, at atmospheric pressure, ensure condensation after exit.

The rate of heat loss from the toluene could of course be calculated in an elementary way from the mass flow rate, the temperature drop and a suitable averaged value of the specific heat. Use of a single value of the specific heat imposes a degree of error. No such error is entailed in calculating this from enthalpies instead. From tables [3], the enthalpy of toluene vapour is $909.0\,\text{kJ}\,\text{kg}^{-1}$ at 130°C (403 K) and $884.1\,\text{kJ}\,\text{kg}^{-1}$ at 112°C (385 K). Hence, for a throughput of $10^{-2}\,\text{kg}\,\text{s}^{-1}$, the rate of heat loss from the toluene is:

$$10^{-2}\,\text{kg}\,\text{s}^{-1} \times [909.0 - 884.1]\,\text{kJ}\,\text{kg}^{-1} = 249\,\text{W}$$

Such information is helpful in energy audits on plant and, possibly, in 'process integration', but it is with safety matters that our discussion is concerned. The heat lost by the toluene will go into the saturated steam with which the pipe is in contact, but this will not rise in temperature. As long as two phases of a pure chemical substance are in phase equilibrium at specified pressure, the temperature cannot change. This follows from the phase rule. This provides for a medium of effectively infinite heat capacity, and heat exchange plant often uses saturated steam in this way. Of course, the heat (in our example from the toluene) cannot 'disappear': though its release does not change the *temperature* of the saturated steam it does add to its *enthalpy*, causing the weight distribution between the two phases – vapour and liquid – to adjust.

Enthalpies such as those used in the calculation, taken from standard tables, are referred to an arbitrary zero. All substances do not have the same enthalpy at absolute zero of temperature, therefore there is no one unifying reference as, by reason of the Third Law of Thermodynamics, there is for entropies.

6.6.5 Vapour pressures

The vapour pressure *of a pure chemical substance* is a fundamental thermodynamic quantity obtainable from tables or from the Clausius–Clapeyron equation where the heat of vaporisation and normal boiling point are known. Vapour pressures of binary mixtures of organic compounds are discussed at length in many classical thermodynamics texts (e.g., [7]). Such discussions feature *inter alia* Raoult's Law, Henry's Law, activity coefficients and azeotropic (constant boiling) mixtures.

For the lighter petroleum fractions, the Reid vapour pressure (RVP) is the usual measure. For a hugely complex mixture there is no such thing as *the* vapour pressure at a particular temperature, that is, the vapour pressure is not single-valued. It depends on the space which the vapour occupies. When such a mixture is placed in a vessel and liquid–vapour equilibrium established, the very composition of the phases depends on how much has been transferred from liquid to vapour state. In the attainment of equilibrium, the liquid becomes less rich in the lighter constituents to an extent which depends on the quantity which has vaporised which, in turn, depends upon the volume occupied by the vapour. In RVP measurements on petroleum fractions the volume into which the vapour expands is standardised so that there is at least consistency in this regard when different samples are examined. These points were touched on in the discussion of flash points in Chapter 3.

Notwithstanding the statement above concerning the Clausius–Clapeyron equation as a route to vapour pressures of pure compounds, it has to be remembered that for precise calculations, especially where the temperature of interest is a long way one side or the other of the normal boiling point, the equation does not, in its simple form, 'deliver the goods'. This is because of the implicit assumption that the heat of vaporisation is a constant, a reasonable approximation over a few tens of degrees but not true in principle. For example, the heat of vaporisation of benzene at 25°C is 435 kJ kg^{-1} and at the boiling point of 80°C it is 395 kJ kg^{-1}. A possible solution to this is to use a mean value of the latent heat across the temperature range of the integration or, less crudely, to incorporate the latent heat as a function of temperature and take it inside the integral.

There are other ways of estimating the vapour pressure of an organic substance, including the Antoine equation [1, 3, 8], which in fact goes back to the 1880s. This states:

$$\log P^* = A - \frac{B}{T + C} \qquad \text{Eq. 6.6}$$

where P^* = vapour pressure in mm Hg, T = temperature in °C and A and B are constants characterising a particular substance. These are obtained from interpolation of experimental vapour pressure/temperature plots. Care must be exercised when using the Antoine equation, or 'Antoine constants' from various sources; for example, in some texts the equation uses Naperian rather than decadic logarithms. An application follows below.

In Chapter 9 the combustion of acetaldehyde (CH_3CHO) will be considered. The Antoine constants for acetaldehyde in the range −45 to +70°C (encompassing the normal boiling point of +20°C) are [8]:

$$A = 6.81089, \ B = 992.0, \ C = 230$$

What is the equilibrium vapour pressure of acetaldehyde at 38°C?

Solution

We apply equation 6.6:

$$\log P^* = A - \frac{B}{T + C} = 3.1093975$$
$$P^* = 1286 \text{ mm Hg} \equiv 1.693 \text{ bar}$$

6.6.6 Other quantities relevant to design

These include thermal conductivity of gases, liquids and solids which are obtainable from handbooks such as [3] and, in today's world, from electronic sources. Where the conducting medium is a fluid – gas or liquid – conduction and convection are both possible, and whether or not convection dominates has to be judged from the Rayleigh number[22].

Specific heats also feature in many design calculations, and for mixtures these can be taken to be additive according to the proportions of the respective constituents. Diffusion coefficients (units $m^2 s^{-1}$) are also available in the literature for many substances – gases and liquids – of interest in chemical processing, and there are correlations for calculation of the diffusion coefficients of mixtures from the properties of the individual components.

6.7 Concluding remarks

The aim of this chapter has been to reintroduce some of the topics from the previous chapters against a background of design practice. The book has been concerned primarily with safety aspects of *operation*, but to step back from *operation* to *design* is clearly helpful and indeed necessary in having a good understanding of process safety.

References

[1] Coulson J.M., Richardson J.F., Sinnott R.K. *Chemical Engineering* Volume 6: Design, First Edition. Pergamon, Oxford (1983)
[2] Geankoplis C.J. *Transport Processes and Unit Operations*, Second Edition. Allyn and Bacon, Boston, MA (1983)
[3] Perry R.H., Green D. *Perry's Chemical Engineers' Handbook*, Sixth Edition. McGraw-Hill, New York (1984)
[4] McCabe W.L., Smith J.C. *Unit Operations of Chemical Engineering*, First Edition. McGraw-Hill, New York (1956)
[5] Atkins P.W. *Physical Chemistry*, First Edition. Oxford University Press, Oxford (1978)
[6] Jones J.C. *The Principles of Thermal Sciences and their Application to Engineering*, First Edition. Whittles Publishing, Caithness, UK and CRC Press, Boca Raton, FL (2000)
[7] Dickerson R.E. *Molecular Thermodynamics*, First Edition. W.A. Benjamin, Menlo Park, CA (1969)
[8] Rousseau R.W., Felder R.M. *Elementary Principles of Chemical Processes*, First Edition. Wiley, New York (1978)

Numerical problems

1. Return to the example in the main text of this chapter in which LPG storage in a horizontal cylindrical container of diameter 2 m is considered. Previously, the design was such that at each end there was a flat plate, bolted to the cylinder. Reconsider the problem with a ellipsoidal dome at each end, for which (retaining the symbols used in the main text):

$$e = \frac{PD}{2f - 0.2P}$$

Comment on your answer.

2. Return to the numerical problem appended to Chapter 4 where vinyl chloride (CH_2CHCl) storage as a liquid under its own vapour pressure is considered. If it is in a spherical vessel of 5 m diameter constructed seamlessly of stainless steel, what must the wall thickness be for the conditions as set out in the earlier question?

Also perform the calculation for the cases where: (a) there is a 'fully radiographed' welded seam in the vessel, and (b) there is a welded seam such that a J value of 0.85 applies.

3. Return to the problem in the main text of Chapter 5 (Section 5.2) where, in a cracking plant, hydrocarbon liquid exits the process at 250°C and is pumped into a vessel for cooling. The pump has a discharge pressure of 12 bar and the liquid flows along the pipe at a speed of $3 \, \mathrm{m\,s^{-1}}$, the required delivery rate being $5.5 \, \mathrm{m^3\,minute^{-1}}$ (1450 US gallon minute^{-1}). Using tables such as those in [1, 3], make a reasoned selection of a suitable stainless steel pipe in terms of schedule and nominal outer diameter.

4. Return to the problem in Chapter 5 of the main text (Section 5.6.2) where ethylene and hydrogen are reacted and the final pressure is approximately 34 bar and the final temperature 305°C. If the reactor is cylindrical, seamless and to be fabricated of 304 stainless steel with internal diameter 5 m, what must the thickness of the curved walls be?

5. For the situation in question 3, calculate the 'most economic diameter' from the equation in the main text. Make a reasoned estimate of your own for the liquid density.

6. Return to the question in the main text for an ethanol–water mixture of specified composition, where the density estimated without regard to volume effects of mixing was calculated and compared with the value from a more rigorous calculation using partial molar volumes. Recalculate the density using Amagat's Law [3], viz.:

$$V_{mix} = \sum_{i=1}^{i=n} x_i V_i$$

where n = number of components
    x_i = mole fraction of component i
    V_i = molar volume of pure liquid i at the mixture temperature
Comment on your answer.

7. A hydrocarbon liquid of dynamic viscosity $3 \times 10^{-4} \mathrm{kg\,m^{-1}s^{-1}}$ and density $700 \mathrm{kg\,m^{-3}}$ enters a pipe of inside diameter 5 cm at a temperature of 47°C. It travels along the pipe at a speed of $4 \mathrm{cm\,s^{-1}}$. It is necessary, in order to prevent an explosion hazard at the next stage of the processing, for the hydrocarbon to have cooled to below its flash point of 38°C on exit. The pipe outside surface is maintained at 30°C. The following equation applies for the convection coefficient for transfer of heat from the hydrocarbon:

$$h = 0.05 \ \mathrm{Re}^{0.8}$$

where h = convection coefficient $(\mathrm{W\,m^{-2}K^{-1}})$
    Re = Reynolds number

What is the minimum length of piping for the necessary cooling to be accomplished under steady conditions? Use a value of $2000 \mathrm{J\,kg^{-1}K^{-1}}$ for the specific heat of the liquid.

8. Return to the calculation in the main text, which considers toluene passing at a mass flow rate of $10^{-2} \mathrm{kg\,s^{-1}}$ along a pipe and cooling from 130°C to 112°C in so doing. Using a value of $1135 \mathrm{J\,kg^{-1}K^{-1}}$ for the specific heat of toluene vapour find the rate of heat loss. Compare your answer with that given in the main text.

9. One of the questions appended to Chapter 4 (question 2) required calculation of the temperature of benzene when its equilibrium vapour pressure is 7 bar. The Clausius–Clapeyron equation was used. Now repeat the calculation using the Antoine method, with the following values of the constants (taken from [8]).

$$A = 6.90565$$
$$B = 1211.033$$
$$C = 220.790$$

Comment on any discrepancy between the two values.

10. In Chapter 3 the flash point of ethanol was given as 13°C. On the basis that this corresponds to half-stoichiometric, calculate the flash point from the Antoine constants for ethanol below. The chemical equation is:

$$C_2H_5OH + 3O_2 \ (+11.3 \ N_2) \rightarrow 2CO_2 + 3H_2O \ (+11.3 \ N_2)$$

Antoine constants [8]:

$$A = 8.04494$$
$$B = 1554.3$$
$$C = 222.65$$

11. Methanol (CH_3OH) is placed in a $5\,m^3$ drum at 25°C, there being air in the space above the liquid surface. When the drum is emptied of liquid a gas/vapour mixture comprising the equilibrium pressure of methanol plus air, at 1 bar total pressure, is left behind and has to be diluted to below the lower flammability limit of methanol, which is 6.7% (volume or molar basis), by purging before the tank can be opened to the atmosphere. The dilution is carried out by injection of previously cooled process gas of molar weight 0.038 kg. What quantity (in kg) of gas is necessary for the purging. The Antoine coefficients for methanol are:

$$A = 7.87863$$
$$B = 1473.11$$
$$C = 230.0$$

12. Butane (C_4H_{10}) has a boiling point of 272.5 K and a heat of vaporisation of $22\,kJ\,mol^{-1}$. It is proposed to store butane as a liquid under its own equilibrium vapour pressure in a cylindrical vessel of diameter 10 m at a site where the ambient temperature is up to 35°C (308 K).

(i) Calculate the pressure which the container must be capable of withstanding.
(ii) If the container is to be cylindrical and made of plain carbon steel for which the design stress at the temperatures of interest is $135\,N\,mm^{-2}$, find the thickness required for the curved walls, using the expression:

$$e = \frac{PD}{2Jf - P}$$

where e is the wall thickness (m), D is the internal diameter of the cylinder (m), P the pressure (MPa) and f is the design stress (MPa) and J the welding factor (no units). Make the assumption that any welded seam in the cylinder has been fully radiographed so that the seam has the same mechanical properties as 'virgin plate'.

(iii) Recalculate the required wall thickness if the welded seam is such that a value of J of 0.85 applies.

13. Gasolines have an upper flammability limit (UFL) of about 7% in air at a total pressure of 1 bar. What value of the vapour pressure would be required for the mixture of air and gasoline vapour in a vented tank to be flammable? Compare your answer with the vapour pressure of iso-octane (the benchmark hydrocarbon for gasolines) at 25°C.

14. Diesels have a lower flammability limit (LFL) of about 0.6% in air at a total pressure of 1 bar. What value of the vapour pressure would be required for the mixture of air and diesel vapour in a vented tank to be at the LFL? Express your answer in Pa.

Endnotes

[20] For a good introduction to pumps and pumping see Geankoplis [2].

[21] This expression has been adapted from the Dittus–Boelter equation [2].

[22] For example, the calculation in Chapter 4 on the storage safety of IPN took the internal heat conduction to be purely conductive. This had been confirmed by reference to the source from which the data for the compound were taken (Oxley *et al.*, Chapter 4 [6]). Nevertheless, generally in applications of thermal ignition theory to fluids the possibility that conduction is being augmented by convection has to be examined.

7

SOME RELEVANT MEASUREMENT PRINCIPLES

7.1 Introduction

In Chapter 6, design principles were discussed in the context of hydrocarbon process safety. Once a plant is 'up and running' there are continual measurements to be made as routine operating procedure, and these we outline and incorporate into selected parts of the previously presented material.

7.2 Flow measurement

7.2.1 The venturi meter and the orifice meter

In Chapter 2 we encountered Bernoulli's equation for incompressible flow as:

$$v_1^2/2 + p_1/\sigma = v_2^2/2 + p_2/\sigma$$

the above form of the equation applying to flow which is horizontal so that there are no potential energy effects. Bernoulli's equation is the basis of two widely used flow metering devices: the venturi meter and the orifice meter, for an excellent introductory discussion of which the reader is referred to Geankoplis [1].

Imagine an incompressible fluid passing along a pipe part way along which there is a throat, where, over a considerable axial length, the pipe diameter narrows from its previous diameter D_1 to a smaller value D_2 and then, equally gradually, widens back to D_1. Continuity (also encountered previously in this book in the calculation of mass flow rates of fluids) gives:

$$v_1 \times \pi D_1^2 / 4 = v_2 \times \pi D_2^2 / 4$$

Substituting from the continuity condition into Bernoulli's equation gives:

$$v_2 = \sqrt{\left[\left(2/\sigma\right)\left(p_1 - p_2\right)\right]/\left[1 - \left(D_2/D_1\right)^4\right]}$$

If there are frictional losses along the throat of the venturi an experimental coefficient can be incorporated to correct. However, for Reynolds numbers of 10^4 or above this is seldom necessary. Use of this equation is illustrated in the shaded area below, which returns to the discussion in section 5.2 where safety aspects of cracking were discussed.

Hydrocarbon liquid at a cracking unit is supplied along a pipe of 3.5 inch (0.089 m) diameter. The rate at which the liquid flows along the pipe is to be determined with a venturi meter. Installed part way along the pipe length is a venturi of throat diameter 2.5 inch (0.064 m). Pressure tappings taken upstream and downstream reveal a pressure difference of 85 kPa across it. The density of the liquid is 900 kg m^{-3}.

It is possible to calculate the speed of the liquid at the orifice from the above equation, inserting $(p_1 - p_2) = 85\,000$ N m^{-2}, $\sigma = 900$ kg m^{-3} and $D_2/D_1 = 0.71$ to give:

$$v_2 = 16\,\text{m s}^{-1}$$

Now although the velocities are not the same at positions 1 and 2 the mass flow rates are. The flow rate is clearly:

16 m s^{-1} × [$\pi(0.089)^2/4$] m^2 × 900 kg m^{-3} = 90 kg s^{-1} ($\approx$2 million US gallons per day)

which is the measured flow rate along the pipe. It is advisable to check the Reynolds number for this rate of flow along the pipe. Using a value for the dynamic viscosity of 0.05 kg m^{-1}s^{-1} the Reynolds number is clearly:

$$(16 \times 0.089 \times 900)/0.05 = 2.6 \times 10^4$$

The venturi meter has the disadvantage of a fixed throat diameter, whereas an orifice meter, which works along the same principles, uses an orifice plate which can be changed, affording a range of diameters. Either venturi meter or orifice meter can be used for compressible flow as well as for incompressible. For compressible flow, however, it is necessary to insert into the equation an expansion factor, the value of which depends on the ratio D_2/D_1 as well as on the pressure difference across the throat or orifice and, for different gases, on the ratio of principal specific heats. From this information, the appropriate value of the expansion factor is obtainable from standard graphs, available from many sources e.g., [2, 3].

The venturi meter has a fixed throat diameter and an orifice meter can be used with one of a set of orifices of a range of sizes. A rotameter works along the same principles but has the advantage that the orifice size can be changed *in situ* and smoothly rather than (as in the case of a orifice meter) in steps.

7.2.2 The weir

This means of measuring flow rates of liquids is for external flow, that is, flow in an open channel. This contrasts with internal flow in tubes and ducts. The weir as a means of measuring flow rate finds particular application to distillation processes.

The weir might extend across the entire width of the channel, alternatively the liquid can pass through a 'notch' before exiting the channel. Analysis according to Bernoulli's equation is closely similar to that in Chapter 2 for liquid exiting an orifice in the side of a vessel. The measurement which leads to a value of the flow rate is the height of the crest of the weir on exit, and there are empirical correlations for the flow rate as a function of this, e.g. [4]:

$$Q = 1.84[B - (0.1 \times nD)]D^{1.5} \qquad \text{Eq. 7.1}$$

for a rectangular notch, where Q = flow rate (m^3s^{-1}) B = notch width (m), D = height of the weir above the base of the notch (m) and n is a factor which depends upon the configuration in the following way:

$n = 0$ if the 'notch' is the full width of the channel

$n = 1$ if the notch is narrower than the channel but has its base at the level of the horizontal surface of the channel

$n = 2$ if the notch is narrower than the channel and has its base above the level of the base of the channel, but is situated half way along the channel width so as to give symmetrical flow

The height of the weir must be measured at a point some distance back from the notch, where it is fully developed, as shown in Figure 7.1, which relates to the case for which $n = 0$. Some of these principles are illustrated in the following calculation.

A hydrocarbon liquid flows along a distillation tray of width 1 m. It exits the tray under gravity and in so doing passes over a weir. The height of the liquid surface above the weir (measured from the surface edge of the weir, upstream of it) is 2.5 cm. On exiting the distillation tray the liquid enters a vessel of capacity 100 m³, spillage from which through overflow would con-

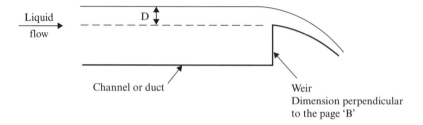

Figure 7.1 *Schematic of a weir for use in measuring liquid flow rates.*

stitute a fire and explosion hazard. For how long can the vessel receive liquid from the distillation tray without there being such a hazard?

Solution
This is clearly a case where the notch is the full tray width, that is, in the above equation $n = 0$, therefore:

$$Q = 1.84BD^{1.5}$$

with symbols as previously defined. Putting $B = 1$ m and $D = 0.025$ m gives:

$$Q = 0.0073\,\text{m}^3\text{s}^{-1}$$

Time taken to fill the vessel = $(100/0.0073)\,\text{s} = 13\,699\,\text{s}$ (3.8 hour)

One would probably build in a safety margin by making the time between emptyings of the tank about 3.25 hour, some 15% lower.

In the chemical industry, weirs are used to convey water *between* processing units and organic liquids *within* processing units. A float gauge can be used to determine the weir height.

7.3 Pressure measurement

The reader has almost certainly encountered the U-tube manometer during his or her early training in physics. This device, and simple adaptations of it, find significant use in the process industries. Such adaptations include the inclined manometer, where the fluid column is at an angle. Either the column height has to be multiplied by the sine of the angle made with the horizontal or, more commonly, the manometer is designed so that the column fits into a slot, affording a fixed angle of say 15°, in which case the column can be calibrated directly. Other variants on the basic U-tube manometer [1] include a two-liquid manometer where, for example, one liquid might be mercury and the other water. This arrangement is particularly suitable for measurement of small pressure difference in gases.

The principles involved in the operation of manometric devices of various sorts are simple and related calculations fairly straightforward. Such a calculation is in the shaded area below.

We return to the situation considered in the previous chapter, a pipe conveying toluene (density 780 kg m^{-3}). A pressure difference of 80 kPa was recorded between two points along the pipe by means of a simple differential manometer, one limb of which was connected to a pressure tapping point upstream of the venturi, the other to a pressure tapping point downstream of the venturi. The liquid in the manometer is mercury, density 13 590 kg m^{-3}, which is totally immiscible with toluene. What would be the difference in heights of the liquid in the two sides of the manometer?

Solution
Initially the two limbs of the manometer are open to the atmosphere, so the heights of fluid are at the same level at each side. Then, the limbs are connected to the respective tapping points, and at each limb an interface forms between the toluene and the manometer fluid. At the high-pressure side (pressure P_2) this interface is a vertical distance 'h' below the level of the interface at the low-pressure side (pressure P_1).

Let the depth (below the toluene-bearing pipe) of the interface at the high-pressure side be 'x' m, then the depth of the interface at the low-pressure side $= (x - h)$ m. For a fluid at rest (which the manometer contents are, after settlement) the pressure at any horizontal level is the same. We apply this to the level 'x' m below the pipe, to give:

$$P_2 + (x \times \sigma_T \times g) = P_1 + \{(x - h) \times \sigma_T \times g\} + h \times \sigma_F \times g$$

where subscript T denotes toluene and subscript F the manometer fluid, and g is the acceleration due to gravity. Rearranging:

$$P_2 - P_1 = hg(\sigma_F - \sigma_T) = 80 \text{ kPa} = 80\,000 \text{ N m}^{-2} \Rightarrow h = 0.636 \text{ m } (636 \text{ mm})$$

Whereas these devices work along hydrostatic principles, the Bourdon gauge, also widely used, is a mechanical device whereby fluid enters a metal tube which is configured in the shape of a letter 'C'. Changes to this shape in response to internal pressure are the basis of calibration of such a gauge.

These have been in use from the earliest days of the hydrocarbon processing industry.

7.4 Temperature measurement

7.4.1 Use of thermocouples

Introduction

Temperature measurement is a vast topic, the subject of numerous texts and of ongoing research and development. Thermocouples have featured several times in this book, and thermocouple thermometry continues to find wide application across many industries.

The principle of operation of a thermocouple is that two wires of dissimilar metals ('thermoelements') are welded to form a tip. *Contrary to what is widely believed, the thermocouple e.m.f. does not develop at the tip.* It develops along the wires where the temperature changes. The thermoelements are extended back to the cold end of the thermocouple where they can be connected to a reference thermocouple which stands in ice or, much more probably, a recorder with the instrumentational equivalent of a cold junction at 0°C. There are still, in 2002, only eight 'letter-designated' thermocouple types[23], that is, thermocouple types of internationally standardised composition and calibration. Three of these are 'noble metal' types, e.g.:

> *Type S:* One wire Pt, the other Pt-10% Rh, which will
> operate at temperatures up to about 1600°C.

The other five are 'base metal' types, e.g.,

> *Type K:* One wire 'chromel' (nickel–chromium) the other 'alumel' (nickel–
> aluminium–manganese). These operate at temperatures up to 1100°C.
> Invented a hundred years ago, type K is the most widely used
> of the eight thermocouple types.

Thermocouple selection and circuitry

In section 5.3, a thermocouple in a reactor of gas and vapour was considered in the context of partial oxidation of hydrocarbons. Let us return to that and adapt it to a design exercise by imagining that we are commissioned to install the thermocouple in the reactor having previously selected the thermocouple type and configuration, and to connect it to a suitable recording system. Several important facets of thermocouple thermometry will be brought out in the context of this exercise.

The temperature in the reactor is a little under 300°C and all of the letter-designated types are therefore suitable in terms of range. Type K is therefore, on these grounds, as good a choice as any and will be adopted here. The user will not expect to have to calibrate the thermocouple; it is supplied so that it conforms to standard e.m.f.s for the thermocouple type. Next comes the choice of configuration: 'bare-wire' or mineral-insulated, metal sheathed (MIMS). These terms require enlargement.

In a so-called bare-wire thermocouple, only the tip is actually bare. The thermoelements as they extend back from the tip are coated with a suitable insulator such as fibre glass (otherwise, of course, they would short on contact). Bare wire protrudes from the insulation at the 'cold end' of each thermoelement, and thus the thermocouple is wired into circuitry. By contrast, in the MIMS configuration the thermoelements are inside a sheath, which is usually made of stainless steel. The space between the thermoelements and the sheath interior surface is packed with a ceramic, usually magnesium oxide powder which is a excellent electrical resistor. The obvious advantage of the metal sheath is that it protects the thermoelements from chemical attack by their environment. MIMS thermocouples come in sheath diameters from 0.5 mm to 1 cm or larger. The smaller sizes are made by drawing down, under heat, MIMS stock previously of wide diameter, and this necessitates annealing afterwards in order to remove strain introduced into the thermoelements during heating. If this strain is not removed it can jeopardise the calibration. A MIMS configuration of 1 cm sheath diameter is a sensible choice for the hydrocarbon example under consideration. At 300°C this has an intrinsic uncertainty of ±2.2 K. Radiation errors such as were identified in the original numerical example will add to this. Figure 7.2 shows a pair of MIMS thermocouples.

Once the type and sheath diameter are decided upon, a MIMS thermocouple can be made to order in terms of its length and the nature of its cold end, whether it is simply 'flexible tails' or a 'terminal head'. Arbitrarily we select one metre for the length with a terminal head, by means of which the thermocouple will be connected to the measuring circuit (see below). A thermocouple supplier would probably regard this information as complete and go ahead and construct the device to these specifications, and in so doing terminate the junction of the thermoelements in such a way that it was separated from the interior of the sheath tip by a few mm of magnesium oxide packing. This is called an *isolated* junction and is likely to be supplied by default if no instructions to the contrary are given. The alternative is the *grounded* junction, where the thermoelements are welded to the inside of the sheath tip; the three metals, two thermoelements and sheath alloy, form a single blob of metal at this point. The grounded junction has the advantage

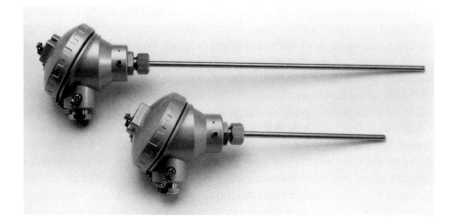

Figure 7.2 *MIMS thermocouples. Reproduced courtesy of Labfacility, UK.*

of faster response time when immersed in a fluid, but in our application response time is fairly unimportant as it is an essentially steady temperature, displaying only small variations, which is being followed. Moreover, the grounded junction type can, under some circumstances, exhibit spurious signals called 'earth loops' [5]. In our application there is no reason to choose grounded junctions.

Thus far then a type K, MIMS 1 cm sheath diameter thermocouple with an isolated junction and a termination head has been specified. There are the following points to be considered when incorporating it into a 'measurement circuit'. First, extension or compensating cable is required to take the thermocouple signal to the recording system from the plant. Extension cable has the same composition as the thermocouple itself, in this case chromel and alumel, but is of less stringent specification so as to have the same thermoelectric properties only up to about 200°C. Compensating cable has a quite different composition from the thermoelements in the thermocouple, but the same thermoelectric properties up to about 80°C. Copper/constantan is the usual compensating cable for type K. Compensating cable is cheaper, but extension cable is more accurate and will be used in this application. Insulator colour codes have to be checked, especially if ANSI (USA), BSI (UK), JIS (Japan) or DIN (Germany) products are being mixed. It should be safe to attach, for example, BSI type K extension cable to an ANSI type K thermocouple provided that the connection is made correctly having regard to the polarities. If the polarity is unwittingly reversed the error accruing will be exactly twice the temperature difference between the cold end of the thermocouple and the terminals of the recording system. Also, although modern recorders are good at rejecting non-genuine signals, it is wise to shield the extension cable in order to eliminate pick-up. This is especially desirable in the presence of heavy-current electrical plant such as compressors and pumps, or if the extension cable passes close to power cables.

Whether the recording system is a potentiometric recorder, giving a continuous printed record, or a PC, is immaterial provided that the following conditions are met. First, the recording device must have *internal cold junction compensation* appropriate to type K. This can be retrofitted if necessary; inexpensive printed circuit boards are obtainable for this. To ensure proper functioning of the cold junction compensation facility, the recording device should be positioned where ambient temperature changes are not severe. In this application that would mean placing it far enough away from the vessel for the recorder not be receiving radiation from its outer walls. Secondly, the recording system must have *thermocouple burnout protection*. If a thermocouple fails by reason of breakage of one of the wires, there might still be an e.m.f. at mV level because of stray electric fields, and the operator could mistake this for a genuine temperature signal. However, if the thermocouple is broken the resistance in the circuit will be extremely high, orders of magnitude higher than that of the thermocouple/extension cable combination before breakage. With thermocouple burnout protection the recording instrument is programmed to respond to such an inordinately high resistance by disregarding the reading, thereby alerting the operator to the thermocouple failure; some sorts of recording system also provide for an audible alarm to come into action in this eventuality.

General advice on thermocouple usage and comments on accuracy

We conclude with a number of practical tips on setting up thermocouple circuits, relevant not only to the design exercise above but also to thermocouple usage generally. One follows from the statement above that the e.m.f. develops not at the tip but along the thermoelements where the temperature changes. It follows from this that particular attention should be paid to the condition of a pre-used thermocouple at points along its length where the temperature gradient is greatest. In terms of our example this means the part of the thermocouple actually passing through the reactor wall, where at one side the temperature is in the neighbourhood of that being measured whilst at the other side the temperature is in the neighbourhood of that of the room. Any part of the sheath length showing symptoms of wear and tear, e.g., by having been bent, should be placed either fully inside or fully outside the reactor and not in the part of the circuit where the temperature gradient is severest.

Electrical connections, for example between the thermocouple terminal head and the extension cable, have to be made and it is usually sufficient for these to be general-purpose electrical connectors. It is true in principle that a temperature gradient across such a connector will introduce an unwanted thermal e.m.f., but in all circumstances relevant to day-to-day chemical processing such e.m.f.s are much too small to have an effect. Non-metallic connectors, or connectors made of alloys to match the thermoelements they are in contact with, are available. These are, however, required only in certain very specialised applications, for example in a physiology laboratory where the temperature response of an organism to a stimulus is being monitored. Thermocouples intended for such applications not only have to be calibrated against a standard to give closer tolerances than those given by the manufacturer but also, before calibration, have to be checked for homogeneity of their thermoelements to ensure that this is sufficiently good to support such calibration.

The contents of the previous paragraph are all of interest, and indeed practical importance, to the specialist in thermoelectric thermometry. A reader of this text is informed of them only so that he or she will have a true appreciation of thermocouple readings and the uncertainty in them. It is extremely difficult, in routine application of thermocouples, to measure even a steady temperature to better than to about ±2–3°C (or more at higher temperatures). This is due chiefly to the calibration uncertainty which increases with usage and which, as we have seen, in a gas- or vapour-phase application might be compounded by radiation effects. The recording device, even when fully 'within spec', also adds a significant fraction of one degree to any uncertainty. Claims of thermocouple readings to within a tenth or even (as the author has seen) a fiftieth of one degree need to be viewed with the utmost caution.

The signal from a thermocouple must go to a high-impedance recording device. This requirement can be assumed to be fulfilled for recorders pre-calibrated for thermocouples with internal cold-junction compensation, but if a general-purpose recorder of basic design is used the impedance should be checked especially if there are very long runs of extension or compensating cable the total resistance of which might not be negligible in comparison with the impedance of the recorder. Lengths of such cable of the order of a kilo-metre are not uncommon in chemical processing plants. Of course, when the e.m.f. is measured there has to be, in principle, a current in the thermocouple circuit. With a suitably high impedance this current will be of the order of μA or even nA, so that no heating effect results. This point is taken up in the next section which deals with resistance thermometry.

The discussion of thermocouples continues with a related numerical example which illustrates the points made above in relation the thermocouple accuracy.

A previously unused MIMS type K thermocouple has an uncertainty of ±2.2°C or 0.75% of the reading in °C, whichever is larger [6]. Such a thermocouple is used to measure the temperature of a stirred tank of hydrocarbon liquid at 150°C. The terminal head of the thermocouple, a few centimetres from the outside wall of the vessel, is at a temperature of 40°C. This connected to extension cable with a nominal tolerance of ±2.5°C, which is electrically shielded. The extension cable conveys the thermocouple signal to a recording device which is set up in an air-conditioned instrument room where the atmospheric temperature is controlled at 18°C. The recording device has internal cold junction compensation appropriate to type K, and this has an uncertainty of 0.8°C. Estimate the error in the thermocouple reading when:

(i) the temperature is read at a digital display
(ii) the temperature is read from a recorded analogue trace where the signal occupies a field of 10 cm width on the recorder chart and there is an uncertainty of 2 mm attributable jointly to the thickness of the printed signal and possible alignment error, and of 0.25 mm attributable jointly to thermal expansion and mechanical stretching of the chart paper[24].

Solution
First, since the thermocouple is immersed in a liquid, which can be assumed to be opaque to thermal radiation at the measured temperature, there are no radiation errors. Also, since the extension cable is shielded there will be assumed to be no errors due to electrical pick-up. The errors in a digital reading are then:

that due to the calibration uncertainties in the thermocouple
that due to the calibration uncertainties in the extension cable
that due to the instrumentational cold junction

We retain the ±2.2°C uncertainty in the thermocouple at 'face value', but it is necessary to incorporate the uncertainty in the extension cable calibration a little more subtly. In the limit where there is no temperature difference between the cold end of the thermocouple and the terminals of the recording device there is no e.m.f. in the extension cable: an uncertainty of 2.5°C in the reading could not possibly, in these circumstances, be caused by the extension cable. The Seebeck coefficient – voltage thermally generated per unit temperature rise – of type K at temperatures between 0 and 200°C is almost constant at $40\,\mu V\,°C^{-1}$. The nominal uncertainty re-expressed in thermal e.m.f. terms is therefore about $100\,\mu V$, and this should be understood as appertaining to the maximum permissible temperature range, actually 200°C. In the example of interest the temperature range experienced by the extension cable is only 22°C, so a reasonable estimate of the accompanying uncertainty is:

$$(22/200) \times 100 \, \mu V = 11 \, \mu V \equiv [(11/40) \times 1°C] = 0.28°C$$

All of the errors can be combined by taking the square root of the sum of their squares to give, for a digital reading:

$$\left(\sqrt{2.2^2 + 0.8^2 + 0.28^2} \right)°C = 2.4°C$$

In the case where an analogue trace is used, the uncertainty due to the thickness of the print and imprecise alignment is:

$$(2/100) \times 150°C = 3°C$$

That due to expansion, contraction and stretching is clearly:

$$(0.25/100) \times 150°C = 0.38°C$$

so for an analogue reading the total uncertainty is:

$$\left(\sqrt{2.2^2 + 0.8^2 + 0.28^2 + 3^2 + 0.38^2} \right)°C = 3.8°C$$

In the hypothetical example above all errors are due to the measurement system: that this is not always so, as the reader is already aware. All other conditions are ideal in the example: no electrical pick-up, no extraneous thermal e.m.f.s and a thermocouple previously unused and therefore within spec. The view stated above that a temperature is never, by routine application of thermocouples, measurable to better than about ±2–3°C is supported.

Non-letter-designated thermocouples

There remain only eight letter-designated thermocouple types as noted previously, and it is 30 years since the most recently letter-designated (Type N) joined the previously existing seven. 'Letter designation' in the sense in which the term is being used requires that for a newly developed thermocouple a case be made to bodies including ANSI for inclusion with the exclusive eight, making it an exclusive nine! The incentive for developers of thermocouples to seek letter designation might not be strong, and there are thermocouples in use which have a letter signifying their specifications but are not letter designated in the formal sense. One of these in Type M, whose thermal e.m.f.s are given in [7]. This has one thermoelement made of molybdenum and another of molybdenum-18% nickel. It can operate in the temperature range –50 to +1410°C, giving it a small advantage in terms of upper limit of use over Type K. There is also the Platinel thermocouple, in which one thermoelement is gold-palladium and the other gold-palladium-platinum [8]. These have been used in the chemical industry [9]. Tolerances for these are about the same as those for letter-designated types, that is of the order of plus or minus one or two degrees. A few other examples of thermocouples identified by letters but lacking formal letter designation exist and details can be found on web sources.

7.4.2 Resistance thermometry

Thermocouples work by thermally generated e.m.f.; the resistance of the wires is totally irrelevant. Indeed, thermocouples of the same type with different wire diameters obviously have different resistances, but this does not affect the calibration. By contrast resistance temperature detectors (RTDs) work on the principle that the electrical resistance of a conductor increases with temperature. RTDs have found widespread use in the chemical processing industry.

Platinum is the usual choice of metal for use in RTDs, in particular the Pt100 sensor. This has a resistance of 100 ohm at 0°C, and at temperatures between that and 850°C the polynomial:

$$R(\theta) = R_o\{1 + A\theta + B\theta^2\} \qquad \text{Eq. 7.2}$$

where $R(\theta)$ = resistance of the platinum at temperature θ°C
$\qquad R_o$ = resistance of the platinum at 0°C
$\qquad A = 3.9083 \times 10^{-3}\,°C^{-1}$
$\qquad B = -5.775 \times 10^{-7}\,°C^{-2}$

applies. The platinum wire can be mounted on a ceramic support, or sheathed like a MIMS thermocouple (Figure 7.3). In principle, the RTD can be connected, by two lead wires, to a resistance-measuring device such as a Wheatstone bridge (not, of course, in the simple classical form that one encounters in elementary physics texts, but incorporated into as instrument which gives temperature readings directly). This two-wire arrangement has the disadvantage that the resistance of the lead wires adds to the reading at the bridge, introducing an inaccuracy. If the leads are made of copper, which has an extremely low resistivity, this effect will broaden the error bars but will not invalidate the temperature measurement. A three-wire arrangement is possible whereby the resistance of the leads is corrected for. Even here the correction will be exact only if the resistances of the three leads are identical

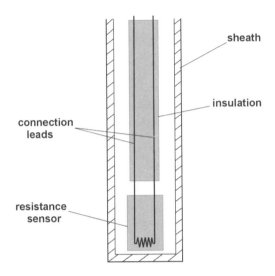

Figure 7.3 *RTD in sheathed configuration.*

to each other. Whilst for most industrial purposes a three-wire configuration is the norm, the most accurate is a four-wire configuration.

Recording devices are often made to be interchangeable for thermocouples or RTDs. The recorder contains both cold-junction compensation for thermocouples and resistance measurement for RTDs, and the user selects one or the other. However, terminals need to be specific to one or the other, there being a two-wire input for thermocouples and usually (as we have seen) a three-wire input for RTDs. The cold-junction for thermocouple applications might itself use resistance thermometry. In this event somewhere close to the terminals there will be a resistance temperature sensor which 'informs' the instrument of the ambient temperature and, therefore, of the thermal e.m.f. compensation required.

As already stated RTDs have found wide application in chemical plant. In some quarters there has been a swing towards them because they have better intrinsic accuracy than thermocouples which, as we have seen, are always at least ±2°C. A Pt100 thermometer has a tolerance of less than 1°C at temperatures up to 400°C.

With thermocouples an e.m.f. is being measured and the impedance can be made as high as one wishes to reduce the current to minuscule values as we have already seen. With RTDs a resistance is being measured, and the current during measurement is not necessarily undetectable in terms of possible heating effects. Such effects, though small, are a possible source of error in RTD measurements.

In the previous section, a calculation based on thermocouple measurement the temperature of a stirred tank of hydrocarbon liquid at 150°C was presented, and here we extend it to RTDs. Whereas a thermocouple can be immersed in a liquid, an RTD has to be protected if so immersed, possibly with a thin glass sheath. However, in measurement of a steady or slowly changing temperature there is no reason to suppose that this introduces any error.

A Pt100 sensor is immersed in a stirred hydrocarbon liquid at 150°C. If the tolerance on the sensor is 0.4 ohm what will be the total uncertainty:

(a) if the sensor is configured in a three-wire arrangement, and the uncertainty due to the recording device is 0.3 ohm. Assume that all three leads have the same electrical resistance.
(b) if the sensor is configured in a two-wire arrangement by means of two copper wires of diameter 1 mm and length 5 m. The uncertainty due to the recording device is 0.3 ohm and the resistivity of copper is 1.6×10^{-6} ohm cm.

Solution
Once the error δR in the resistance is estimated, that in the temperature ($\delta\theta$) can be determined as:

$$\delta\theta = \delta R/(dR/d\theta)$$

Now $dR/d\theta = AR_o + 2BR_o\theta$ (symbols as in equation 7.2) giving:

$$dR/d\theta = 0.374\,\Omega°C^{-1} \quad \text{at } \theta = 150°C$$

For the three-wire configuration:

$$\delta R = \sqrt{0.4^2 + 0.3^2} = 0.50\,\Omega \;\Rightarrow\; \delta\theta = 1.3°C$$

Now with the two-wire configuration there is also the resistance of the copper to be accounted for:

$$R_{copper} = \{2 \times 1.6 \times 10^{-6} \times [500/(\pi \times 0.05^2)]\} = 0.20\,\Omega$$

$$\delta R = \sqrt{0.4^2 + 0.3^2 + 0.2^2} = 0.44\,\Omega \;\Rightarrow\; \delta\theta = 1.4°C$$

The device, even in the two-wire configuration, therefore performs in a superior way to thermocouples, the resistance of the leads in this application adding only a fraction of one degree to the error. Moreover, the intrinsic uncertainty given on the RTD is for a Class B Pt100 device. With the superior Class A device the tolerance is 0.23 ohm at 250°C. Temperature errors in the above example for the three- and two-wire configurations with this are, respectively, 1.0 and 1.1°C.

It must be remembered that heat transfer effects on temperature measurements apply to RTDs as well as to thermocouples. For example if an RTD is measuring a gas or vapour temperature and has a 'view' of a colder surface there will be a radiation error, as with thermocouples or indeed any thermometric device.

7.4.3 Measurement of cryogenic temperatures

The cryogen of most interest to the specialist in hydrocarbon technology is likely to be liquefied natural gas (LNG). Its boiling point, approximated to that of pure methane, is −161°C (112 K). Thermocouples including type K can be used to measure such temperatures, but not in the straightforward way in which they are used to measure the temperatures for which they are calibrated and to which the standard tables apply, 0–1250°C in the case of type K. Mention was made earlier in the chapter of the need to anneal thermocouples to remove strain from the thermoelements which would otherwise affect the thermal e.m.f. To immerse a thermocouple into a cryogen reintroduces such strain and takes it 'out of spec' in terms of its conventional measurement range. Hence a thermocouple for use at cryogenic temperatures has to be set aside for such use and specially calibrated, perhaps using liquid nitrogen (77 K) as one calibration point. Type K thermocouples can, under these circumstances, measure down to −200°C (73 K) with a tolerance of about ±4°C. A thermocouple so calibrated would not be 'in spec' in the conventional measuring range.

A Pt100 RTD can also be used down to −200°C, though a different resistance–temperature correlation applies from that given previously for such devices working in a much higher temperature range. A related numerical example follows.

Consider LNG at 112 K (−161°C) passing along a pipe. The resistance–temperature correlation for a Pt100 device at such temperatures is, retaining the notation in equation 7.2:

$$R(\theta) = R_o\{1 + A\theta + B\theta^2 + C(\theta - 100)\theta^3\}$$

where A and B have the same values as previously and C has value $-4.183 \times 10^{-12}\,°C^{-4}$.

At the temperatures of interest, a Class A Pt100 RTD has a tolerance of 0.20 ohm. How accurately can the temperature of the LNG be measured by this means?

Solution
Differentiating the polynomial:

$$dR/d\theta = AR_o + 2BR_o\theta + 3R_o\theta^2 C(\theta - 100) + R_o C\theta^3$$

Substituting A, B and C, with $\theta = -161°C$, gives:

$$dR/d\theta = 0.413\,\Omega\,°C^{-1}$$

Hence for a resistance uncertainty of 0.2 ohm the temperature uncertainty is:

$$(0.2/0.4)°C = 0.5°C$$

The RTD displays a very significant accuracy advantage over thermocouples in the above example. For cryogenic applications the RTD also has the advantage that no special calibration is needed, as it is for thermocouples operating at such temperatures.

7.5 Fire protection of sensitive measurement instruments

Thermocouples and RTDs are used not only in temperature measurement at chemical plants but also in control. For the thermocouple and RTD circuitry and the devices to which their signals go to be affected by fire could have serious consequences for the plant. Such a fire would probably not originate at the scene of chemical processing and might be electrical in origin. The fighting of such a fire would be on a much smaller scale than the fighting of a well developed hydrocarbon fire, and fire protection for instruments is best provided by carbon dioxide extinguishers. There are several manufacturers of carbon dioxide extinguishers who supply refineries with them to protect the instrumentation. A related calculation follows.

An electronic device at a refinery approximates to a cube of 50 cm side. It becomes overheated, its surfaces reaching a temperature of 700°C. At what rate will it radiate heat? What would be the required amount of carbon dioxide in an extinguisher with a discharge time of one minute to hold the temperature of the malfunctioning device at 700°C for that period of time whilst measures were taken to isolate it from the power supply. Use a value of $1000 \, J \, kg^{-1} \, °C^{-1}$ for the specific heat of CO_2.

Solution
Taking the surfaces to be 'black':

$$\text{Rate of heat release by the device} =$$
$$5.7 \times 10^{-8} \, W \, m^{-2} K^{-4} \times 6 \times 0.52 \, m^2 \times 9734 \, K^4 = 77 \, kW$$

and this is the rate at which the carbon dioxide must remove heat to maintain the quasi-steady conditions required.

Calling the mass flow rate m, rate of uptake of heat by the CO_2 (discharge temperature say 25°C)
$$= m \, kg \, s^{-1} \times 1000 \, J \, kg^{-1} \, °C^{-1} \times (700 - 25) = 77\,000 \, W$$
$$\downarrow$$
$$m = 0.1 \, kg \, s^{-1}$$

$0.1 kg \, s^{-1}$ for one minute requires 6 kg.

The amount of carbon dioxide and the discharge time are both representative of extinguishers available.

7.6 Concluding remarks

Though measurements of various quantities have featured in this chapter, the emphasis has been on temperature measurement. Temperature measurement devices are of course also used in temperature control, and a good understanding of thermometric devices is helpful in understanding control devices. Several of the principles of thermoelectric thermometry covered in the main text also feature in the appended numerical examples.

References

[1] Geankoplis C.J. *Transport Processes and Unit Operations*, Second Edition. Allyn and Bacon, Boston, MA (1983)
[2] Perry R.H., Green D. *Perry's Chemical Engineers' Handbook*, Sixth Edition. McGraw-Hill, New York (1984)
[3] Streeter V.L. *Fluid Mechanics*, any available edition. McGraw-Hill, New York
[4] Coulson J.M., Richardson J.F. *Chemical Engineering Volume 1 Fluid Flow, Heat Transfer and Mass Transfer*, Fourth Edition. Pergamon (1990)
[5] Jones J.C. A combustion scientist's view of thermocouple temperature measurement. Seminar on Advanced Sensors and Instrumentation Systems for

Combustion Processes, pp 11/1–11/4 Institution of Electrical Engineers, London (2000)

[6] Bentley R. *Theory and Practice of Thermoelectric Thermometry*. Springer Verlag, Singapore (1998)

[7] http://www.pyromation.com/downloads/data/emfm_f.pdf

[8] http://digital.library.unt.edu/ark:/67531/metadc13234/m1/22/

[9] http://www.catalysts.basf.com/p02/USWeb-Internet/catalysts/e/function/conversions:/publish/content/microsites/catalysts/prods-inds/temp-sens/BF-8219_Enclad-0209A.pdf

Proprietary literature published by Labfacility, UK, has also been drawn on, with permission, in this chapter.

Numerical examples

1. In the previous chapter we considered the flow of toluene vapour, and its eventual condensation, in the context of a design calculation. Such a process might, for example, represent the separation of toluene from BTX (benzene, toluene, xylenes) which frequently occur together. Now consider the liquid toluene, which exits the condensation vessel and is passed along a horizontal 4-inch (10.2 cm) pipe on its way to a chemical reactor. A venturi, of throat diameter 1 inch (2.54 cm), is put in position to measure the flow rate. Pressure tappings up- and downstream of the venturi give a pressure difference of 80 kPa. The liquid density is $780 \, kg \, m^{-3}$. Calculate the flow rate of the toluene.

2. A distillate of density $800 \, kg \, m^{-3}$ flows along an open channel and over a weir having a rectangular notch of width 25 cm. Liquid flow faster than $10 \, kg \, s^{-1}$ is too rapid, there being a risk of flooding further along the process. Suggest at what weir height the alarm should be given that the flow is too rapid.

3. The Antoine constants (see Chapter 6) for methanol are:
$$A = 7.87863$$
$$B = 1473.11$$
$$C = 230.0$$
Methanol is stored in a tank at 40°C. A manometer, with mercury as the internal liquid and open to the atmosphere at one side, is attached to the vessel in order to measure the pressure in the space above the liquid surface. What will be the difference in heights of liquid between the two limbs of the manometer:

(a) If the vessel was previously evacuated, so that the vapour phase contains methanol vapour only?
(b) If the space contains, in addition to the methanol vapour, air at a partial pressure of 1 bar?

Take the surrounding atmosphere to be at exactly 1 bar.

4. Return to problem 9 at the end of Chapter 5, where hydrocarbon liquid after cracking is required to cool from 250°C to 90°C by means of a heat exchanger. It is desired to record the exit temperature of the hydrocarbon liquid by means of a type J (iron/constantan) thermocouple the intrinsic uncertainty in which is ±2.2 K when new. The cold end of the thermocouple is at 40°C and is connected by extension cable to an instrument, having cold junction compensation, which gives a digital reading every 30 s. The recorder terminals are at 27°C. There is a vapour explosion, involving plant damage but no death or injury to persons, downstream of the heat exchanger and in thelegal follow-up to the incident the exit temperature of the liquid during the previous 5 minutes is ascertained by examining the digital readout from the recorder. The temperature during this period was recorded as being initially 85.5°C rising, in the half-minute before the explosion, to 87.7°C. Examine this range in the light of thermocouple errors and estimate by how much, if at all, the true liquid temperature might have exceeded 90°C:

(a) if the tolerance on a new thermocouple is assumed to apply.
(b) if the thermocouple has been in use for a long period without a calibration check and, consequently, it is suggested that the intrinsic uncertainty might be as high as ±4.5°C.

The tolerance on type J extension cable at its maximum operating temperature of 200°C is 2°C. Make a reasonable estimate of your own for the error due to the cold-junction compensation. What would be the uncertainty in measurement of the temperature of the liquid cooled to 90°C with an RTD the uncertainty in which is 0.4 ohm?

5. Return to the calculation in the main text, in which a type K thermocouple is used to measure the temperature of a hydrocarbon liquid. Further details are as follows. The software by means of which the instrument converts the e.m.f. to a temperature draws on e.m.f.–temperature information for ANSI (US) specification thermocouples, and initially ANSI products – thermocouple and extension cable – are used. There is then renewal of the thermocouple though not of the extension cable (which seldom needs renewing). A thermocouple of BSI (UK) colour code is substituted for the original one. Predict the effect if any when:

(a) the replacement thermocouple is connected to the instrument with correct polarity via the ANSI compensating cable also correctly connected.

(b) the replacement thermocouple is connected to the instrument correctly in terms of its own polarity but in such a way that the chromel lead of the thermocouple is unwittingly connected to the alumel lead of the extension cable and (obviously) vice versa.

6. In refrigeration of a hydrocarbon storage tank, the evaporator temperature is −18°C. It is desired to monitor this temperature, as a check on correct functioning of the refrigerator, by means of an immersed type K thermocouple. This is connected to another type K thermocouple the tip of which is welded to a block of metal which can be taken to be at a constant +18°C and which forms in effect the reference junction. According to standard tables for type K, with the reference junction at 0°C the e.m.f. at −18°C is −0.701 mV and at +18°C is 0.718 mV. The measuring circuit goes to a recorder. What is the expected reading at the recorder if nothing is amiss at the evaporator of the refrigerator?

Repeat the calculation using the information that the Seebeck coefficient of type K in the temperature range of interest is $39.5\,\mu V\,°C^{-1}$.

Which of the two calculations is, in principle, more precise?

7. If in the above question a Pt100 RTD were used what would the resistance be?

8. It is desired to measure continually the temperature of liquid ethane in equilibrium with its vapour at 1 bar, that is, at its normal boiling point which is −88°C. With a platinum RTD what would be the resistance?

The Seebeck coefficient of Type K at 0°C is $39.5\,\mu V\,°C^{-1}$ as stated above. Use that to estimate the e.m.f. at the boiling point of ethane, with the cold junction at ice temperature.

Go online to http://www.thermometricscorp.com/PDFs/Thermocouple-Charts/Type-K-Thermocouple-Chart-C.pdf or equivalent and find the temperature in the tables corresponding to that e.m.f. and comment on the result.

Comment on the possible relative merits of bare-wire or MIMS configuration for the thermocouple.

9. A thermocouple reads a gas temperature of at 300°C (573K). If the thermocouple tip is a black body and the walls are at 297°C (570K) what convection coefficient at the tip is necessary for the radiation effect not to exceed 1K?

10. Return to the question of fire extinguishment in section 7.5. It often happens that exit of carbon dioxide from a pressurised extinguisher is accompanied by some solid formation. What percentage of the amount of CO_2 calculated would need to be in the solid phase for the effectiveness time to be extended by one second. The heat of sublimation of carbon dioxide is $571 \, kJ \, kg^{-1}$.

11. Saturated water vapour at 6 bar is in use at a plant. Imagine that it is desired to check its temperature with an Pt100 RTD the tolerance on which is 0.4 ohm. Calculate the resistance of the device at that temperature and the error in the temperature measurement if there is no error due to the leads. Steam tables will need to be consulted.

Endnotes

[23] The most recent to attain 'letter designation' was the type N (nicrosil-nisil), in 1983.

[24] The two effects referred to in relation to recorded traces are usually discussed in the user manuals provided with such recording instruments.

8

Offshore oil and gas production

8.1 Introduction

The 'oil industry' dates from about three-quarters of the way through the nineteenth-century, as we saw in Chapter 1. The *offshore* oil and gas industry dates from 1945 [1, 2], when the first offshore oil installation, in the Gulf of Mexico, began operation. There had been very limited offshore production in shallow waters off California in the 1890s, but this was at the time that the Texan oil fields were rapidly expanding and there was little incentive to develop offshore fields anywhere. There was also extension offshore of fields in Borneo previously worked onshore in the 1930s; British oil companies were involved in this. The year 1945 however is therefore correctly seen as the year of commencement of modern offshore oil and gas production. Over that time sea depths and well depths have increased greatly. Current limits are about 3000 m sea depth and >10 000 m well depth.

At the present time there is offshore oil/gas production off the coasts of about 75 countries. Aberdeen, where the author has lived and worked for many years, is adjacent to the British sector of the North Sea oil and gas fields and is closely associated with them. In promotional contexts, the city is sometimes called the 'oil capital of Europe'! Figure 8.1 shows the Beryl Alpha platform in the North Sea. It is operated by Mobil.

Many residents of Aberdeen are employed in the industry, sometimes as offshore workers, sometimes in related jobs onshore, for example at Cruden Bay, a few miles up the coast, where oil is brought in from offshore, and at St. Fergus where the gas is brought in. There are also, of course, many engineering firms which service the offshore industry and numerous consultancy practices which specialise in offshore matters. As well as the North Sea and the Gulf of Mexico, which have already been mentioned, major scenes of offshore activity include the Bass Strait, off the coast of south eastern Australia, the coast of Honshu, Japan (as mentioned in Chapter

Figure 8.1 *Beryl Alpha platform in the British sector of the North Sea. Reproduced courtesy of the Health and Safety Executive, UK.*

1), off the Malay Peninsula and off some of the many islands comprising the nation of Indonesia.

Any production platform would be an 'accident waiting to happen' were it not for the safety practices which have been developed over the years since offshore activity began. In spite of the hazards peculiar to offshore activity, major accidents offshore oil and gas production[25] have been few, and this is an indication of the high standards which have been maintained. A whole branch of safety engineering has evolved in response to the needs of the industry. Relevant disciplines include structural engineering, fire and explosion science, statistics and probability, occupational psychology and law.

Production platforms effect the separation of the well contents as summarised in Figure 8.2.

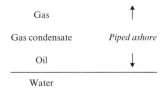

Figure 8.2 *Schematic of the separation effected by an offshore production platform.*

A good deal of the water, arising from aquifers in the geological formation, also finds its way ashore. To dump it straight back into the sea would obviously be environmentally unsound; it has to be cleansed, and this can be done onshore by means of separator tanks. (Oil pollution of the sea is touched on in Chapter 2.) Other ways in which such water can be processed is by ground disposal or by re-injection into the well. The dominant practice in the North Sea is however settlement tanks and return to the sea. Such water finds its way back into the sea after cleansing, and retains some of its original organic contaminants. Discharge limits, typically 40 mg of oil per litre of water (≈40 p.p.m., weight basis), apply. Dilution of the contaminants by a factor of about 10^6 is realised a kilometre or so from the discharge point.

8.2 Some features of an offshore platform

An offshore platform is uniquely hazardous in that persons are miles out to sea and surrounded by huge quantities of powerfully combustible material. Components bearing hydrocarbon inventory include those briefly described in Table 8.1. To enter all of the configurational details of a production plat-form into a software package for risk assessment requires weeks of effort by a trained analyst. Some parts of a platform not bearing inventory are described in Table 8.2.

The note in the penultimate row of Table 8.2 requires some qualification. It is true of many current platforms, but in the design of a new platform it would be difficult to reconcile the accommodation of off-duty personnel actually on the production platform with the concept of *inherent safety*.

8.3 The role of structural components in platform safety

A *truss* has a role in supporting modules containing inventory and/or persons (including the temporary refuge). A particular truss will have an endurance time in the event of flame impingement, beyond which it will fail. Some walls and solid floors have a role in fire protection as well as in structural support.

Table 8.1 *Some key hydrocarbon-bearing facilities at an offshore platform*

Component	Description of function and supplementary comments
Wellhead	Interface of the well pipe work and the platform pipe work Oil and gas pass from the wellhead to separators *q. v.* Periodic maintenance work required on a wellhead already in service. Possible blowout (loss of containment) during this
Separators	Well fluid separated into oil and gas. Gas further processed by separation of condensate by compression and cooling Fire and explosion hazards created by accidental loss of containment of inventory
Gas compressor	Compression of gas, after condensate removal, for conveyance ashore along the gas pipeline *q. v.*
Gas or oil riser	Pipe work by means of which inventory enters or leaves a platform. May be rigid or flexible
Oil and gas pipelines	Conveyance ashore of oil and gas. A powerful pump required for the oil. Condensate mixed with the oil stream and separated onshore. Gas pipelines operate at pressures in the neighbourhood of 140 bar. Regular cleaning of pipelines, especially the oil pipeline, by means of a device called a 'pig' which, having been launched at the 'pig launcher' on the platform, travels along the pipeline and scrapes waxy deposit from the pipe interior surface. 'Pigs' might also contain detectors, signals from which give information on the condition of the pipeline. There will also be considerable lengths of pipe work on the platform, bearing oil or gas

Table 8.2 *Some parts of a platform not bearing hydrocarbon inventory*

Component	Function
General utilities areas	Stores, ventilation, air conditioning, production of fresh water from seawater, etc. Common for fire water protection system pipe work to pass through such areas
Accommodation	Housing of personnel. Often located on the production platform, but can be a separate platform
Temporary refuge area	For mustering of personnel to await evacuation by lifeboat or helicopter in an emergency. A suitable accommodation module might double up as a temporary refuge

These are assigned, in risk assessment, an endurance time of flame contact and a maximum explosion overpressure which can be withstood. Such a wall or floor is sometimes, in ORA, referred to as a *real divider*. By contrast, a *null divider* is a wall or floor having no protective function: nil flame impingement endurance time and unable to withstand any overpressure (though its presence might impede the drift of smoke). *A grated floor* has no resistance to fire or to smoke, but might have the safety benefit, in the event of explosion, of relieving the overpressure.

8.4 Background to offshore accidents

It is of course possible for an accident at an offshore installation to be initiated by structural factors, for example failure of a load-bearing member through corrosion, severe weather or collision of a vessel. More commonly the *initiating event* is leakage of hydrocarbon, ignition and escalation with the potential to lead to destruction of the platform and death of its occupants. Of course, not all leaks are followed by ignition, but there is a legal requirement that all leaks are formally recorded and documented. We know from Chapter 3 that the type of combustion depends on the hydrocarbon inventory and the conditions; for example, gas leaking through a small orifice will give rise to a jet fire. If the jet fire torches a pipe or vessel itself containing hydrocarbon inventory and causes it to fail, this is the commencement of *escalation*.

Consequence analysis, which has a section to itself later in this chapter, is concerned with the effects of particular types of combustion behaviour on previously unaffected components. For example, given that a jet fire has resulted from the initiating event, will it be long enough to impinge upon adjacent hydrocarbon-bearing pipes or vessels? Will it be in the right direction to do so? The basic equations for jet fires, and those for the other sorts of combustion phenomenology discussed in Chapter 3, are utilised in consequence analysis.

8.5 Measures taken in the event of an initial leak

One such is operation of emergency shutdown (ESD) valves to isolate hydrocarbon inventory from the fire. Another is engagement of fire water protection systems (FWPS). Blowdown valves divert gas inventory from the fire. Persons muster at the temporary refuge (TR) (see Table 8.2) for evacuation by lifeboat or helicopter. All of this is subject to reliability limitations, as the calculation in the following shaded area shows.

At an offshore installation two emergency response devices are required to operate in the event of leakage at a particular part of the platform: a shut-down valve with a reliability of 0.95 and a water spray with a reliability of 0.97. Express all of the possibilities and show that their sum is unity.

valve and spray both work: $0.95 \times 0.97 = 0.9215$

valve works, spray fails: $0.95 \times 0.03 = 0.0285$

valve fails, spray works: $0.05 \times 0.97 = 0.0485$

both fail: $0.03 \times 0.05 = 0.0015$

Total $= 1.0000$

Also installed at platforms for emergency response are 'deluge systems' for reticulation of sea water around the platform for the cooling of plant and structures. These can be driven by electricity. There is also a 'blowdown system', diversion of gas to a flare. The following calculation illustrates the potential benefits of deluge systems.

During a fire at an offshore platform an enclosure bounded by walls, floors and the outside surfaces of certain plant and pipe work becomes heated to 600°C (873 K). If evacuees previously delayed by the effects of smoke have to pass through this part of the platform on their way to the temporary refuge, and to do so takes 7 s, what percentage of such evacuees will experience fatal burns?

Imagine that the part of the platform under consideration was served with a deluge system to cool the surfaces. To what temperature will the deluge need to reduce the surface temperature of the enclosure in order for there to be no fatalities during evacuation? The probit equation for fatal injury by exposure to thermal radiation is:

$$Y = -36.38 + 2.65 \ln\{tI^{4/3}\}$$

where I is the thermal radiation intensity ($W\,m^{-2}$), referred to the area of the receiver, and t the exposure time (s).

Solution
Taking the enclosure inside surface to be 'black':

$$\text{radiative flux} = 5.7 \times 10^{-8} \times 873^4 W\,m^{-2} = 33\,108\ W\,m^{-2}$$

This is the flux at the source. That at the receiver – in this case a human being – will be less than this by a factor between 2 and π (Lees, *op. cit.*). Using the factor of 2, then substituting into the probit equation:

$$I = 16554 \, \mathrm{W \, m^{-2}} \implies Y = 3.1$$

Converting the probit to a percentage:

3% of the evacuees are expected to experience fatal injuries

Negligible fatality rate for Y less than ≈ 2. Putting $Y = 2$ and retaining the 7 s as the time for movement gives:

$$I = 12\,123 \, \mathrm{W \, m^{-2}}$$

$$T = [2 \times 12\,123/5.7 \times 10^{-8}]^{0.25} = 807 \, \mathrm{K} \; (534°\mathrm{C})$$

Hence, elimination of fatal injuries would result if the deluge cooled the surfaces by about 70°C. Hence only quite moderate cooling has a major effect on hazards.

The likelihood of non-fatal injuries is examined in one of the appended numerical questions.

Computer packages are available to predict, for particular platforms, the consequences of certain initiating events in terms of possible escalation paths. Input includes the dimensions and configuration of the pipe work, the distribution of quantities of hydrocarbon (distinguishing between oil, gas and two-phase inventory), the time-averaged populations of the various parts of the platform and the reliabilities of all of the emergency response devices.

8.6 Background on frequencies and probabilities

The simple calculation in the previous section illustrates the use of probabilities. Input to programs for ORA also includes frequencies of particular initial or consequential events. Returning to the numerical problem in the previous section, in which the probabilities of correct functioning of two ESD valves was considered, only probabilities were of interest. In ORA probabilities and frequencies have to be correctly applied together. A simple example is provided below.

At an offshore installation the frequency of leakage of natural gas from pipe work is estimated as 10^{-6} per year per metre length of pipe work. The probability of ignition is 0.2. Consider leakage from a particular section of the pipe work of length 100 m. The ESD valves at either side of such a leak each have a 99% probability of correct functioning. Vertically above the pipe is a truss, failure of which would result in collapse of one module of the platform. The fire rating of the truss is such that it can withstand a jet fire resulting from the quantity of gas isolated between the two ESD valves. If however one or both ESD valves do not work, therefore isolation does not take effect, the jet fire will be of long enough duration to cause the truss to fail. Calculate the frequency with which there will be a jet fire resulting in module collapse.

Solution

$$\text{Frequency of leak} = 100 \times 10^{-6} \text{ per year} = 1 \times 10^{-4} \text{ per year}$$

$$\text{frequency of leak with ignition} = 2 \times 10^{-5} \text{ per year}$$

$$\text{frequency of a jet fire in the direction of the truss}$$
$$= (1/6) \times 2 \times 10^{-5} \text{ per year} = 3 \times 10^{-6} \text{ per year}$$

$$\text{probability that both ESD valves will operate} = 0.99^2$$

$$\text{probability that one or both will not} = 1 - 0.99^2 = 0.0199$$

$$\text{frequency of a fire sufficient to cause module collapse}$$
$$= 0.0199 \times 3 \times 10^{-6} \text{ per year} = 7 \times 10^{-8} \text{ per year}$$

Let us examine the above calculation in detail. The frequency of leakage of natural gas from pipe work, given as 10^{-6} per year per metre of pipe work, will have been based on records. Recall the comment made previously of the statutory requirement to report all leakages, however minor. These can be statistically processed in order to provide this sort of information. The value of 10^{-6} will not be a 'hard number' and is likely to be based partly on statistical data across the industry and partly on more precise information appertaining to the platform features. Such figures are constantly subject to review.

As we have seen, once gas leakage has occurred it is not inevitable that ignition will occur. The likelihood that it will depends partly on the rate of leakage: the more rapid the rate of leakage the greater the probability of ignition. In the above example the probability that, given leakage, ignition will ensue is given as 0.2, so the frequency of leakage with ignition can be calculated as a simple product, having units appropriate to frequency. In terms of impingement on to the truss, this depends on the direction of the jet fire. A common approximation is that a jet fire will be in one of six orthogonal directions (vertically up, vertically down, horizontal N, S, E and W) and that all potential targets can be taken to be in one of these six directions. Hence, in calculating the frequency with which there will be leakage, ignition and impingement on to a particular truss a directional factor of one-sixth has to be incorporated. The probability that one or other of the valves will not operate is calculated from the reliability information given and this is used to complete the calculation. In such calculations, since the 'bottom line' has to be a frequency clearly only one frequency can be used as the input. Probabilities have no units, so to multiply a frequency by one or more probabilities does not affect the units. It is important not to confuse probabilities with frequencies.

Note that in this calculation it is assumed that if the jet fire occurs and is in the right direction it will be long enough to affect the truss. More commonly in programs for ORA the length of the jet fire is calculated by means of a correlation such as the one presented in Chapter 3, and this is compared with the distance from the source of the jet fire to a particular component of the platform which might be affected. This brings us back to consequence analysis. Finally we should note that the 'bottom line' of the calculation – a frequency of $\approx 10^{-7}$ per year – is very low, and it is doubtful whether there would be need to be further negotiation or discussion of the possible incident which is the subject of the analysis.

8.7 Consequence analysis

8.7.1 Jet fires

A jet fire, like the other sorts of combustion behaviour of interest, can be the result of the initiating event or of escalation. A jet fire is very powerfully heat-releasing. If there is impingement on to a structural component of a hydrocarbon-bearing pipe or vessel, there can be failure. Moreover, fatalities can easily be caused directly by a jet fire. It is sometimes assumed in ORA that a quarter of the persons occupying the part of the platform where a jet fire occurs will die as a result of burns.

Figure 8.3 shows a segment of an escalation path for an incident initiated by natural gas leakage, structured as five rows and three columns. The length of pipe work under consideration in the first row of the figure is unspecified, and the frequency of 10^{-2} per year for the pipe work under consideration, served by a single set of ESD valves, is much higher than that in the numerical example in the previous section. The second row contains calculations linking frequencies and probabilities like those performed earlier in this chapter. Early in the escalation path is an application of the equations encountered in Chapter 3 for jet fire length. The fatalities calculated in row 3 will be included in the total fatalities. Fatalities calculated at particular stages of an escalation path can be non-integers. It was stated above that some programs assign a 25% death rate to persons at the part of the platform where a jet fire occurs, so if there are seven such persons a program so set up will attribute 1.75 deaths to the jet fire. Deaths from each point where a fatality is possible are summed and, if not instructed to do so earlier, a program will cease running

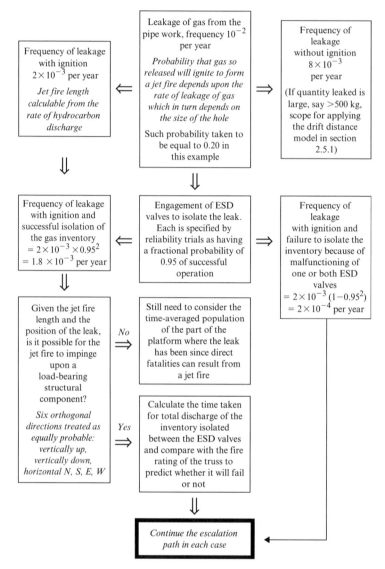

Figure 8.3 *Simplified escalation path for a gas leak at an offshore platform.*

when the number of fatalities equals the time-averaged population of the platform.

Another facet to consequence analysis for jet fires is the effect of the resulting heat flux on structures. The well known [3, 4] 'semi-infinite solid' model can be utilised. In a semi-infinite solid, one dimension is much smaller than the other two so that conduction is entirely across the small dimension. An example would be a thin sheet of metal receiving heat flux at one side. The model has been solved, and algebraic or graphical solutions made available, for three sets of conditions: at time zero, a step change in the surface temperature; at time zero, constant heat flux applied at the surface; at time zero, convection with constant coefficient applied at the surface. For each of these, the whole of the solid is taken to be at a uniform temperature initially, that is, before application of the flux.

The model enables temperatures at particular depths from the surface to be calculated at different times. In ORA however, where a failure temperature is assigned to a component or to a structural member, it is sufficient to calculate the time taken for the surface to reach the failure temperature. As far as this is in error, it is on the safe side, and the equation for surface temperature as a function of time, for any of the three forms of the analysis, is quite simple. The calculation below illustrates the example of the semi-infinite solid to consequence analysis.

Imagine that a jet fire occurs within 'sight' of a plane wall. The geometrical relationship between flame and wall is such that the wall receives $8\,\mathrm{kW\,m^{-2}}$ from the jet fire. The wall is made of an insulating material of thermal diffusivity (α) $3 \times 10^{-7}\,\mathrm{m^2 s^{-1}}$, thermal conductivity ($k$) $0.1\,\mathrm{W\,m^{-1}K^{-1}}$. How long will it take for the surface of the wall to rise from its initial value of 20°C to 600°C?

The thermal diffusivity (usual symbol α) $= k/c\sigma$ where c is the heat capacity and σ the density, units are $\mathrm{m^2 s^{-1}}$.

Solution
Invoke the 'semi-infinite solid' model. This means a plane wall having two dimensions much larger than the remaining one. If heat is applied uniformly to one of the large faces heat transfer within the solid will be entirely across the small dimension. Clearly, the form of the model which applies here is that in which a constant flux is applied at time zero.

For this, for heat flux q' ($\mathrm{W\,m^{-2}}$), the surface temperature at time t given by [3]:

$$q_1 = A\sigma\epsilon F_{\text{ground-sky}}(T_{\text{cover}}^4 - T_{\text{sky}}^4)$$

where $T_s(t)$ = surface temperature at time t and T_i = initial temperature of the solid

In the example under consideration

$$T_s = 600°C, \; T_i = 20°C, \; \alpha = 3 \times 10^{-7} m^2 s^{-1}$$

$$k = 0.1 \, W \, m^{-2} K^{-1}$$

$$q' = 8 \, kW \, m^{-2}$$

$$\Downarrow$$

$$t = 138 \, s$$

8.7.2 Pool fires

In ORA, it is usually considered that if a pool fire begins in an enclosed part of a platform, about 10% of the occupants will die from burns. However, in escalation a pool fire is less powerful than a jet fire. Some trusses and other structural members have fire ratings such that they can withstand pool fire impingement for very long periods. This is because the equilibrium temperature resulting from the balance of heat transferred from the pool fire and to the surroundings is below failure temperature.

8.7.3 Fireballs

In ORA the view is usually taken that a fireball will create a significant overpressure if its volume, approximated to a sphere, is larger than the enclosure in which the fireball occurs, therefore if it is not confined at all there can be no overpressure. The calculation below illustrates these ideas.

A vessel containing hydrocarbon stands in part of an offshore installation bounded by a grated floor supporting the vessel and another grated floor vertically above and parallel. The two floors are supported by common vertical steel posts in a rectangular arrangement such that at any one horizontal level the structure forms a rectangle 100 m by 65 m. The vertical areas between the posts are occupied by walls, also supporting electrical cables and water pipes, forming an enclosure of rectangular cross-section.

The vessel contains, at any one time, 5 tonne of hydrocarbon inventory. How far apart must the two grated floors be in order to ensure that if there is leakage and fireball behaviour the walls are not subjected to overpressure?

Solution
If the vessel contains 5 tonne of hydrocarbon, the diameter of a fireball resulting from its failure is:

$$5.25\{5000\}^{0.314} = 76 \, m$$

Approximating to a sphere, the maximum volume is $231\,187\,\mathrm{m}^3$. The volume bounding the vessel must exceed this if there is to be no overpressure. Hence the minimum separation of the grated floors is:

$$231\,187/(100 \times 65)\,\mathrm{m} = 36\,\mathrm{m}$$

This is a convenient point at which to discuss ALARP (As Low As Reasonably Practicable) ideas, though of course they could equally well have been introduced in the context of jet fires or of pool fires, nor of course, are ALARP principles applied only to offshore operations.

An extension of the example above can be used as a framework for the introduction of ALARP. Imagine that the grated floors are actually 25 m apart, so that if the 5 tonne of hydrocarbon inventory leaked catastrophically and ignited there would indeed be an overpressure, which could threaten the electrical cables and water pipes. We have to estimate the frequency with which there is leakage and ignition. Imagine that the frequency of leakage is 8×10^{-5} per year and the probability of ignition 0.4. The frequency of a fireball, and accompanying damage to cables and pipes, is then:

$$3 \times 10^{-5} \text{ per year}$$

and a judgement has to be made as to whether this is ALARP. This requires a value for the number of deaths resulting from the course of events to which the above frequency applies. A fireball in that part of the platform could, by reason of its overpressure, cause loss of the electricity supply to other parts of the platform and failure of hydrocarbon-handling components (e.g., pumps) which rely on this supply. The number of deaths resulting if events took their full course in this sort of way could be estimated. Let us suppose for illustrative purposes that the number of deaths so estimated is 10.

In ALARP the number of deaths per year is fitted to an equation of the form:

$$N\phi(N) = 10^{-n} \qquad\qquad \text{Eq. 8.1}$$

where $\phi(N)$ is the frequency of an occurrence resulting in N deaths. In the present example, $N = 10$ and $\phi(N) = 3 \times 10^{-5}$ per year, giving:

$$N\phi(N) = 3 \times 10^{-4} \text{ per year} = 10^{-3.52} \text{ per year}$$

signifying one death every 3333 years. One way of judging the acceptability of risk is for this equation to be plotted in log-log form for $n = 1$ and $n = 4$, giving two parallel lines. The former signifies one death in 10 years, 100 deaths in 1000 years and is at the high end of the risk which can be regarded as acceptable. The latter signifies one death in 10 000 years, 100 deaths in 1 000 000 years, and is viewed as involving negligible risk. The area in between is where negotiation is possible as to the acceptability of risk, and clearly our example falls in this area.

Returning to the original context of the problem, the risk could be reduced very significantly by widening the gap between the horizontal grated floors, say from its original 25 m to 45 m, giving a comfortable margin of safety whereby it could be assumed that a fireball would not have an overpressure. In this event there would still be a risk to the wall bearing cables and pipes, but a smaller one due to heat only in the absence of a blast. It might be that, from formal analysis, it was concluded that to widen this gap to 45 m would increase n by an order of magnitude from 3.5 to 4.5. If this improvement could be effected at reasonable cost, as it perhaps could if the two floors had no structural role to play, it would be required. If, on the other hand, such a modification would require major structural work at vast expense, the value of 3.5 would be accepted as ALARP. We return to ALARP[26] in Chapter 9. HAZOP (Chapter 4) and ALARP are sometimes used together.

8.7.4 Smoke

Smoke inhalation is often fatal and smoke impairs visibility during evacuation. Fires at offshore installations are non-premixed as we have seen and as such tend to be rather smoky. Figure 8.4 shows intentional flaring of gas at a platform. Note the abundance of smoke in the burnt gas.

In an accident offshore where there has to be evacuation, for the temporary refuge to be impacted by smoke is particularly undesirable. In a hydrocarbon accident, whether or not there is significant smoke depends upon the total quantity being burnt. Smoke contains carbon particles, unburnt volatiles, nitrogen and oxygen, oxides of carbon and polyaromatic hydrocarbons (PAH). In smoke movement there is dilution, according to [5]:

$$\mathrm{d}N/\mathrm{d}t = -GN^2 \qquad\qquad \text{Eq. 8.2}$$

where N = number of particles per cm^3
$\quad\quad t$ = time (s)
$\quad\quad G$ = a constant characteristic of the burning fuel, often in the neighbourhood of $10^{-9}\,\mathrm{cm^3\,s^{-1}}$

Integrating between the limits $N = N_0$ at $t = 0$ and $N = N$ at $t = t$ gives:

$$N = N_0/\{1 + GtN_0\}$$

Reduction of visibility by smoke is according to:

$$I/I_0 = 10^{-kL} \qquad\qquad \text{Eq. 8.3}$$

where I is the intensity at a distance L (path length, units m), I_0 is the intensity at $L = 0$ and k (m^{-1}) the extinction coefficient.

Figure 8.4 *Flaring of gas at an offshore platform. Reproduced courtesy of the Health and Safety Executive, UK.*

$k = k_m m$ where k_m = the specific extinction coefficient (units $m^2 g^{-1}$) and m is the mass concentration of smoke (units $g m^{-3}$)

typically, $k_m \approx 5 m^2 g^{-1}$ and m is in the range 0.05 to $3 g m^{-3}$

This is applied in the assessment of illumination signs at offshore installations, in particular EXIT signs to direct persons to the TR in an emergency. The author has had recent experience in the assessment of cold-cathode fluorescence tubes as an alternative to light-emitting diodes in EXIT signs for use at offshore installations [6].

The maximum distance S at which an illuminated sign can be seen is given by:

$$kS = 8$$

The maximum distance S at which a light-reflecting sign can be seen is given by:

$$kS = 3$$

A numerical example follows.

At a fire at an offshore installation there is a pool fire which releases smoke with 'G' value 5×10^{-9} cm^{3}s^{-1}. At release from the fire the particle concentration is 10^7 cm^{-3}. If the smoke travels in the direction of the TR and takes 4 minutes in reaching it, by what factor will it have become diluted on reaching the TR?

$$\text{dilution by a factor } N_o/N = 1 + GtN_o = 13$$

The elevated temperature of smoke causes it to have a lower density than air at ambient temperature and hence buoyancy. The smoke behaviour is also influenced by wind speed. An expression sometimes used in ORA is:

$$\tan \theta = 5/u$$

where θ = plume angle (radians) and u = wind speed (m s^{-1}). For wind speed zero, $\tan \theta$ = infinity, $\theta = 90°$, and there is a vertical plume.

8.8 Construction of escalation paths

In this section we continue the discussion in the second part of section 8.7.1, and incorporate some of the principles of consequence analysis from the intervening sections, especially that concerned with smoke. Figure 8.3 comprised a segment of an escalation path. That began where all escalation paths begin: at the initiating event, which in this case was leakage of natural gas from a pipe. A segment of an escalation path might of course 'take up the story' beyond the initiating event, and in the following example we begin at row three on Figure 8.3. Here it is confirmed that there is a jet fire, and going right from there the path bifurcates depending on whether the jet fire is in the right direction to impinge upon a load-bearing structural component. We now go beyond here to determine whether smoke will impede evacuation to the TR. This is shown as Figure 8.5, where the lightly shaded areas represent parts brought from Figure 8.3.

Beyond the bifurcation referred to, in considering smoke the programme brings the two parts of the second column together, since whether or not smoke is released is independent of whether the jet fire is impinging upon a truss. It depends only on the total rate at which hydrocarbon is being burnt on the platform. In making a judgement as to whether smoke occurs, the software refers back to the leakage rate, and hence burning rate, of methane. Generally speaking if this is in excess of 5 kg s^{-1} there will be enough smoke for there to be possible visibility impairment. This value is arbitrary and any preferred value can be selected by the user. As shown in Figure 8.5, the program then predicts what wind conditions would cause the smoke to impact the TR, how long the smoke under such conditions would take to reach the

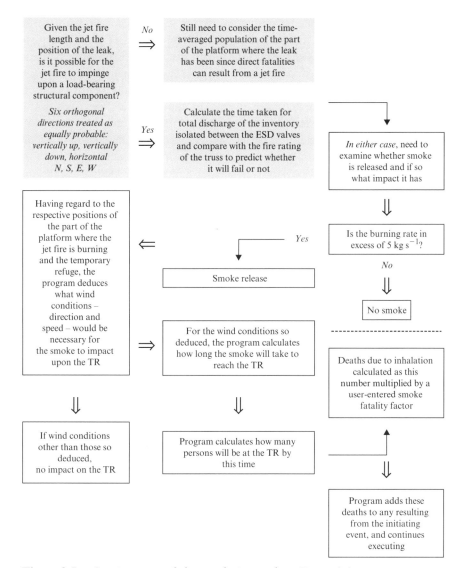

Figure 8.5 *Continuation of the escalation path in Figure 8.3.*

TR and how many deaths from inhalation there would be. Whereas we have considered, in the segment below, the TR as the smoke target, a full escalation path would also consider its possible impact on helidecks and lifeboat launching points. For it to impact upon either of these might prevent evacuation from the platform.

The reader is hopefully becoming aware that for given conditions there is no one way of charting an escalation path. Many degrees of detail are possible and many variations on the basic ideas can be conceived and implemented. Moreover, user-entered numbers such as the rate of burning necessary to produce significant smoke and the smoke inhalation fatality factor (both of which feature in Figure 8.5) are obviously subject to a very significant uncertainty. There is clearly scope for development and refinement of methods of charting escalation paths. Mention has been made a number of times of lifeboat evacuation. Figure 8.6 shows, for a production platform in the North Sea, a suspended 'survival capsule' and also the lifeboat mustering points.

8.9 Offshore accident case studies

On the Ekofisk platform in the Norwegian sector of the North Sea there was a fire in 1975 which resulted from failure of an oil riser. The accident cost six of the platform personnel their lives. Although the fire took two hours to extinguish, structural damage to the platform was minor. The 1988 Piper Alpha disaster in the British sector of the North Sea resulted in 167 deaths and is the worst offshore accident to date. Also very grave was the accident in the North Sea in March 1980. Overturning of an installation during high winds led to 123 deaths.

Of course, accidents, including fatal ones, occur at offshore installations which have no connection with the unique circumstances there. A distinction has to be drawn between lives lost through the type of incident under discussion in this chapter – fire and consequent platform loss or damage – and lives lost in other ways, e.g., 'slips, trips and falls', lifting operations, falling objects. Such accidents occur in all workplaces (and indeed in homes and places of recreation) and have no particular association with offshore oil and gas production. The author has obtained the following figures for offshore installations which report to the UK Health and Safety Executive (HSE).

1997–98 Three fatalities: one due to a falling object, the other due to the falling overboard of a person and an attempted rescue
1996/97 two fatalities
1995/96 one fatality
1994/95 one fatality
1993/94 one fatality

To the slightly dated information immediately above will be added details of some more recent examples of incidents at scenes of offshore oil and gas production, taken from [7]. At Bohai Bay offshore China oil production by

Figure 8.6 *Suspended survival capsule and lifeboat mustering points at an offshore platform in the UK sector of the North Sea. Reproduced courtesy of Shell International.*

means of an FPSO – floating production, storage and offloading unit – was taking place in 2011 when the mooring of the FPSO failed in severe weather. In the same year an FPSO in service in the North Sea experienced a similar fate: loss of mooring in a storm. The FPSO, called *Gryphon*, was secured by tugs and there was no loss of oil.

The overturning of the AHTS – anchor handling, tug and supply vessel – *Bourbon Dolphin* in the North Sea in April 2007 [8] resulted in eight deaths. The sense of tragedy at the time was exacerbated by the fact that one of the dead was a boy doing work experience.

Also to be noted is the *Montara* (Timor Sea) drilling rig explosion in 2009. This accident involved leakage into the sea of 30 000 barrels over a 74 day period as well as associated gas. There were no fatalities. It occurred only a few months before the much more serious Macondo spill at the Gulf Coast, of which the author has written in full elsewhere [9].

8.10 Other matters relating to offshore safety

After Piper Alpha the offshore industries in the United States and Europe (primarily the UK) took different approaches safety in offshore activity. In the UK as a result of the Cullen Report, more importance was attached to 'safety cases'. This was so in sectors of the North Sea other than the UK one. The philosophy of a safety case is that arguments are used in advance which would have been used in defence had there been failure.

In the US the standard

API RP 75: Recommended Practice for Development of a Safety and Environmental Management Program for Offshore Operations and Facilities

was widely applied over a very long period. API RP 75 is concerned with the following inter alia: well drilling, production well maintenance, pipelines, structural reliability of platforms and supports. Such standards are seen as 'prescriptive', that is, they set out procedures for particular operations. By contrast safety cases are based more on the application of safety principles than on instructions and practices.

Since the 2010 Gulf of Mexico (GoM) spill, drilling in US waters has required a safety case. A structure ('template') for such a safety case has been drawn up by the International Association of Drilling Contractors (IADC). The number (>5000) of platforms in the GoM precludes a safety case for each, which in any case would involve needless duplication as many platforms closely resemble each other in structure and design. Notwithstanding the GoM spill the use of API standards has been declared successful, there having been nothing in US waters on the same scale over the previous 41 years (the 'Santa Barbara blowout'). At the Montara incident in the Timor Sea outlined above there had been a safety case. We note in conclusion of this discussion that the safety case concept is not restricted to offshore oil and gas production, nor indeed did it originate with it. The idea began when nuclear facilities first came into being in the UK.

Control of Substances Hazardous to Health (COSHH) Regulations 2002 involve issue of 'essential sheets' for the numerous operations within its scope. Substances concerned with exploration and production which come within the remit of COSHH include drilling muds ('Essential sheet' OCE9), chemicals for well maintenance ('Essential sheet' OCE7) and products of 'pigging' (pipeline cleaning): ('Essential sheet' OCE25; see Table 8.3).

Table 8.3 *COSHH applications to offshore oil and gas production*

Background on the substances	COSSH recommendations include:
Drilling muds: ('Essential sheet' OCE9)	Water based muds to be preferred over oil-based ones on safety grounds, otherwise attention to the oil toxicity necessary.
Most drilling muds aqueous. Bentonite a very common ingredient of aqueous muds, formula: $(Na,Ca)\,Al,Mg)_6(Si_4O_{10})_3(OH)6nH_2O$.	'Shaking' of muds to improve homogenisation before use in a ventilated area of specified
Chemicals for well maintenance ('Essential sheet' OCE7)	Protection from diesel exhaust gases. Respiratory protective equipment (RPE) when any mists or vapours are generated, and in particular during the handling of corrosion inhibitors.
Diesel motors used for injection. A wide range of chemical agents from simple mineral acids to corrosion inhibitors and biocides (recall SRB).	Residues from all operations to be classified as 'hazardous waste'.
Products of 'pigging' (pipeline cleaning): ('Essential sheet' OCE25)	BTX expected to be in the residue, and precautions against poisoning from it required.
Organic residues present at the conclusion of the journey of the 'pig'.	Hydrogen sulphide present, especially from a pipeline which has also conveyed gas, and here again precautions necessary.

The ancillary operations below (Table 8.4) are also the subject of 'essential sheets'.

These examples are intended to give a reader a degree of insight into the formal requirements of offshore safety. After the Macondo spill the EU proposed a directive to review and reorganise offshore safety in its waters. (the Proposed EU Regulation of Offshore Safety 2011). Its introduction was accompanied by a statement that 'the likelihood of a major offshore accident in European waters remains unacceptably high'. EU centralisation of offshore

Table 8.4 *Examples of COSHH applications to support roles at offshore platforms.*

Laundry	Avoidance of stockpiles of soiled clothing (which can be a spontaneous heating hazard). A fire hazard in subsequent tumble drying if not all of the oily substance is removed.
	Close attention to tumble drier filter dust (avoidance of a dust explosion).
Insulation removal	Confirm absence of asbestos before commencing. Such information in an asbestos register. Wetting of the old insulation to keep the fibre release down.
Preparing surfaces for painting	Enclosure around the scene of old paint removal. Vacuum cleaner of specified performance for removal of debris within an enclosure.

regulation recommended instead regulation by member countries individually. The statement in inverted commas is NOT equivalent to saying that that the North Sea is unsafe. However, some parts of it have higher risks than others, and some installations in the North Sea have been selected as good examples of safety practices. Responses to the directive by some major interested parties are summarised in Table 8.5. It has drawn on references 10–15.

An obviously relevant question is 'Why should the approach to safety at European oil and gas fields be radically changed because there has been an accident in the Gulf of Mexico?' and this has probably stimulated the negativity evident in the table. Revisions of the Regulation are likely to follow.

8.11 Concluding remarks

The topic of offshore safety is vast and complex. This chapter will, hopefully, have given the reader a basic appreciation of the framework within which engineering calculations concerning offshore safety are carried out as well as familiarising him or her with the 'language' of the subject.

References

[1] *Our Industry Petroleum*, Fourth Edition. British Petroleum, London (1970)

Table 8.5 *Summary of responses to the Proposed EU Regulation of Offshore Safety 2011*

Party	Views expressed
Norwegian Oil Industry Association, an employer's organisation	'...nothing in the Commission's proposal that would reduce the safety in any of the countries affected. As a minimum standard regulation, it only acts to raise the standards across Europe...'
The Bellona Foundation, an independent international environmental body based in Norway	'...welcomes the proposal, but would like to see it strengthened.'
Government of Norway	'...refused to enter a constructive debate on the proposal and on the contrary claimed this year that the legislation would not be applicable on the Norwegian continental shelf.'
European Parliament	'...mixed messages, with reports by two rapporteurs backing a strong legislative response to protect against a major oil or gas accident, whilst another rejected several positive measures.'
A representative of SNR Denton, a multi-national law firm with a strong involvement in oil, writing in *Petroleum Review*	'The UK regime is undoubtedly fit for purpose as a broadly effective offshore safety and environmental regulatory regime.' '...there would be a case for overarching legislation at EU level in certain areas, for example clarity on environmental liability following an offshore disaster.' '...it is difficult to find much positive commentary on the proposal amongst UK stakeholders.'
Oil & Gas UK, a not-for-profit representative body for the UK offshore oil and gas industry	'While Oil & Gas UK will always support proper moves to improve safety standards, the Commission's proposal to dismantle the UK's exemplary safety regime is likely to have exactly the opposite effect. Regulation 'poorly drafted and clumsy.'
IndustriAll, European Trade Union	'Strong scepticism.'

[2] *Encyclopaedia of Technology*, Seventh Edition. McGraw-Hill, New York (1992)

[3] Holman J.P. *Heat Transfer*, SI Metric Edition. McGraw-Hill, New York (1989)

[4] Jones J.C. *The Principles of Thermal Sciences and their Application to Engineering*, Whittles Publishing, Caithness, UK and CRC Press, Boca Raton, FL (2000)

[5] Mulholland G.W. Smoke Production and Properties in *Handbook of Fire Protection Engineering*, Second Edition. SFPE, Boston, MA (1995)

[6] Jones J.C., Kotzias I. and McBrien C. Cold-cathode fluorescence tubes as an alternative to light-emitting diodes in the illumination of EXIT signs *Journal of Fire Sciences* **19** 255–257 (2001)

[7] Jones J.C. *Dictionary of Oil and Gas Production*, Whittles 288pp. Publishing, Caithness (2012)

[8] Jones J.C. 'On the capsizing of the Bourbon Dolphin in April 2007' *Journal of Loss Prevention in the Process Industries,* **20** 286 (2007)

[9] Jones J.C. *The 2010 Gulf Coast Oil Spill*, Ventus Publishing, Fredericksberg (2010)

[10] http://www.bellona.org/articles/articles_2012/1346073068.06

[11] http://www.bellona.org/articles/articles_2012/1345705600.84

[12] http://www.bellona.org/articles/articles_2012/1341998761.14

[13] http://www.snrdenton.com/news__insights/in_the_media/2011-12-09-offshore-regulation.aspx

[14] http://www.oilandgasuk.co.uk/aboutus/aboutus.cfm

[15] http://tgwuoffshore.org.uk/Documents/Offshore_reg_industriAll_position_final.pdf

Numerical questions

1. A fire in a gas export riser is the initiating event in an offshore accident. Isolation of hydrocarbon from the riser necessitates successful operations of two ESDs, A and B, each with a reliability of 0.95. Express all of the possible outcomes and show that their sum is unity.

2. 100 m³ of crude oil at an offshore facility is isolated by successful operation of the ESD valves. Making reasonable estimates of your own for the quantities involved, calculate the heat released in the event of leakage and ignition of this quantity of crude oil.

3. For a gas pipe at an offshore platform the frequency of leakage is estimated as 5×10^{-3} per year, and the probability of consequent ignition as 0.25. In the event of leakage two ESD valves, each with a reliability of 0.98, come into operation. If both of these work, the inventory isolated between them is insufficient for any jet fire resulting from the leak to impinge upon a truss nearby long enough to cause it to fail. If one or other ESD valve does fail, there is enough inventory for a jet fire to torch the truss and cause it to fail. Estimate the frequency with which there will be leakage and failure of the truss. If such failure would be expected to result in five deaths, determine whether the risk is acceptable, unacceptable or of a value such that ALARP principles apply.

4. At a part of an offshore platform where crude oil is conveyed through pipes, the time-averaged population is five persons. The frequency of leakage of crude oil is 1×10^{-3} per year and the probability of ignition to form a pool fire 0.35. Calculate the frequency with which a fatality is expected in this situation and comment on it. Refer to the main text for a value of the probability of death from pool fire exposure.

5. The following information relates to the initiating event at an offshore accident:

Occurrence	Probability or frequency	Cumulative probability
Leak of gas from the gas export riser	0.1 per year	–
Ignition with no overpressure	0.02	v_1
Both emergency shutdown valves work. In which case hydrocarbon not prevented from escaping by the shutdown valves will, if ignited, form a jet fire	0.81	v_2
Vertically downwards orientation of such a jet fire	0.166	v_3
Fire water pump will work	0.98	v_4
The deluge sets, each with a reliability of 0.9, will both function	0.81	v_5

Calculate all of the cumulative frequencies up to and including v_5. Explain the meaning of the value of v_5. The time averaged population of the part of the platform impinged upon by the jet fire is 0.2, and the fatality rate for a jet fire is 0.25. How many deaths are there as a result of the initiating event?

6. The post-combustion gas from a localised hydrocarbon fire drifts towards a wall and transfers heat to it with a convection coefficient of $5\,W\,m^{-2}K^{-1}$. The gas is at 600°C on arrival at the wall, which is initially at 20°C. Calculate the wall surface temperature after 1 hour. The thermal diffusivity of the wall material is 4×10^{-7} m^{2}s^{-1} and the thermal conductivity $1.5\,W\,m^{-1}K^{-1}$.

N.B. For a semi-infinite solid with constant convection coefficient at the surface:

$$\frac{T_s(t) - T_i}{T_F - T_i} = 1 - \exp(h^2 \alpha t / k^2)\,\mathrm{erfc}\left[h\sqrt{\alpha t}/k\right]$$

where T_F is the fluid temperature, h is the convection coefficient ($W\,m^{-2}K^{-1}$), other symbols as previously defined.

7. Reconsider the question on smoke movement in the main text by assuming that smoke uniformly having a mass concentration of $0.1\,g\,m^{-3}$ fills the space between the site of the pool fire and the TR. From what distance will an illuminated sign at the TR be visible under these conditions?

8. Crude oil from a pipe at an offshore vessel has leaked and there has been ignition resulting in a pool fire. Determine the maximum pool area for the burning not to be accompanied by a smoke hazard.

9. Return to the problem in the main text where evacuees from a platform where fire has spread experience radiative flux of $16554\,W\,m^{-2}$ for $7\,s$. What percentage of the evacuees would suffer first-degree burns? If the deluge sets operate and the surroundings are consequently cooled to the extent calculated previously, what will be the percentage of evacuees with first-degree burns? The probit equation of first-degree burns from radiative flux is:

$$Y = -39.83 + 3.02\ln\{tI^{4/3}\}$$

10. Sulphate-reducing bacteria in the sea are an agent in the souring of oil and gas, diminishing its value once produced [7]. The process is conversion of SO_4^{2-} to S^{2-}. At a particular scene of oil activity the produced water has a sulphate concentration of $100\,mg$ per litre. If conversion to sulphide by SRB is total what will be the concentration of the sulphide in p.p.m. molar basis?

11. A crude oil not having been stabilised (that is, the lowest boiling constituents not removed to eliminate explosion hazards) will burn as a fireball if there is catastrophic leakage. At an offshore site a sphere of such crude oil is positioned in a module of capacity $1000\,m^3$ which also contains utilities failure of which would have serious consequences. What is the maximum amount of the oil which can be held in a spherical container consistently with there being no blast damage to the utilities in the event of a fireball? Choose a density value of your own for the crude oil and the following expression for fireball diameter D:

$$D(m) = 5.25(M/kg)^{0.314}$$

where M is the quantity leaked.

12. Using the equation for ALARP given in the main text:

$$N\phi(N) = 10^{-n}$$

what value of n would correspond to (a) one death every 150 years (b) one death every 1200 years and (c) one death every half a million years?

Endnotes

[25] See comment in section 8.9 concerning accidents occurring offshore but they are no more frequent that in equivalent situations in other work environments, i.e. 'slips, trips and falls'.

[26] One sometimes encounters the equivalent term ALARA: as low as reasonably achievable.

9

HAZARDS ASSOCIATED WITH PARTICULAR HYDROCARBON PRODUCTS

9.1 Introduction

This chapter will review certain hydrocarbons or groups of hydrocarbons in terms of hazards associated with them. Some such hydrocarbons have already featured in the book, but further points of practical interest will be added in this chapter. We start with crude oil.

9.2 Crude oil

Sometimes crude oil has to be stripped of its lowest boiling components before shipping and handling in a process known as crude oil stabilisation. The crude oil so stabilised has a Reid Vapour Pressure not exceeding 10 p.s.i. and ethane is often present in major quantities in the vapour stripped off. Scenes of crude oil stabilisation include the Forties Field in the North Sea. Crude oils so stabilised if necessary have flash points in the neighbourhood of 125°C [1]. As we saw in Chapter 8, spillage of crude oil at an offshore platform will, if there is ignition, result in a pool fire. An accident during pipe transportation of crude oil occurred in the Middle East in 1990. It is believed to have been caused by the failure of a feed line in the crude pre-heating part of the refinery. There was one fatality, two injuries and substantial fire damage. Sulphur in crude oil can cause corrosion to pipes and vessels, and eventual failure. In Bantry Bay, Eire in 1979 there was a pool fire involving 40 000 tonne of crude oil. Fifty lives were lost. The crude oil leaked from a tanker on to the sea where it ignited.

In Chapter 3 the point was made that a pool fire always has a mass transfer rate, from pool to fire, of about $0.1\ kg\,m^{-2}\,s^{-1}$. This is a less satisfactory approximation for petroleum fractions than for pure compounds, since in the former the lighter constituents will evaporate before the heavier. This is even more so for crude oil, which contains hydrocarbons from about C_4 to C_{25}.

A further hazard associated with crude oil is, of course, its spillage from ships, with resulting contamination of the coastline. There have been several major spills around the coast of Britain [2] including the Torry Canyon accident of 1967, when 110 000 tons of crude oil were released into the sea off the Cornish coast. There have been other spills including one off the coast of East Anglia in 1978, involving 5000 tons of hydrocarbon. Highly serious though such incidents are, they account for only a few per cent of the total oil pollution of the sea. Most of the oil in the sea comes from two other sources: routine cleaning of the tanks of ships, and transfer to the sea by natural agencies including rivers of hydrocarbon from land having previously been either disposed of or accidentally released there [2]. The non-destructive disposal of hydrocarbons on land is briefly discussed in Chapter 11.

9.3 Natural gas

9.3.1 Background

Interestingly, a great deal was known about methane hazards long before natural gas became a widely used fuel. This is because of leakage of methane in coal mines (and other sorts of mine, including salt mines) which, through-out the last several centuries, has been the cause of a huge number of fatal accidents. Methane in mines is referred to as 'firedamp'. Equally interestingly, a natural gas reserve at Chi liu ching in China, which still features in compil-ations of gas reserves in that country, is known to have been producing gas since 211 B.C. [3].

As we saw in Chapter 1, in the USA utilisation of natural gas commenced in the late nineteenth century. The dominant constituent of natural gas from almost all known sources is methane, with small amounts of ethane, propane, butane, and inerts (N_2, CO_2). The calorific value is about 37 MJ m^{-3} ($\equiv$55 MJ kg^{-1}), cf. bituminous coal $\approx$30 MJ kg^{-1}, petroleum products $\approx$45 MJ kg^{-1}. The flammability limits of natural gas are 5% to 15%.

Methane, being C_1, is the least reactive alkane. It has a relatively high mini-mum ignition energy (about 300 μJ) and being less dense than air (density of methane relative to that of air = 0.56) disperses readily when leaked, thereby dropping fairly quickly to sub-flammable proportions. One consequence of the low reactivity is a low flame speed. A methane–air flame propagates at about a third of the speed of a hydrogen–air flame. Consistent with this is the view [4] that methane clouds when ignited always display flash fire behaviour to the exclusion of v.c.e. behaviour, though there are those in the profession who dissent from this view. Of course, v.c.e.s and flash fires relate to large quantities, tonnes or tens thereof. There is no doubt that there can be a significant overpressure when there is a confined 'gas explosion' involving smaller amounts.

Natural gas finds many applications as a general-purpose fuel in homes and industry. It can also be used as feedstock for conversion to liquid fuels by two methods: via synthesis gas and by partial oxidation. Since the first edition of this book there has been enormous growth not only in tight gas as mentioned in Chapter 1 but also in methane from coal bed methane (CBM) as a source of natural gas. There is major activity in CBM in countries including Australia, Canada and Indonesia. There is one important way in which CBM differs from gas from conventional reserves or from tight gas: CBM yields no condensate (see section 9.5).

Much more speculative, but featuring widely in 'beyond oil' discussions, is methane from natural gas hydrates. Such hydrates consist of methane enclosed in an ice structure; in the terminology of structural chemistry natural gas hydrates are clathrates. Natural gas hydrates occur at continental shelves and on the sea floor, and the amount of methane they contain exceeds by three orders of magnitude or more the known conventional gas reserves of the world. There has been very little if any production as yet.

The worked example below relates to conveyance of natural gas along a pipe.

Natural gas (assume to be pure methane, molecular weight $16\,\mathrm{g\,mol^{-1}}$) flows along a pipe of $0.5\,\mathrm{m}$ inner diameter at a speed of $3\,\mathrm{m\,s^{-1}}$. The pressure of gas in the pipe is $1\,\mathrm{bar}$ and the temperature $300\,\mathrm{K}$. Calculate: (a) the mass flow rate, (b) the molar flow rate, (c) the volumetric flow rate, (d) the Reynolds number (Re).

$$\text{gas constant (usual symbol R)} = 8.314\,\mathrm{J\,K^{-1}\,mol^{-1}}$$
$$\text{dynamic viscosity } (\mu) \text{ of methane at } 300\,\mathrm{K} = 1.1 \times 10^{-5}\,\mathrm{kg\,m^{-1}\,s^{-1}}$$

Solution
(a) mass flow rate (m^*) calculable from:
$$m^* = \sigma a u$$
where σ = density ($\mathrm{kg\,m^{-3}}$), a = cross-sectional area ($\mathrm{m^2}$), u = velocity ($\mathrm{m\,s^{-1}}$)

Density from the Ideal Gas Law:
$$PV = nRT \Rightarrow n/V = P/RT \Rightarrow \sigma = m_r P/RT$$
P = pressure ($\mathrm{N\,m^{-2}}$), V = volume ($\mathrm{m^3}$), n = number of moles, T = temperature (K), m_r = molecular mass ($\mathrm{kg\,mol^{-1}}$).

Hence, $\sigma = [(0.016 \times 10^5)/(8.314 \times 300)]\,\mathrm{kg\,m^{-3}} = 0.64\,\mathrm{kg\,m^{-3}}$
therefore $m^* = 0.64 \times \{\pi \times 0.25^2\} \times 3\,\mathrm{kg\,s^{-1}} = 0.38\,\mathrm{kg\,s^{-1}}$

(b) molar flow rate = $(0.38/0.016)$ mol s^{-1} = 23.8 mol s^{-1}
(c) volume occupied by a mole at 300 K, 1 bar = RT/P

$$= 0.025\,m^3$$

hence volumetric flow rate = $0.025 \times 23.8\,m^3s^{-1} = 0.59\,m^3s^{-1}$

(d) Re = $ud\sigma/\mu$

$$= (3 \times 0.5 \times 0.64)/1.1 \times 10^{-5}$$
$$= 8.7 \times 10^4$$

This calculation, presented here because of its natural gas context, would have belonged equally well to Chapter 2. In the solution the continuity condition is invoked in order to obtain the mass flow rate, and the Reynolds number obtained straightforwardly from the information given. This would be of use in choosing a convection coefficient if heat transfer to or from the methane were taking place, as in the calculation in section 6.6.3. Even on its own, the Reynolds number conveys one piece of very important information: that the flow is turbulent, the value of Re being in excess of 2500. The first of the numerical problems at the end of this chapter, for the reader to attempt, is a continuation of the above.

9.3.2 Case studies

Melbourne, Australia, 1998 [5]

Natural gas safety practices were put to a most severe test in Melbourne in 1998 when, after a refinery accident, the whole city of approximately 3 million people was without gas supply for a period and restoration of supply had to be implemented. The accident occurred at a facility receiving oil and gas from offshore oil and gas fields in the Bass Strait. Separation of the gas and oil takes place onshore. The gas from this facility accounts for 80% of the residential and commercial needs of Victoria (population about 4 million). Crude oil, once separated from the gas, is refined at the site.

Shortly after midday on a weekday, several explosions in rapid succession occurred at the facility, and there were consequent fires including fireballs that could be seen 60 km away. There were two fatalities – both employees – and eight other persons were hospitalised. Fire fighters required 48 hours to extinguish the fires. Emergency measures taken included evacuation of the facility, of homes within a 5 km radius and of an adjacent car manufacturing plant. Two LPG-bearing vessels, each containing 0.2 million litre, had become engulfed by flames and were at risk of exploding violently. Isolation valves were used where possible, and there was controlled burning of inventory not so isolated. There was also shutdown of the entire gas supply system to the State of Victoria.

As well as the deaths and injuries from the accident itself, there was major disruption to the entire activity of the State of Victoria for several weeks because of the loss of the gas supply. Supply from neighbouring New South Wales met initially only 2% of the state's needs, eventually rising to 20%. There was closure of many large and small businesses, and loss of gas supply to homes and to industrial and agricultural properties. Losses during the earlier period of the emergency were estimated at about $A50 000 000 per day. Many employees were forced to take annual leave or unpaid leave. There were visits to all homes and commercial premises to turn off the gas supply and, under emergency legislation, fines for refusing to allow the gas to be turned off.

Panic buying of electrical heating and cooking appliances and of LPG cylinders ensued, and warnings not to connect LPG to plant designed for natural gas were sometimes ignored, with resulting further hazards. A serious accident was caused by LPG leakage at a restaurant. Unauthorised installation of gas cylinders took place in certain hospitals. By a week after the accident, the losses to Victoria were $A100 000 000 per day. Industries in neighbouring states were also affected by disruption of supply of components from Victoria. There were more temporary layoffs and compulsory unpaid leave.

Efforts to get the gas supply working continued round the clock. By about 10 days after the accident many commercial firms had had their gas supply restored and people were returning to work. However, start-up had to be phased so that a sudden surge in amount drawn, sufficient to cause a major pressure drop and consequent drawing of air into the pipes, did not occur. About two days later, phased start-up of domestic gas supply began. The fire service was 'on alert' during this period. This gas start-up 10–12 days after the accident was one of the biggest operations of its kind ever to have been attempted, and it was completed without a major accident. Information given by radio, TV and the press throughout the crisis greatly helped people to cope. The value of the media in management of such a crisis became very clear.

Mexico, 2012

A very recent example of a natural gas accident is that in Tamaulipas, Mexico in September 2012, which resulted in 26 deaths. Mexico, it must be remembered, has for a long time been very productive of hydrocarbons. Like Nigeria (an OPEC country, which Mexico is not), Mexico has problems with theft of fuel, a serious enough matter in itself but with enormous safety consequences: will thieves so abstracting fuels act with a regard to safety, other than their own? The accident was at a site run by the Mexican company Pemex, a matter of days after there was an accident involving non-fatal injuries at another Pemex facility. Fairly obviously, in the fatal September explosion there had been overpressure due to acceleration of the combustion by turbulence.

9.3.3 Liquefied natural gas (LNG)

9.3.3.1 Nature of LNG

This is a cryogenic substance (b.pt. of methane $-161°C$). It is often stored without any refrigeration in holes excavated in the ground, and can be transported by land or sea. Such storage and transportation always involves some evaporative losses. The worked example below illustrates this.

A $20 \times 20 \times 20$ m cubic container made of material of thickness 50 cm and thermal conductivity $0.05\,W\,m^{-1}K^{-1}$ is used to store LNG. At the outside surface of the tank the convection coefficient is $5\,W\,m^{-2}K^{-1}$ and at the inside surface the convection coefficient is $25\,W\,m^{-2}K^{-1}$. The tank contents are at $-160°C$ and the outside temperature is $20°C$. Calculate the percentage loss per day of a payload of LNG initially occupying 80% of the volume of the tank. Use a value of $424\,kg\,m^{-3}$ for the density of the LNG and a value of $535\,kJ\,kg^{-1}$ for its heat of vaporisation.

Solution
Initial amount of LNG $= (20)^3 \times 0.8\,m^3 = 6400\,m^3 \equiv 2714$ tonne.
Thermal resistance of $1\,m^2$ of tank

$$= [(0.5/0.05) + (1/5) + (1/25)]\ K\,W^{-1}m^{-2}$$
$$= 10.2\,K\,W^{-1}m^{-2}$$

rate of heat transfer per $m^2 = [20-(-160)]/10.2 = 17.6\,W$
total heat transfer $= 17.6 \times 6 \times 20 \times 20 = 42\,kW$
rate of evaporation $= \{42\,000\,W/(535\,000)\,J\,kg^{-1}\}$

$$= 0.08\,kg\,s^{-1}$$
$$= 6.8\ \text{tonne day}^{-1}$$
$$0.25\%\ \text{of the contents per day}$$

Such losses can of course be reduced by refrigeration. A reader requiring clarification of the expressions for thermal resistance in the above calculation is encouraged to consult one of the many basic texts in which they are found.

9.3.3.2 Hazards in storage and land transportation of LNG and necessary precautions

Hazards with LNG include stratification. Over time methane, the lightest hydrocarbon compound present, is lost by evaporation more than proportionately, creating a liquid which is richer in the minor heavier components such as ethane than the parent gas. This can cause density gradients and mechanical instability leading to rollover, which results in rapid evaporation. The saturated vapour pressure of the liquid at its boiling point is of course 1 bar, therefore LNG is not required to be stored under pressure.

In stationary storage situations, a safety pressure valve can be set to open at only just above atmospheric pressure. When a depleted vessel is being purged with an inert gas the effluent must be conducted to a flare. In road transportation of LNG there must be no leakage: leaked methane might find its way to the engine of the vehicle and interfere with the vehicle's own fuel intake and 'engine management'. Consequently pressure relief valves are set high in such situations, 3–4 bar. Road, rail and even barges have been used to transport LNG from the storage tank to the consumer.

Whereas with LPG (see section 9.4) there is a risk that electrostatic effects will provide an ignition source on leaking, this is less true of LNG. Nevertheless, vessels containing LNG and pipes conveying it, e.g., from a stationary tank to a mobile one, should be earthed.

The low temperatures involved in storing and transporting LNG necessitate special materials for vessels and pipes. Ordinary mild steel becomes brittle at cryogenic temperatures. Materials suitable for LNG containment include copper, aluminium alloys, some stainless steels (especially those high in nickel/chromium). In designing plant for LNG the contraction which results from the low temperatures has to be 'designed in', and if different materials are interfaced, their compatibility in this respect has to be considered. If (as is usually the case[27]) the LNG is vaporised by heat exchange, e.g., with sea water, copper and/or aluminium tubes and pipes have to be used. Metallic elements have higher thermal conductivities than alloys, e.g., copper 386 in SI units, stainless steel about 15. If a cryogenic substance such as LNG was heat-exchanged using steel pipes the temperature gradient across the pipe wall would be sufficient for mechanical damage.

As well as conventional heat transfer plant, there is a further means of effecting conversion of LNG to the gaseous state. This is the Submerged Combustion Vaporiser (SCV), in which product gas from the burning of previously evaporated methane is used to heat water with which the LNG is heat exchanged. Depending on the load this can be used concurrently with conventional heat exchange using unheated seawater.

In LNG spillage on water, e.g., during shipment, there can be the additional hazard of non-chemical explosions due to contact with the water. The mechanism of this is not entirely clear, but probably involves one or both of:

• violent boiling of the methane and the development of pockets of high pressure

- formation of ice, encapsulation of LNG and subsequent evaporation and bursting of the ice

As with organic liquids and vapours at ordinary temperatures, it is the *vapour* that burns. It is quite standard practice for tanks of LNG to have electrical devices (for example, for pumping) on their inside surfaces. LNG is an excellent insulator, and electrical resistances of metallic conductors are very low at LNG temperatures.

Table 9.1 gives brief details of accidents involving LNG storage and transport facilities. Incidents in LNG production are discussed after the table.

An LNG production facility will consist of multiple 'LNG trains', in which progressive operations are carried out starting with removal of anything from the parent gas which falls in the condensate/NGL category and is therefore a very saleable by-product. The LNG production facility at Skikda, Algeria, has six such trains, three of which were destroyed in an

Table 9.1   *Accidents involving LNG*

Location and date	Details
Cleveland, Ohio, USA, 1944	Escape of LNG and spread through the city streets. 128 deaths
La Spezia, Italy, 1971	Rollover, discharge of 2000 tonnes of LNG
Staten Island, USA, 1973	An insulated LNG storage tank had been out of use for almost a year when repairs to the plastic liner were attempted. A fire started to spread through the foam insulation and all 40 workers were killed
Nevada USA, 1987	Ignition of a cloud of LNG created by intentional release of LNG spillage at a USA Department of Energy test site. The cloud had been expected to disperse as in previous such tests
Manchester, UK, 1993	Rollover due to the development of temperature gradients. Loss of LNG, no injury or damage
Spain, 2004	Explosion of a road tanker carrying LNG. One fatality, two injuries
Nigeria, 2005	Loss of containment from a pipe bearing LNG. Fire spread over a wide area

explosion in January 2004. Compression processes along a 'train' require work in the thermodynamic sense of that term, usually at an LNG train obtained from a gas turbine. At Skikda steam turbines were in use, one of which exploded. As with the Staten Island incident (Table 9.1), one might argue that LNG *per se* was not the origin of the mishap. One of the appended numerical examples relates to Skikda.

It was an exploding gas turbine which was the origin of an incident at an LNG train in Trinidad in the same year. Evacuation of the entire facility was necessitated but there were no injuries. Trinidad and Tobago is rich in natural gas from offshore fields but lacking a pipeline structure for export, and infrastructure sharing with Venezuela holds little promise. The gas is mainly converted to methanol for export as such; BP and Methanex are both active in T&T. The alternative to that is LNG production.

Below is a numerical example appertaining to LNG transportation through a pipe.

A heavily lagged copper pipe of 1 m diameter is used to convey LNG at its boiling point of 111 K ($-162°C$) which travels at $0.1\,\mathrm{m\,s^{-1}}$ along the pipe. Determine the mass flow rate of the LNG.

The lagging comprises a material of thermal conductivity $0.01\,\mathrm{W\,m^{-1}\,K^{-1}}$ which is mounted concentrically with the pipe so as to have an outer diameter of 3 m. The thermal resistance of the pipe is negligible in comparison with that of the lagging, and the pipe walls maintain a temperature uniformly 30 K warmer than the LNG. If the temperature of the outside surface of the lagging is 15°C, how much of the LNG will evaporate per km travelled along the pipe?

properties of the LNG:
density $425\,\mathrm{kg\,m^{-3}}$
heat of vaporisation $512\,\mathrm{kJ\,kg^{-1}}$

Solution
Mass flow rate (m^*) calculable from:

$$m^* = \sigma a u$$

where σ = density ($\mathrm{kg\,m^{-3}}$), a = cross-sectional area ($\mathrm{m^2}$), u = velocity ($\mathrm{m\,s^{-1}}$)

$$m^* = 0.1 \times \pi\,(0.5^2) \times 425\,\mathrm{kg\,s^{-1}} = 33\,\mathrm{kg\,s^{-1}}$$

rate of transfer of heat to the LNG $= \Delta T / R_{\mathrm{th}}$

where ΔT = temperature difference between outside and inside surfaces of the lagging

R_{th} = thermal resistance of the lagging

For a long (length $\gg$ radius) hollow cylinder [6]:

$$R_{th} = (1/2\pi Lk) \ln [r_o/r_i]$$

where k = thermal conductivity (W m^{-1} K^{-1})
L = length (m)
r_o = outer radius
r_i = inner radius

$\Downarrow$

$R_{th} = 0.0175$ K W^{-1} for a 1 km length

Rate of passage of heat through the walls for a 1 km length
$$= \{15 - (-162)\}/0.0175 \text{ W}$$

$$= 10.1 \text{ kW}$$

Amount of LNG evaporated $= 10.1 \times 10^3/(512 \times 10^3)$ kg

$= 0.020$ kg (20 g, a very small amount, indicating that the lagging is effective)

An alternative to simple lagging of a pipe bearing LNG is for there to be two concentric pipes with the annulus between them evacuated. Currently such devices are available which have a vacuum of the order of 10^{-6} torr in the annulus.

The standard BS 7777-2:1993 which is for cryogenic liquids including LNG requires that a cylindrical tank be vertically orientated. It will therefore receive direct solar flux across its projected area which is πr^2 where r is the radius. If the same cylinder was sited horizontally it would receive solar flux across an area rL, where L is the length. So for the vertical orientation to be advantageous requires:

$$\pi r^2 < rL$$

i.e. the length greater than about three times the radius, which is almost certain to be so as one would not expect a squat cylinder to be used in such an application.

9.3.3.3 Ocean transportation

A great deal of LNG is transported, in specially designed and constructed vessels, by sea, e.g., from Indonesia to Japan, from Libya to USA. One such vessel is called the *Methane Princess*.

Tanks for the LNG can be integral to the ship's structure, the outer layer of the tank being the ship's wall, with balsa wood as insulation. Alternatively, tanks may be prismatic, conforming to the profile of the ship's hull but self-supporting. Tanks may also be spherical; these have the advantage of good accessibility but their use leaves considerable proportions of the ship's internal volume unoccupied. Figure 9.1 shows such a vessel. The fact that LNG and balsa wood are both less dense than water means that most of the cargo is carried above the water line.

In Chapter 1 mention was made of the vulnerability of Japanese industry through lack of indigenous fuels. At the present time this is largely overcome by importation of LNG from countries including Indonesia. Indonesia was for a considerable period the world's largest exporter of LNG, having recently been surpassed by Malaysia. This is because of the facility in Bintulu, Malaysia. Australia's 'north west shelf' gas production has enormously increased over the last few years, and a great deal of this becomes LNG for markets in the Asia-Pacific region.

9.3.3.4 Combustion phenomenology

Catastrophic leak of LNG will, if there is ignition, result in a fireball. LNG having spilt on to a surface and subsequently ignited will burn as a pool fire. As noted in the Appendix, in the Canvey study radiative fluxes of about $140 \, kW \, m^{-2}$ were predicted for LNG pool fires. Using this and configurational details, it was estimated that a tank of hydrocarbon nearby would, in the event of such a pool fire, receive heat at a rate of 30 MW. There have been

Figure 9.1 *LNG carrier. Reproduced courtesy of Shell International.*

a number of experimental investigations of LNG pool fires, and these are summarised by Mudan and Croce [7].

The fatal accident involving LNG in Cleveland in 1944 is referred to in Table 9.1. Notwithstanding the modern view that methane does not display v.c.e. behaviour, it was reported in the follow-up to the Cleveland accident that there had been some confined v.c.e.s both close to the gas facility where the escape had occurred and in buildings which the methane, once evaporated, had entered. 'Gas explosions' are known to have occurred not only in buildings but in sewers. The LNG at Cleveland had been stored in several tanks, failure of one of which initiated the accident. Twenty minutes later another tank failed, and the contents of this ignited as a pool fire.

9.3.4 Compressed natural gas (CNG)

This is gas stored as such at ordinary temperatures in vessels capable of withstanding high pressures. It is an alternative to gasoline in powering spark ignition engines. It finds extensive application to motor vehicles in New Zealand, with gas brought ashore from the Maui field off the west coast of the North Island.

9.4 Liquefied petroleum gas (LPG)

9.4.1 Nature of LPG

As we have already seen, LPG is usually principally propane. In refining it comes over before gasoline. Its critical temperature is such that it can be liquefied by application of pressure at room temperature. It is a quite widely used fuel; applications include motor vehicles. If leaked catastrophically, and ignited, it displays BLEVE behaviour, discussed fairly fully in Chapter 3. If it leaks through an orifice and ignites it will burn as a jet fire. Figure 9.2 shows a group of LPG containers.

9.4.2 Examples of risk assessment for LPG transportation

LPG is frequently moved about, by road, ship, rail or pipeline, and such movements require a formal risk assessment. LPG is piped around refineries in pressure piping, and below is a fairly simple risk analysis for LPG release caused by pipe failure which has been adapted from a case study in [8].

At a site where LPG is in use, the frequency of leakage and consequent ignition of a vertically orientated LPG-bearing pipe is estimated as 4×10^{-3} per year. The site approximates to a circle of 0.5 km diameter and

Figure 9.2 *LPG storage tanks. Reproduced courtesy of Shell International.*

the only other hydrocarbon it contains is a portable tank of diesel which, at any one time, can be anywhere within the site. Calculate the frequency with which there will be a leakage, ignition and impingement on to the tank of diesel. Approximate the flame resulting from ignition of leaked LPG to being horizontal with a plume angle of 20° creating an arc of radius 70 m which remains entirely within the site.

Solution

Area swept out by the flame on ignition = $\pi(70)^2$ (20/360) m^2 = 855 m^2.

Total area of the site = $\pi(250)^2$ = 196375 m^2

Probability that at any one time the tank of diesel will within the area enveloped by the flame = [855/196375] = 0.0044

Frequency of leakage, ignition and flame impingement on to the tank

= 0.0044 × 4 × 10^{-3} per year = 1.7 × 10^{-5} per year

In Chapter 8, ALARP principles were explained against a background of an offshore example. Here they will be reiterated against a background of LPG transportation on land. It is often necessary to transport LPG by land.

It might be, for example, that an enterprise such as a glass factory, which has a very large energy requirement (and which in 'days gone by' might well have relied on producer gas), wishes to select LPG as its primary fuel, necessitating deliveries in suitable railcars on a weekly basis. In risk assessment, the route between where the LPG is produced and where it is required has to be examined. The populations of points along the way at times of transportation have to be estimated, and in that way a frequency of fatality calculated and submitted for judgement as to whether it is acceptable. These ideas are implemented in the partial escalation path in Figure 9.3.

The risk analyst assigned to this task will need to identify an initiating event for a serious accident; the obvious one is derailment, overturning and leakage. A frequency is assigned to the first of these and probabilities to the second and third, as shown in Figure 9.3.

With reference to this escalation path, consider what we shall call scenario 1, summarised below:

the train derails
an LPG-bearing car overturns
the contents leak
they ignite and BLEVE

From Figure 9.3 it is clear that the frequency of this is:

$$\phi \times \eta_1 \eta_2 \eta_3 \text{ per year}$$

Scenario 2 is summarised below:

the train derails
an LPG-bearing car overturns
the contents leak without ignition

and the frequency of this is:

$$\phi \times \eta_1 \eta_2 [1 - (\eta_3 + \eta_4)] \text{ per year}$$

and frequencies for other scenarios, possibly involving overturning of more than one LPG-bearing car, can be estimated. Each possible scenario will have, as its bottom line in an escalation path such as that shown in simplified form in Figure 9.3, a number of deaths per year. These can be summed across all of the scenarios to give an aggregate number of deaths. The procedure then follows the same lines as that shown for the offshore example in Chapter 8.

Imagine that in our LPG transportation the frequency of a death is 10^{-5} per year ($n = 5$). According to the criteria expressed in Chapter 8, this is an acceptable risk and transportation can go ahead. Now imagine that, instead,

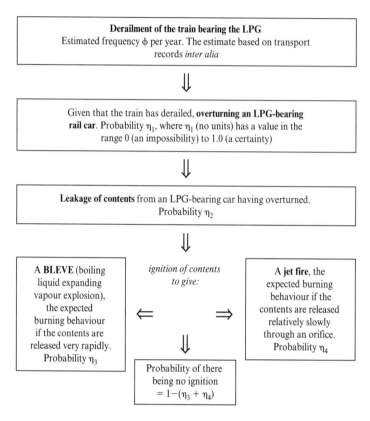

Figure 9.3 *Partial escalation path for LPG leakage during rail transportation.*

the frequency was $10^{-0.5}$ per year. This is too high and the transportation cannot be authorised. Imagine now that the frequency is 10^{-3}. This is in between the two reference values of *n*, and the onus will be on the proposers of the transportation to satisfy the authorities that this is ALARP. Some modification to the transportation procedure, such as an alternative route taking the consignment of LPG through fewer highly populated places, might be identified whereby this frequency could be brought down to $10^{-3.5}$. If this modification can be brought about at moderate cost it will be required: if it cannot, the risk embodied in a frequency of 10^{-3} per year will be accepted as ALARP.

9.4.3 Combustion phenomenology and case studies

BLEVE behaviour and jet fire behaviour were discussed in Chapter 3. Note that generic equations for fireballs, for example for the maximum diameter, are frequently applied without distinction to BLEVEs. It is also possible [7] for spilt LPG to burn as a pool fire, on land or on water.

The worst ever disaster involving LPG was in Illinois a little over 30 years ago, and is fully described elsewhere [9]. It involved derailment of a train pulling several LPG-bearing cars. BLEVE and jet fire behaviour were both observed. Indeed, an LPG-bearing car released its contents through a small hole caused by the impact, and burnt as a jet fire which 'torched' a previously unaffected LPG-bearing car, eventually causing it to burst open and its contents to BLEVE. There have been more recent LPG accidents in Sydney, Australia and in Iowa, USA. In August 2012 a crash involving a road tanker of LPG in India resulted in an appalling death toll of 20 persons. The tanker collided not with another vehicle, but with a barrier separating the right-hand lanes of a highway from the left. There was fire spread to buildings.

Storage of LPG from the point of view of design principles for vessels was discussed in Chapter 6. Parts of the discussion there will be revisited below in our treatment of the effects of solar radiation on stored or transported LPG. We saw in Chapter 4 that solar flux incident on a tank of hydrocarbon will increase its temperature and stimulate evaporation. Lees *op. cit.* mentions that current codes of practice are for a vessel to be 'finished in white', and that the codes should be modified in the direction of stringency if this is not the case. The code gives a temperature range for liquid offtake from a propane vessel as −18 to +38°C. At either temperature, the working pressure must not exceed 14.5 bar g.

9.5 Natural gas condensate

This can occur with associated or with non-associated gas, and with tight gas but not with coal bed methane. It comprises simple hydrocarbons up to C_5 and after refining can be put to automotive or aviation use. Whilst there are 'condensate refineries' set up as such, a conventional refinery can straightforwardly be adapted for condensate. Movement of condensate from its initial scene of extraction from natural gas to a scene of storage or processing might require its stabilisation by removal of ethane. Densities are in the range 600 to 750 kg m^{-3}. Even condensate so stabilised will have a Reid Vapour Pressure in excess of 2 bar, which whilst well superatmospheric is not sufficient for BLEVE behaviour. Ethane alone can BLEVE. In October 2009 two young people were killed in an explosion of stored condensate in rural Mississippi [10]. The storage facility contained fourteen barrels of condensate, and the absence of fences around the facility and of warning signs was noted in the follow-up.

Natural gas is sold on a heat – not a quantity – basis, unlike crude oil which is sold on a volumetric basis. (Coal is sold on a weight basis.) The presence of ethane and other higher hydrocarbons in natural gas makes for complications with pricing and this is addressed in one of the numerical examples at the end of the chapter.

9.6 Oxygenated hydrocarbons

9.6.1 Introduction

We saw in Chapter 5 how oxygenates such as cyclohexanone can be manufactured by partial oxidation. Routes to oxygenated hydrocarbons other than via the parent hydrocarbon include:

- fermentation processes
- pyrolysis of cellulosic materials

Breakdown products of wood include methanol (CH_3OH), sometimes called 'wood alcohol'.

There are very many oxygenated hydrocarbons of industrial importance. Some have featured in earlier parts of this book, for example in the discussion of flash points in Chapter 3 and in a number of case studies, for example Ludwigshafen.

9.6.2 Combustion characteristics

Oxygenated compounds obviously have lower calorific values than the parent hydrocarbons because they are already partly oxidised, e.g., $27 MJ kg^{-1}$ for acetaldehyde (CH_3CHO) compared with $52 MJ kg^{-1}$ for the parent hydrocarbon, ethane. However, oygenates are often very reactive.

In the shaded area below is a calculation of the adiabatic flame temperature of acetaldehyde.

Acetaldehyde burns according to:

$$CH_3CHO + 2.5O_2 (+ 9.4N_2) \rightarrow 2CO_2 + 2H_2O (+ 9.4N_2)$$

net heat of combustion $1105 kJ mol^{-1}$

specific heat of water vapour $= 43 J °C^{-1} mol^{-1}$

specific heat of carbon dioxide $= 57 J °C^{-1} mol^{-1}$

specific heat of nitrogen $= 32 J °C^{-1} mol^{-1}$

Specific heats are values about half way along the expected range from room temperature to flame temperature

When a mole of acetaldehyde is burnt, the reaction products clearly have a heat capacity:

$$[(2 \times 57) + (2 \times 43) + (9.4 \times 32)] \, J\,°C^{-1} = 501 \, J\,°C^{-1}$$

$$\text{temperature rise on burning} = (1105000/501) \, K = 2206 \, K$$

$$\text{actual temperature (for reactants at 298 K)} = 2504 \, K$$

It is clear from this example that in spite of the lower heat of combustion the adiabatic flame temperature is the same as a typical one for a hydrocarbon. Less oxygen is required to burn a mole of acetaldehyde than to burn a mole of the parent alkane, acetaldehyde being already partly oxidised. As a result of this, the amount of nitrogen in the post-combustion gas is also lower than for the C_2 hydrocarbon. These two factors – smaller heat of combustion and lower heat capacity of the burnt gas – to a good approximation cancel each other out. The calculation below also relates to an oxygenated hydrocarbon, whilst invoking some principles of thermal radiation not previously encountered in this book.

According to a source cited by Thomas [11], for heat balance purposes (e.g., in weather forecasting) the sky can be treated as a black body at a temperature 5 to 20 K below that of the earth's surface at the place to which the heat balance relates.

At a petrochemical plant an outdoor tank, mounted in the ground, is used for organic compound storage, and this has a polished metal lid. A deep layer of a highly insulating substance is attached across the entire area of the lower surface of the lid. A convection coefficient of $10 \, W\,m^{-2}\,K^{-1}$ applies to heat transfer between the atmosphere and the upper surface of the lid.

The plant is set up in a place where the climate is hot and during a typical day of usage the ambient temperature reaches 40°C. The oxygenated compound methyl formate (CH_3OOCH, boiling point 31.5°C) is also present at the site in a refrigerated storage facility. Predict whether, in the event of leakage of the methyl formate on to the vessel lid nearby, the lid would be hot enough to act as a 'heating element' in boiling of the methyl formate. Assume that thermal conditions are approximately steady and take the sky temperature to be 300 K.

Solution
Anything on the earth's surface receives solar flux of about $1400 \, W\,m^{-2}$ in daylight, as we saw in an earlier chapter. The information that the lid is 'polished' is a cue to the reader to assume that it reflects most of the solar radiation incident upon it.

Heat once received by the earth is redistributed in a number of ways. For example, in the situation under discussion, having been absorbed by the earth's surface this heat is in part transferred by convection to the air. If we treat a point on the earth's surface as being warmer than the sky, the heat which has been received from that direction in the first place has to be understood as now being transferable by modes of heat transfer other than radiation. The following thermal influences therefore operate:

(i) Radiation of heat from the cover of the tank to the surroundings (rate q_1)

(ii) Transfer of heat from the atmosphere to the tank cover by convection (rate q_2, convection coefficient $h = 10 \, \mathrm{W \, m^{-2} \, K^{-1}}$)

In formulating an expression for q_1, we have to take the view factor into account, viz.:

$$q_1 = A\sigma\epsilon F_{\text{ground-sky}}(T_{\text{cover}}^4 - T_{\text{sky}}^4)$$

where A is the area (m²) of the radiating source, σ is the Stefan–Boltzmann constant, ε is the emissivity and $F_{\text{ground-sky}}$ the appropriate view factor; the subscripts on the temperatures are self-explanatory. Because of the expanse of the sky in comparison with the lid, the sky will 'look black' to the lid and we assume:

$$\varepsilon = 1$$

Now to obtain a value for the view factor, we treat the tank cover and the sky as parallel surfaces of unequal area. For 2 surfaces [10] there have to be $2^2 = 4$ view factors, easily characterised as follows:

$$F_{\text{ground-ground}} = 0$$

since the ground, being flat, does not 'see itself', likewise the sky which is being treated as a flat emissive surface, hence:

$$F_{\text{sky-sky}} = 0$$

Now $F_{\text{sky-ground}}$ can also be taken to be zero; the sky is so expansive in comparison with the cover of the tank that any radiation from it has an infinitesimal likelihood of striking the cover, therefore, from [10]:

$$\sum F = 1.0 \Rightarrow F_{\text{ground-sky}} = 1$$

So the equation for radiation becomes:

$$q_1 = A\sigma(T_{\text{cover}}^4 - T_{\text{sky}}^4)$$

Now under steady conditions:

$$q_1 = q_2$$

We therefore have two equations to combine and solve. These are:

$$q_1 = A\sigma \left(T_{cover}^4 - T_{sky}^4 \right) \text{(see above)}$$

$$q_2 = hA(T_{ambient} - T_{cover})$$

Combining these, having regard to the fact that conditions are steady, and inserting the given value for h ($10\,\text{Wm}^{-2}\text{K}^{-1}$), the value for the Stefan–Boltzmann constant ($5.7 \times 10^{-8}\,\text{W}\,\text{m}^{-2}\text{K}^{-1}$), a value of 313 K for $T_{ambient}$ and a value of 300 K for the sky temperature gives:

$$5.7 \times 10^{-9} T_{cover}^4 + T_{cover} = 359.17$$

The above equation has to be solved by trial and error, yielding a solution of 308 K (35°C). The methyl formate would indeed boil if brought into contact with the surface.

The above calculation requires qualification, since in the original formulation cited radiation towards the sky is at night so the point on the earth's surface to which the heat balance relates receives no solar flux. Application of the same approach in daylight requires that the point – in the case of the calculation above the vessel lid – shall be a non-gray body. In thermal radiation a gray body, though less emissive at any one temperature than a black one, acts equivalently towards thermal radiation of all wavelengths. In the above question there is the requirement that the lid, being polished metal, reflects most of the incident solar radiation which is therefore lost from the lid. By contrast, it acts as a black body when itself radiating at low temperature. This is non-gray body behaviour, and it is not at all uncommon for practical materials to display such behaviour, especially when the incident and emitted radiation are widely separated in wavelength.

It is counterintuitive that the lid can settle to a steady temperature below that of the atmosphere, but non-gray body behaviour on the part of the lid and black body behaviour on the part of the sky together allow for this. This has recently been discussed in a short contribution to the research literature [12], where its relevance to LNG tankers experiencing widely different weather conditions during a voyage is emphasised.

9.7 Organic peroxides

9.7.1 Introduction

A simple example of an organic peroxide is diethyl peroxide: $C_2H_5OOC_2H_5$. Organic peroxides are extremely reactive, not by oxidation but by explosive decomposition. They can be formed in side reactions in partial oxidation of hydrocarbons, and if undetected constitute a considerable hazard. The one possible exception to the general rule that hydrocarbon accidents in industry display deflagration rather than detonation behaviour is the decomposition of organic peroxides.

9.7.2 Case studies and related calculations

In Los Angeles in 1974 there was a huge explosion involving stored mixed peroxides awaiting transportation. There had been a similar incident in New York about 20 years earlier. A simplifying factor in the prediction of thermal stability or otherwise of peroxides is that the mechanism of decomposition is, in comparison with that of oxidation of a hydrocarbon, fairly simple therefore incorporation of the kinetics of reaction into classical models for thermal ignition is straightforward. Of course, an unstable substance in storage will not necessarily explode. Whether it does depends on the quantity, the ambient temperature and, in principle, the shape of the assembly. If conditions are such that there will be an explosion, this will not be immediate; indeed, 'induction times' can be long enough to prevent anything untoward in short-term storage, though reliance on this is not necessarily acceptable practice.

Some of the principles of ignition theory can be applied to such substances, for example, to prediction of times to ignition. In fact, in the Los Angeles accident a significant factor was that the storage was unplanned, the purchaser having requested a delay in delivery. Clearly in such circumstances it is useful to predict times to ignition, and this can be done in the way shown in the numerical example below.

A semi-solid effluent substance from a processing plant contains significant amounts of unstable organic compounds including a variety of peroxides. A small 1 m³, approximately cubic, lump of the substance when tested in an oven is shown to display thermal instability at oven temperatures above 80°C. The sample is placed in the oven at room temperature, and rises to 80°C because of heating by the oven. The time taken between uniform attainment of 80°C by the sample and incipient explosion is 2 hours. In the event that a 3 m × 3 m × 3 m pile of the substance awaiting disposal was at its critical condition, how long would it take to explode?

Solution

An approximate rule which has found wide if not always uncritical application is [13]:

$$\text{time to explosion} \propto \text{sample dimension}^{2.0}$$

where the sample dimension is for a sphere, for example, the radius, or for a cube the half-width.

So calling the time taken for the $3\,m \times 3\,m \times 3\,m$ to explode τ hours:

$$\tau/2 = (1.5^2/0.5^2) \implies \tau = 18 \text{ hours}$$

In interpreting this calculation one must bear in mind that it is only valid when conditions for the large pile are, like those for the small pile in the oven test, such that the ambient temperature exceeds only slightly the critical ambient temperature. If the former significantly exceeds the latter, explosion will be accelerated.

Another 'spontaneously unstable' substance, mentioned in the Appendix as having been present at Canvey Island at the time of the study, is ammonium nitrate. Its uses are as a fertiliser, as an ingredient of high explosives and (more rarely) as one component of certain propellants. In view of its having been noted in the Canvey study and because this substance has caused two accidents each of which involved hundreds of fatalities, brief coverage of it here is not inappropriate even though the substance is not, of course, a hydrocarbon. An ammonium nitrate explosion in Germany[28] in 1921 claimed 561 lives, and one on board a ship berthed at a US port in 1947 claimed 522 lives (Lees, *op. cit.*). There has also been a fatal ammonium nitrate explosion in Belgium. Ammonium nitrate is made by neutralising nitric acid with ammonia, itself made from synthesis gas (see Chapter 5). Like organic peroxides, ammonium nitrate requires no external oxidant for explosive behaviour.

References

[1] Kanury A.M. 'Ignition of liquid fuels' in *Handbook of Fire Protection Engineering*, Second Edition. SFPE, Boston, MA (1995)

[2] Perry A.H. *Environmental Hazards in the British Isles.* George Allen and Unwin, London (1981)

[3] Jones J.C. 'The development of fuel production and utilisation in the Far East during the 20th Century' *Energy World* Issue No. 303: 18–19 (2002)

[4] Marshall V.C. *Major Chemical Hazards.* Ellis Horwood, Chichester, UK (1987)

[5] *National Fire Protection Association Journal*, March/April 1999 pp. 64–69.

[6] Holman J.P. *Heat Transfer* any available edition, McGraw-Hill, New York

[7] Mudan K.S., Croce P.A. Fire Hazard Calculations for Large Open Hydrocarbon Fires in *Handbook of Fire Protection Engineering*, Second Edition. SFPE, Boston, MA (1995)

[8] Lees F.P. *Loss Prevention in the Process Industries*, First Edition. Butterworth, London (1980)

[9] Jones J.C. *Combustion Science: Principles and Practice*, First Edition. Millennium Books, Sydney (1993)

[10] http://www.icis.com/Articles/2010/10/18/9402406/extra-ethane-poses-obstacle-in-us-marcellus-shale-development.html

[11] Thomas L.C. *Heat Transfer*, First Edition. Prentice-Hall, Englewood Cliffs, NJ (1992)

[12] Jones J.C. Radiation approximations for estimating the cover temperature of outdoor storage vessels *International Journal on Engineering Performance-based Fire Codes* **11** 110–111(2002)

[13] Bowes P.C. *Self-Heating; Evaluating and Controlling the Hazards*, First Edition. Elsevier, Amsterdam (1984)

[14] Oxley C.J., Smith J.L., Rogers E., Ye W. Heat-release behaviour of fuel additives *Energy and Fuels* **15** 1194–1199 (2001)

[15] Bradley J.N., Barnard J.A. *Flame and Combustion*, Second Edition. Chapman and Hall, London (1984) p. 47

[16] http://bulktransporter.com/management/shippers/oilfield_hazards_0101/

Numerical problems

1. Return to the first worked example in the main text of this chapter. There is a breakage of the pipe bearing methane such that there is a full diameter exposed and consequent leakage at the previously calculated flow rate. The leak continues for two minutes until emergency action has taken effect. Over how large a volume will the leaked methane need to have dispersed uniformly in order for the lower flammability limit – 5% volume or molar basis – not to be exceeded?

Identify a factor, relating to the respective properties of air and methane, which might impose some inaccuracy upon this calculation.

2. A tank of LNG is temporarily uncovered and receives solar flux of $700\,W\,m^{-2}$ at its liquid surface. If this state of affairs continues for one hour, calculate the amount of loss if the liquid surface area is $400\,m^2$. The heat of vaporisation is $535\,kJ\,kg^{-1}$. *From MSc (Safety and Reliability) Examination, University of Aberdeen, 1997.*

3. As noted in the main text, LNG can be evaporated by heat exchange with sea water. Sea water for such a purpose is supplied at $40\,kg\,s^{-1}$ and drops 12 K during exchange. If the process is 80% efficient what will be the rate of evaporation of the LNG. The latent heat of evaporation is $512\,kJ\,kg^{-1}$. Take the heat capacity of the sea water as $4200\,J\,kg^{-1}K^{-1}$ and express your answer in tonnes per day.

4. In the Canvey study it was estimated that an LNG pool fire would supply 30 MW of heat to a nearby vessel.

This vessel should be taken to hold hydrocarbon liquid approximating in composition to pure toluene. The initial temperature is 288 K. The vessel has a safety valve set to open at a vapour pressure of 6 bar, and in the event of a pool fire there needs to be at least 30 minutes for emergency measures to take effect before the safety valve opens, i.e., a margin of safety of 30 minutes before the vessel starts to discharge its contents. It is necessary therefore for the vessel contents to provide 'thermal ballast'.

The ullage space in the vessel is small in comparison with the volume occupied by liquid. Neglecting the thermal capacity of the metal vessel wall in comparison with that of the vessel contents, determine the minimum amount of toluene required for the 30-minute response time to be provided for.

Heat of vaporisation of toluene $= 34.5\,kJ\,mol^{-1}$; boiling point of toluene $= 111°C$; and specific heat of toluene liquid $= 2000\,J\,kg^{-1}K^{-1}$.

5. We are told in the first Canvey report that a wall in the vicinity of a source supplying $140\,kW\,m^{-2}$ of heat flux from an LNG pool fire required 42 minutes to attain a steady temperature of 650 K.

Consider a steel wall receiving $140\,kW\,m^{-2}$. How long would it take for the surface temperature to rise from 298 K to 650 K if the entire flux were intercepted by the wall? Given the time actually taken, what proportion of the total heat flux must the wall have been receiving?

Use the following expression for a 'semi-infinite solid' receiving flux q $W\,m^{-2}$:

$$T(t) - T_i = \frac{2q\sqrt{(\alpha t/\pi)}}{k}$$

where $T(t)$ = surface temperature at time t
    α = thermal diffusivity $(m^2 s^{-1})$
    k = thermal conductivity $(W\,m^{-1}K^{-1})$
      thermal conductivity of the steel $= 18\,W\,m^{-1}K^{-1}$
      thermal diffusivity of the steel $= 5 \times 10^{-6}\,m^2 s^{-1}$

6. In a particular analysis of LPG land transportation, derailment of a tank car bearing LPG is estimated to have a frequency of 5×10^{-3} per year. The tank cars are in pairs, the pairs separated by other sorts of cars not containing hydrocarbon inventory, and if one tank car derails there is a 60% probability that the other one in the pair will also derail. The probability of leakage of the contents of any car derailed is 0.2 and of subsequent ignition to form a jet fire 0.25.

Calculate the frequency of a course of events such that one LPG car derails, its contents leak and burn as a jet fire. Its neighbour also derails but does not leak its contents. The jet fire from the first car torches the second, causing eventual failure and catastrophic release. State any assumptions or approximations made in the calculation.

7. According to Lees *op. cit.* the course of events at the fatal 1947 ammonium nitrate explosion referred to in the main text was as follows. A vessel in the harbour was carrying 2300 tons of ammonium nitrate. At 8:00 a.m. fire was observed on this ship, and the ammonium nitrate it was bearing exploded at 9:15 a.m., i.e., 1.25 hours later. A nearby ship, also carrying ammonium nitrate, caught fire at 6:00 p.m. At 1:10 a.m. i.e., 7.2 hours later, there was an ammonium nitrate explosion on the second ship. Using the correlation for explosion induction time as a function of sample size given in the main text, obtain a *very rough estimate* of the quantity of ammonium nitrate which the second ship was carrying.

8. An outdoor tank, mounted in the ground, is used for organic compound storage. The metal lid for use with this tank has an upper surface is polished to make it reflective. To the lower surface of the lid across its entire area is attached a deep layer of a highly insulating substance. A convection coefficient of $20\,W\,m^{-2}\,K^{-1}$ applies to heat transfer between the atmosphere and the upper surface of the lid. The local climate is hot, and during a typical day of usage the ambient temperature reaches 40°C. A hydrocarbon mixture with an initial boiling point of 36°C is present at the site. For most of the daylight period thermal conditions are approximately steady.

Invoking as part of your answer the relevant principles of thermal radiation and also invoking the 'non-gray body model' given in the main text, determine by calculation whether the tank lid would, in the event that there was leakage of the hydrocarbon mixture on to it, be warm enough to cause the hydrocarbon to start to boil. State any assumptions or approximations made.

9. Ethyl hexyl nitrate ($C_8H_{17}NO_3$) is used as a cetane enhancer for compression ignition engines, and is frequently transported and stored. According to recent studies [14] a long[29] cylindrical container, radius 0.91 m, of the material will, from assembly under marginally supercritical conditions, take 21 days to ignite. How long will a container of 2.76 m radius take to ignite under equivalent conditions?

10. Imagine a 'frozen earth' storage situation for LNG; that is, the LNG is in a hole dug in the ground with a thin, reflective cover over it. The air with which the outside layer of the cover is in contact is at temperature 273 K (0°C) and the convection coefficient from air to cover is $10\,W\,m^{-2}K^{-1}$. The space between the underneath side of the cover and the surface of the liquid is occupied by still air and can be taken to be insulating. The upper surface of the cover has, at the long wavelengths that it emits by reason of its own temperature, an emissivity of 0.9. Taking conditions to be thermally steady, calculate the temperature of the cover.

11. Repeat question 10 with an air temperature of 318 K (45°C) and a sky temperature of 303 K (30°C) whilst retaining the values of h and ε.

12. Barnard and Bradley [15] give the following values for the properties of ammonium nitrate (symbols as previously defined).

$$A = 6 \times 10^{13}\,s^{-1}$$
$$E = 170\,kJ\,mol^{-1}$$
$$Q = 4.7\,MJ\,kg^{-1}$$
$$\sigma = 1750\,kg\,m^{-3}$$
$$k = 0.126\,W\,m^{-1}K^{-1}$$

(Note that the density and the thermal conductivity will depend on the particle size.)

Return to the coverage of FK theory in Chapter 4. For a cubic assembly δ_{crit} is 2.57 with r the *half*-width. What size of cubic container filled with ammonium nitrate would be critical at 318 K?

13. Return to question 3 and imagine that instead of the conventional plant an SCV is used, the water being heated to 65°C and dropping to 30°C during the exchange. On the basis of simple heat balance what will be the evaporation rate of the LNG? What factors would need to be evaluated to achieve the calculated result?

14. At Marcellus – a major US tight gas deposit – there is an abundance of ethane in the gas. Condensation in the pipeline conveying the gas once produced has to be avoided. The way in which the amount of ethane is estimated is from the calorific value of the gas. A document on this [16] says:

> *When ethane is mixed in with natural gas it can substantially increase the heat of the natural gas stream.*

If the gas composed of methane and ethane, no other hydrocarbons and no inerts, has a calorific value of 1150 BTU per cubic foot what is the proportion of ethane, volume or molar basis?

15. With reference to the steam device at Skikda which exploded whilst in use at an LNG train, it was operating at a rate of 100 000 horse power. The author is unaware of whether the steam was saturated or superheated. On the hypothesis that the steam was saturated at 1 bar with dryness fraction unity, generated from liquid water at 25°C, and condensed after turbine passage to vapour at 30°C of dryness fraction 0.68, calculate the efficiency of the device and the rate of steam usage. It will be necessary to consult steam tables.

16. In Chapter 3 the equation:

$$Q = AP \sqrt{\{(M\gamma /RT)[2/(\gamma + 1)]^{(\gamma + 1)/(\gamma - 1)}\}}$$

Q = mass flow rate of gas ($kg\,s^{-1}$), A = discharge area (m^2), P = upstream pressure ($N\,m^{-2}$), M = molecular weight ($kg\,mol^{-1}$), R = gas constant = 8.314 $J\,K^{-1}mol^{-1}$, T = gas temperature (K) and γ = ratio of principal specific heats. was given and applied to leakage of methane. The equation is obviously not 'substance specific' and can be applied with perfect rigour to other gases provided that flow is sonic.

Consider a container of LPG, approximating in composition to pure propane, at 25°C. Its pressure will be 9 bar ($\equiv 9 \times 10^5\,N\,m^{-2}$). Imagine that a hole the size of a UK five pence coin (diameter 9 mm) develops. What will be the leakage rate of propane through the hole. γ for propane is 1.136.

Endnotes

[27] Increasingly frequently, the evaporation is 'process integrated' so as to provide some cooling.

[28] This accident occurred in the town of Oppau, across the Rhine from Ludwigshafen, the scene of two fatal chemical plant accidents in the 1940s.

[29] In the context of thermal ignition a 'long' cylinder is one whose diameter is very small in comparison with its axial length. This means that radial conduction dominates to the exclusion of longitudinal.

IO

Toxicity hazards

10.1 Introduction

Although the primary focus of this book has been fire and explosion hazards, passing reference has been made in the earlier chapters to toxic hazards. This has usually been from the point of view of engineering solutions to potential toxic release. This chapter comprises a suitable synthesis of facts and principles apropos of toxicity *per se*. The approach will be to discuss in turn several toxic substances or groups thereof most of which have featured previously in this book. Hydrocarbons themselves, though varying in toxicity, are usually regarded collectively as asphyxiants rather than poisons. Of course, in fire and explosion science hydrocarbons can often be lumped together, their combined effect being expressible simply as the amount of heat released on combustion. Such an approach will not suffice when toxicity is under discussion.

This chapter will therefore be concerned partly with substances which though not themselves hydrocarbons are widely used, for various purposes, in the hydrocarbon industry. There will also be a discussion of certain hydrocarbon derivatives known to be very toxic.

10.2 Chlorine

10.2.1 Introduction

Chlorine is manufactured by electrolysis of brine, and in the hydrocarbon industry finds application in the chlorination of organic substances. The deadly effects of exposure to chlorine were mentioned in Chapter 5, and chlorine always features centrally in any broadly based literature account of toxic gases, e.g., that in Marshall *op. cit.* Chlorine has of course been used in warfare, and much of what is known about the effects of the gas on humans has been gleaned from records of chlorine attacks in war.

10.2.2   Threshold limit values and trends in fatality through exposure

Lees *op. cit.* gives the threshold limit value (TLV) of chlorine as 1 p.p.m. and the odour threshold as 1 p.p.m. According to the same source, 4 p.p.m. is the maximum concentration inhalable for an hour without causing physiological damage, and 1000 p.p.m. is fatal after a few inhalations. Simple ideal gas equations are used in applying such standards and examining particular circumstances for conformity to them. A simple example follows in the shaded area below.

A TLV of 1 p.p.m. for chlorine was given in the previous paragraph. A different source to that cited there gives this as 3 mg m^{-3} at 15°C (288 K), 1 bar. Examine the two for correspondence.

Solution
Invoking the ideal gas equation:

$$PV = nRT$$

where P = pressure (N m^{-2}), V = volume (m^3), n = quantity (mol), T = temperature (K) and R is the universal gas constant (8.314 J K^{-1}mol^{-1}).

Therefore 1 m^3 of gas at the conditions specified contains:

$$(1 \times 10^5 \text{N m}^{-2} \times 1 \text{m}^3)/(8.314 \text{J K}^{-1}\text{mol}^{-1} \times 288 \text{ K}) = 41.8 \text{ mol}$$

Molecular weight of chlorine = 71 g, therefore 3 mg comprises:

$$(3 \times 10^{-3}/71) \text{ mol} = 4.2 \times 10^{-5} \text{mol}$$

p.p.m. chlorine at the TLV = $(4.2 \times 10^{-5}/41.8) \times 10^6 = 1$ p.p.m

The two statements of the TLV therefore correspond exactly.

An important parameter in quantifying the toxic potency of respective chemical agents is the LD$_{50}$, defined as 'the dose which will kill 50% of an exposed population of 'designated animals'. The LD$_{50}$ for chlorine when inhaled by persons is 300 to 500 p.p.m. for 30 minutes.

Also important in quantitative treatments of the effects of toxic substances are vulnerability models. These have their basis in consequence analysis, and are conceptually similar to the construction of escalation paths for fires, considered in an earlier chapter. However, whereas the latter have as their 'bottom line' a number of fatalities, vulnerability models lead to a simple equation for extent of injury as a function of the injury factor, and these can be expressed as probit equations such as were encountered in Chapter 3 when fire and explosion effects were under discussion.

For chlorine lethality the probit equation (Lees, *op. cit.*) is:

$$Y = -8.29 + 0.92 \ln\{C^2 t_i\} \qquad\qquad \text{Eq. 10.1}$$

where C is the concentration in p.p.m. and t_i is the exposure time in minutes. We can at least examine this for consistency with the information previously presented, for example the LD_{50} for chlorine. A 50% rate of death corresponds to a probit of 5.0, giving:

$$14.4 = \ln\{C^2 t_i\}$$

If exposure is for an half an hour, the p.p.m. corresponding to the LD_{50} is 250 p.p.m., broadly consistent with the trends expressed qualitatively previously. Such a level would cause intolerable discomfort very rapidly; 1 p.p.m. can be detected by the human olfactory function.

10.2.3 Chlorine leakage case studies

There have been many fatal accidental leakages of chlorine, the worst of which was in Rumania in 1939 when 60 people were killed.

10.3 Ammonia

Ammonia is the other highly toxic gas which occurs widely in the chemical industry. It may be present as an end product, as a reagent or as a refrigerant. Its presence at Canvey Island at the time of the Canvey study is noted in the Appendix.

It is significantly less toxic than chlorine and a TLV of 50 p.p.m. applies. It has been the subject of a probit equation, the parameters being $k_1 = -30.57$, $k_2 = +1.385$, and the exponent of C is 2.75. There have been a significant number of fatal accidents through ammonia leakage including one in South Africa in the 1970s when 18 people died. A numerical example follows.

At the 1973 ammonia accident in South Africa (Marshall, *op. cit.*) fatalities occurred over an approximately circular area having an estimated population, at the time of the leak, of 250. From these data and using information in the previous paragraph, determine the concentration of ammonia at the site if inhalation was over a 30 minute period.

Solution
18 persons died, the percentage therefore being $(18/250) \times 100 = 7.2$. This corresponds to a probit (Y) of 3.52. Now the probit equation is:
$$Y = -30.57 + 1.385 \ln\{C^{2.75} t_i\} = 3.52$$

$$\Downarrow$$

$$C = 2239 \text{ p.p.m.} \equiv 0.22\%$$

The calculation has, of course, oversimplified the situation by disregarding the fact that there was a concentration profile, not a level concentration. Nevertheless, the result gives a 'feel' for toxic levels of ammonia; an equivalent number of fatalities would of course occur at much lower concentrations of chlorine. This contrast is further addressed in one of the appended numerical problems.

10.4 Hydrogen fluoride

10.4.1 Introduction

This substance (which features in the Canvey study) is used anhydrous or as a solution in water, in which case the name 'hydrofluoric acid' is preferred. The boiling point of hydrogen fluoride is 19°C. Less is known about hydrogen fluoride in terms of toxic levels than is known for ammonia and chlorine[30]. The critical temperature is 230°C, so the substance can be stored or transported as liquid under its own vapour pressure. As well as being a toxic hazard, hydrogen fluoride constitutes a corrosive hazard.

10.4.2 Toxicity

A concentration of $500\,\mathrm{mg\,m^{-3}}$ for one hour involves a significant risk of death (Lees, *op. cit.*). The probit equation is:

$$Y = -26.36 + 2.854\ln L \qquad\qquad \text{Eq. 10.2}$$

where $L = Ct_i$ and, as previously, t_i = exposure time in minutes and C = concentration in p.p.m. In the calculation below this is examined in view of the previous statement that $500\,\mathrm{mg\,m^{-3}}$ for one hour is lethal.

A group of persons experience $500\,\mathrm{mg\,m^{-3}}$ of hydrogen fluoride for one hour. Predict what percentage of the group will be fatally affected. Molecular weight of $HF = 20\,\mathrm{g\,mol^{-1}}$.

First convert the units to p.p.m.:
$$500\,\mathrm{mg\,m^{-3}} = 0.025\,\mathrm{mol\,m^{-3}} = (0.025/40) \times 10^6\,\mathrm{p.p.m.} = 625\,\mathrm{p.p.m.}$$
Substituting into the probit equation:
$$Y = -26.36 + 2.854\ln\{625 \times 60\} = 3.70$$
A probit of 3.70 converts to 10%.

10.4.3 A case study

There were 63 fatalities when a leakage of HF occurred in Belgium in 1930 (Lees, *op. cit.*).

10.5 Selected hydrocarbon derivatives

10.5.1 Introduction

Certain hydrocarbon derivatives, some of which have featured previously in this book, are now reviewed in terms of their toxicity hazards.

10.5.2 Methyl isocyanate: CH₃NCO

The 1984 Bhopal accident, the worst industrial toxic release ever, involved methyl isocyanate. Details of Bhopal[31] can be found elsewhere; the coverage here will be concerned with the nature of the substance, its manufacture and control measures necessary in its usage.

Methyl isocyanate (boiling point 39°C), an intermediate in the manufacture of a major insecticide, is produced by the action of phosgene (itself a very toxic material) on methylamine; the reaction is exothermic. The toxicity of methyl isocyanate far exceeds even that of chlorine. The applicable probit equation (Lees, *op. cit.*) is:

$$Y = -5.642 + 1.637 \ln(C^{0.653} t_i) \qquad \text{Eq. 10.3}$$

symbols as previously defined. The recommended limit is 0.013 p.p.m. Insertion into equation 10.3 of 1 p.p.m. for C and 2.67 (corresponding to a percentage of 1.0) for Y gives $t = 160$ minutes; that is, according to this equation, if a population were to be exposed to 1 p.p.m. of methyl isocyanate for just under three hours, 1% of them would die. Marshall *op. cit.* gives an IDLH (immediately dangerous to life and health) figure of 20 p.p.m. The prediction of the probit equation is that such a level of methyl isocyanate for 30 minutes would cause death to 3% of those exposed.

10.5.3 Benzene, toluene, xylenes (BTX)

These substances are conveniently discussed together, because of their chemical similarity and also because they often occur together. Indeed, they are so grouped in Chapter 4. They are to be found not only in petroleum processing but also in coal processing, for example as a by-product of coking where they occur amongst the tars and oils. Sometimes enterprises involved in coking, e.g., steel manufacturers, have a business arrangement with petrochemical companies whereby BTX is passed along to them.

Benzene itself, very large amounts of which have been used in the petrochemical industry since its inception, was for many years thought to be almost harmless. By the 1960s and 1970s there was evidence that individuals having been exposed to benzene had an increased risk of suffering from certain illnesses including leukaemia. This was followed by a tightening of regulations relating to benzene, and by 1976 a limit of 1 p.p.m. applied in the USA (Lees *op. cit.*). In the event of leakage the danger to persons is that very small amounts will be inhaled for long periods without there being any discomfort or even perception.

Toluene has the advantage, from the toxic hazards point of view, of a very low odour threshold, <1 p.p.m. By contrast xylenes[32], depending on the proportions of the respective xylenes (and having regard to the fact that the odour threshold of a particular substance is not a 'hard number' but varies between individuals) have an odour threshold up to 40 p.p.m. Toluene resembles benzene in terms of its toxicity but does not have the long-term effects of benzene previously noted. The toxicity of xylenes relative to that of toluene has been the subject of some debate [1].

Simple numerical relationships exist [2] for the calculation of occupational exposure limits (OELs) of mixtures of hydrocarbons from knowledge of the OEL of each component. If a hydrocarbon liquid mixture contains n components, the OEL for the mixture if its vapour enters the atmosphere is given by:

$$1 / OEL_{mixture} = \sum \varphi_n / OEL_n \qquad \text{Eq. 10.4}$$

where φ_n denotes the fraction by weight of the nth component. In the calculation below this is applied to BTX. The OELs are all in units of $mg\,m^{-3}$.

A mixture of BTX contains 30% benzene, 65% toluene, balance xylenes (all weight basis). What will be the OEL in a situation where such a mixture is in use. Use the following values [3] for the OELs (units $mg\,m^{-3}$) of the components.

$$benzene = 16$$
$$toluene = 188$$
$$xylenes = 434$$

Solution
Substituting into equation 10.4:

$$(1/OEL_{mixture}) = (0.3/16) + (0.65/188) + (0.05/434)$$

$$\Downarrow$$

$$OEL_{mixture} = 45\,mg\,m^{-3}$$

For more complex mixtures [3], groups of hydrocarbons rather than individual compounds are assigned OELs for the purpose of substitution into equation 10.4. Interestingly, aromatics other than BTX are assigned a value of $500\,mg\,m^{-3}$. The fact that this is an order of magnitude higher than that calculated for the BTX mixture in the above calculation is an indication not only of the particular toxicity of benzene but also of the high volatility of BTX.

It is not inappropriate to mention in this section styrene ($C_6H_5CH{=}CH_2$), which is manufactured from ethylbenzene and has an OEL [3] of 100 p.p.m. It is used in vast quantities in the manufacture of polystyrene. It has a very low odour threshold, <1 p.p.m. Styrene is the subject of one of the appended numerical problems.

10.5.4 Vinyl chloride, $CH_2{=}CHCl$

10.5.4.1 Introduction

The polymerisation of vinyl chloride (coverage of which, from the process safety point of view, is in Chapter 5) was first carried out on a laboratory scale in the second half of the nineteenth century. W.L. Semon, working for the B.F. Goodrich Company, discovered in about 1930 that polyvinyl chloride (PVC) was a potentially very useful product, and the company set about a development programme [4]. In particular, PVC was seen as an alternative to rubber in many applications.

B.F. Goodrich consequently became and remained leaders in PVC manufacture. In 1974 it was found that three long-standing employees of the company had died of a very rare form of liver cancer. The three deceased had been employed in plant cleaning at the PVC facility. Subsequent investigation revealed that, worldwide, this same very rare form of liver cancer had accounted for the deaths of about 25 other persons who had worked in PVC manufacture. Permitted exposure levels in the USA had previously been as high as 500 p.p.m., and those who had died had probably inhaled 1000 p.p.m. for considerable periods. In response to the official report by B.F. Goodrich, the permitted exposure level was lowered by an order of magnitude to 50 p.p.m. and new PVC facilities in the UK are required (Lees, *op. cit.*) to aim for 10 if not 5 p.p.m. The odour threshold of vinyl chloride is above 2000 p.p.m.

10.5.4.2 Precautions in manufacture and processing [5]

We saw in Chapter 5 that PVC is manufactured in batch reactors and that the extent of conversion is 80–90%. This means that considerable amounts of the monomer remain after manufacture, some of it in the suspension water. Moreover, the newly manufactured polymer retains some of the monomer, posing a hazard in subsequent processing and usage. PVC straight from the manufacturing plant is wet and has to be allowed to dry and, after natural drying, will contain 10–1000 p.p.m. (weight basis) of monomer. The extent of retention depends on the porosity of the polymer, which varies with manufacturing conditions affording different 'grades' of PVC. However, all grades respond favourably to milling, which increases the surface area; amounts of residual VCM as low as 1 p.p.m. are obtainable by milling.

10.5.4.3 Case study

There was a major release of vinyl chloride, previously stored as a liquid under its own very high vapour pressure, in Melbourne, Australia in 1982. Immediate emergency response was, of course, concerned with preventing ignition; as we saw in Chapter 3, vinyl chloride can display BLEVE behaviour. There was no ignition in the 1982 Melbourne incident. Very considerable quantities of VCM were, however, released into the atmosphere, and the company responsible was fined.

10.5.5 Acrylonitrile (CH_2=CHCN)

This compound is the cyano analogue of vinyl chloride and is an important ingredient in the manufacture of acrylic fibres. Its normal boiling point is 77°C. It is manufactured by the action of oxygen and ammonia on propylene. It is harmful in a number of ways, and an exposure limit of 2 p.p.m. applies. It is stored in tanks at atmospheric pressure, and small amounts of additives are used to prevent unwanted polymerisation in storage. The numerical example below relates to acrylonitrile storage.

In a vessel used to store acrylonitrile out of doors where the temperature is 5°C, there is a space above the liquid surface of volume 50 litre (0.05 m³). This contains air and an equilibrium quantity of acrylonitrile, total pressure 1 bar. The contents of the space are accidentally discharged into the surrounding atmosphere, and it is estimated that occupants of the nearest housing area will experience, in the short term, a concentration of acrylonitrile equal to that in 50 litre space diluted, because of dispersion, by a factor of 10^3. What level of acrylonitrile, in p.p.m., will the occupants of the housing area be exposed to? The heat of vaporisation of acrylonitrile is 31 kJ mol⁻¹

Solution
The Clausius–Clapeyron equation, with symbols as defined previously, is:

$$\frac{d(\ln P)}{d(1/T)} = -\frac{\Delta H_{vap}}{R}$$

Now for acrylonitrile, the normal boiling point is 77°C (350 K) and the heat of vaporisation is 31 kJ mol⁻¹, therefore:

$$\int_{1\,bar}^{P*} d(\ln P) = -(\Delta H_{vap}/R) \int_{350\,K}^{278\,K} d(1/T)$$

where $P*$ is the equilibrium vapour pressure of acrylonitrile at 298 K. Integrating and substituting:

$$P* = 0.063\,bar$$

$$PV = nRT \Rightarrow n = 0.14\,mol$$

total number of moles of gas in 50 litre at 278 K

$$= [(0.05 \times 1 \times 10^5)/(8.314 \times 278)] = 2.16$$

p.p.m. of acrylonitrile $= [(0.14/2.16) \times 10^6] = 65\,000$

p.p.m. after leakage and dispersion $= 65$

The further numerical example, below, relates to seepage of small quantities of acrylonitrile into a populated area over long periods.

Houses have been built on a previously derelict site. The builders were unaware that several drums of acrylonitrile had, years previously, been illegally buried at the site. There is leakage from them at a rate such that occupants of certain of the houses experience a perpetual background level of 0.04 p.p.m. of acrylonitrile (which is a very long way below the odour threshold). For how long will an individual need to be resident in one of these houses for him or her to have: (a) a 1% probability; and (b) a 10% probability of death from long-term effects of the chemical? The probit equation for acrylonitrile is:

$$Y = -29.42 + 3.008 \ln(C^{1.43}t_i)$$

Solution
(a) The probability that an individual will, over a long period, have a 1% chance of being fatally affected will be taken to be equivalent to the prob-ability of one death in a hundred persons hypothetically resident in the contaminated houses.
Substituting into the probit equation:

$$Y = 2.67 = -29.42 + 3.008 \ln(0.04^{1.43}t_i) \Rightarrow t = 8.2 \text{ years}$$

(b)   Similarly, for a 10% probability:

$$Y = 3.72 = -29.42 + 3.008 \ln(0.04^{1.43}t_i) \Rightarrow t = 11.6 \text{ years}$$

A significant quantity of acrylonitrile was accidentally released at Coode Island, Victoria, Australia in 1990. There were no deaths or injuries.

10.5.6   Fully halogenated organic compounds

10.5.6.1 Introduction

These include carbon tetrachloride (CCl_4), bromochlorodifluoromethane ($CClBrF_2$), both of which have found application in fire extinguishment. Although carbon tetrachloride usage for some applications ceased in the 1970s because of proven toxicity hazards, the chemical is manufactured in large quantities in the manufacture of chlorofluorocarbons (CFCs) as well as in the manufacture of certain pharmaceuticals and pesticides, where it is used as a solvent. This substance will therefore be discussed in the following section.

10.5.6.2 Carbon tetrachloride (CCl_4)

This has a boiling point of 77°C. It is itself very dangerous and a further hazard is that the vapour, when released into moist air, can form phosgene ($COCl_2$, a.k.a. carbonyl chloride) which is a highly toxic gas having, indeed, been used as such in warfare (Marshall *op. cit.*). The reaction of carbon tetrachloride with air to form phosgene requires sunlight. These facts are consolidated in a worked example below.

At a pesticide plant where carbon tetrachloride is used as a solvent there is an accidental leak as a result of which, for a period, the atmosphere contains 25 p.p.m. of carbon tetrachloride, and this is not initially detected. The conversion of carbon tetrachloride to phosgene takes place with an extent of reaction of 10%, and the plant area is occupied by its complement of 80 persons for two-and-a-half hours afterwards. Estimate how many fatalities from phosgene poisoning will result. The probit equation for phosgene (Lees, *op. cit.*) is:

$$Y = -27.2 + 5.1 \ln\{Ct_i\}$$

(symbols as previously defined)

Solution
Clearly the stoichiometry is such that:

1 molar or volume unit of $CCl_4 \rightarrow$ 1 molar or volume unit of $COCl_2$

Hence if there are 25 p.p.m. of carbon tetrachloride and the reaction goes to 10% completion, there are 2.5 p.p.m. of phosgene.

Inserting $C = 2.5$ p.p.m. and $t_i = 150$ minutes into the probit equation gives:

$$Y = 3.0 \equiv 2\% \text{ fatalities, so } 1\text{–}2 \text{ deaths}$$

10.5.6.3 Bromochlorodifluoromethane (a.k.a. BCF or Halon 1211)

This substance has been widely used in fire extinguishers for many years. Its boiling point is −4°C, so it is stored in extinguishers as a liquefied gas. Its effectiveness in extinguishment is due to its ability to release halogen atoms, and their ability to react with reactive intermediates such as the hydrogen atoms which would otherwise have accelerated combustion.

A difficulty with such chemicals is that their use in fires, where water vapour is always present to some extent, leads to the formation of the hydrogen halides HCl, HBr and HF. These are toxic to humans and exacerbate the toxicity of any post-combustion gas in which they occur. That is why different lethal concentrations apply according to whether the BCF is undecomposed or decomposed.

Consequently there is a good deal of current research activity into 'halon substitutes' [6]. One such is a solid propellant incorporating a suitable organic matrix as fuel and potassium bromate as oxidant. If this is activated by contact with an accidental fire, it will release water and carbon dioxide, good extinguishing gases which are not harmful to the environment. Also potassium and bromide ions released dissolve in water to form an aerosol which is itself a good extinguishing agent. Technologies such as these are currently focused on aircraft fuel tanks but there is clear potential for their extension to the hydrocarbon industry.

10.5.7 *Toxicity of combustion products in hydrocarbon fires*[33]

10.5.7.1 Introduction

Earlier parts of this chapter were concerned with toxicity hazards due to industrial organic chemicals which are either starting materials, intermediates or final products. A further important facet is the toxicity of smoke. Fire and explosion hazards of hydrocarbons have been covered in some detail in previous chapters. To the thermal hazards discussed there must be added toxic hazards due to smoke. This final major section of Chapter 10 therefore brings together the toxicity theme of this chapter and the accidental fires theme of earlier ones. Research programmes into smoke toxicity have in fact tended to be focused on building fires rather than on fires at chemical plants.

10.5.7.2 Effects of carbon monoxide

In a hydrocarbon fire such as those discussed earlier in the context of case studies, there will always be significant amounts of carbon monoxide. Purser *op. cit.* emphasises that whereas for some toxic substances the effect of exposure is a function of the inhaled concentration only, for others – including carbon monoxide – the effect depends upon the difference between the concentrations inside and outside the body.

In the simpler case, the dose W required for a specified effect, e.g., loss of consciousness, is expressible as the simple product of concentration and time, so that a particular value of W in p.p.m. minute can be assigned to such effects. This is known to toxicologists as Haber's rule and can be expressed:

$$W = Ct$$

where C is the inhaled concentration, and t the inhalation time.

For carbon monoxide at higher concentrations (>1000 p.p.m.) and shorter times Haber's rule suffices, with W for loss of consciousness having a value 27 000 p.p.m. minute. Hence at an inhaled concentration of carbon monoxide of 5000 p.p.m., the time taken for loss of consciousness would be just over 5 minutes. At lower concentrations the simple Haber rule tends to underestimate the time required for occurrence of the specified effect to which W relates. For example, for an inhaled concentration of 1500 p.p.m. the Haber rule would predict:

$$(27 000/1500) \text{ minutes} = 18 \text{ minutes}$$

for loss of consciousness to occur, whereas the time actually required would be about 30 minutes. For lower concentrations more advanced models are available which contain a parameter which incorporates, amongst other quantities, the total blood volume.

In a fire, the proportion of the fuel carbon which forms, on combustion, carbon monoxide rather than carbon dioxide can be estimated from empirical correlations if the respective rates of fuel and air supply to the fire are known. Otherwise, if this information is not available but it is nevertheless clear that the fire is fuel-rich, it can be assumed that carbon monoxide forms to the extent of 0.2 g per g fuel burned [7]. The calculation below shows how use can be made of this in fire protection engineering.

Liquid hydrocarbon inventory in indoor storage leaks and forms a pool of area equivalent to that of a circle of 1 m radius. The fire burns for 5 minutes without flashing over. If the CO so formed is taken to occupy uniformly the room volume of 6000 m³, for how long will an individual need to remain in the room after the fire to experience loss of consciousness?

Solution

We know from Chapter 3 that a pool fire burns at about $0.1 \, kg \, m^{-2} s^{-1}$, so the quantity of fuel burnt is:

$$0.1 \times \pi \times 1 \times 5 \times 60 \times 1000 \, g = 94\,260 \, g$$

The amount of CO is then:

$$94\,620 \times 0.2 \, g \equiv (18\,852/28) \, mol = 673 \, mol$$

Since the room is large and the fire is both localised and fairly short-lived, the 'room temperature' will be taken to be in the neighbourhood of 300 K, at which temperature, and at atmospheric pressure, $1 \, m^3$ of a gas contains roughly 40 mol, from which:

$$\text{p.p.m. of CO} = \{673/(40 \times 6000)\} \times 10^6 = 2805 \, \text{p.p.m.}$$

Utilising Haber's rule as it applies to loss of consciousness due to CO inhalation:

$$27\,000 \, \text{p.p.m. minute} = 2805 \, \text{p.p.m.} \times t$$

where t is the required time. Solving:

$$t = 10 \, \text{minutes}$$

Carbon monoxide can be detected by semiconductor devices such as those which were outlined in Chapter 2 in the discussion of hydrocarbon leakage, or by electrochemical gas detector devices in which the carbon monoxide is oxidised to carbon dioxide in a galvanic cell the e.m.f. from which is the basis of the measurement signal [9].

10.5.7.3 Incapacitation of fire victims by carbon dioxide

Carbon dioxide is toxic only in concentrations in excess of 5%, but its presence signifies elemental oxygen depletion which is itself a harmful factor. A decline in concentration of ambient oxygen causes acceleration of breathing which means that toxic gases accompanying the carbon dioxide, principally carbon monoxide, enter the body at an enhanced rate. This potential for interaction with other, more intrinsically harmful, constituents of post-combustion gases is the most serious threat posed by carbon dioxide to fire victims. In fires where significant quantities of hydrogen cyanide (HCN) occur in the post-combustion gases, its uptake is also stimulated by carbon dioxide.

Toxicity by combustion products is only marginally within the scope of this text. This section is consequently brief, but is supplemented by a reference [8] to a comprehensive treatment of the subject, in particular in relation to carbon monoxide poisoning.

10.6 Control of major accident hazards (COMAH)

This came into force in 1999 with the remit of ensuring that businesses 'take all necessary measures to prevent major accidents involving dangerous substances, limit the consequences to people and the environment of any major accidents which do occur'. The competent authority for ensuring compliance are the Health and Safety Executive (HSE), the Environment Agency (EA) and the Scottish Environmental Protection Agency (SEPA). What follows has drawn on [10].

The list of substances to which COMAH applies, given in [10], is both long and varied. It ranges from poisonous substances such as antimony pentoxide, which is used in insecticide manufacture, to elemental halogens (Cl_2, F_2, Br_2) and a group of organic substances known to be carcinogenic including bis(chloromethyl) ether. Phosgene also features in the list and, very importantly, distillates of crude oil. COMAH applies wherever such substances exist, whether at the scene of manufacture, the site of storage or place of usage. Each of these is termed, without distinction, an 'operator' in the terminology of COMAH. A formal Major Accident Prevention Policy Document will be required from the operator, and this will need to address issues including identification of hazards due to the materials on site, the assignment of duties relevant to safety to personnel and analysis of processes and procedures for hazards. Adaptation of practices to new circumstances, for example newly installed plant, and preparedness for emergencies will feature in the Major Accident Prevention Policy Document and there will be a plan for monitoring activity for compliance with the 'prevention policy' in the report.

When COMAH was set up a number of previous case studies had 'been identified as illustrating the importance of the technical assessment criteria in preventing, controlling or mitigating major accidents'. These are given in [11] and a selection of them follows as Table 10.1. Some contents of the table can be related to previous parts of the book, for example the ASTM standard for joining metal pipes (row 4) and IP standards for the layout of LPG tanks (row 5).

10.7 Classification and signage

The chart following the numerical exercises summarises the US system of classification of 'hazardous substances' including flammable ones, providing overlap between the contents of this chapter and that of previous ones. They can be compared with the signs in Figure 5.1.

Table 10.1 *Selection of accident case studies*

Location and date	Details
Flixborough UK, 1974	Already studied in detail
Feyzin France, 1966	Loss of containment of propane during removal of a contaminant aqueous layer from an LPG tanks. The cloud of gas drifted 160m before it was ignited by heat from a car. 18 deaths, 81 injuries
Feyzin France, 1966	Loss of containment of propane during removal of a contaminant aqueous layer from an LPG tanks. The cloud of gas drifted 160m before it was ignited by heat from a car. 18 deaths, 81 injuries
Cheshire England, 1994	Release of a miscellany of chemical agents including ethyl chloride (C_2H_5Cl, strictly chloroethane), hydrogen chloride and aluminium chloride catalyst from a reactor. Pipe work believed to have failed at a flange. Manual isolation valves difficult to access
Mexico City Mexico, 1984	500 deaths and excessive damage at site containing LPG tanks, several of which 'BLEVEd'. One contributing factor amongst others was positioning of the vessels
TX USA, 1987	Hydrogen fluoride (a catalyst in alkylation processes) at a refinery released when pipes connected to a container of it was impacted by a piece of plant dropped from a crane. Major environmental damage
Not disclosed	Observation of sparks due to static during transfer of an aqueous liquid to a storage tank. No serious consequences as the liquid was not flammable, and seen as a near miss
Dalmeny Scotland, 1987	Three men inside an 'empty' crude oil storage tank, cleaning it, when one of them lit a cigarette and the residual vapour ignited, killing one of the men. The men were contractors, and the matter of supervision of contractors was raised
Stanlow England, 1990	Explosion at a fluoraromatics plant. The relief valve pressure at the reactor was set at 5 bar, yet the pressure was in the range 60 to 80 bar at the time of failure. The escaped contents burnt as a fireball

10.8 Concluding remarks

A number of the most notable toxic hazards in the petrochemical industry have been covered with sufficient detail for a reader to have awareness of their danger and also background on how hazards from such chemical agents are quantified. A second facet of toxic hazards is the danger to life and health by the inhalation of the vapour products of accidental combustion, and this has also been covered as has classification. The toxicity theme is carried through to the next chapter, where safe disposal of unwanted hydrocarbons is considered.

References

[1] 'Solvents in Common Use: Health Risks to Workers' Royal Society of Chemistry, London (1988)

[2] 'Occupational Exposure Limits 1997' HMSO, London (1997)

[3] 'Occupational Exposure Limits: Criteria Document Summaries' First Edition, HMSO, London

[4] Seymour R.B. (Ed.) *History of Polymer Science and Technology*, Marcel Dekker, New York (1982)

[5] Burgess A.R. (Ed.) *Manufacture and Processing of PVC*, Applied Science Publishers, London (1982)

[6] Olander D.E., Baker J.B., Schall M. Paper presented at the 33rd International Conference on Fire Safety, Columbus, OH, July 2001

[7] Gottuck D.T., Roby R.J. 'Effect of Combustion Conditions on Species Production' in *Handbook of Fire Protection Engineering*, Second Edition. SFPE, Boston, MA (1995)

[8] Hirschler M.M. (Ed.) *Carbon Monoxide and Human Lethality*, Elsevier, New York (1993)

[9] Jones J.C. *Dictionary of Fire Protection Engineering*, 297 pp. Whittles Publishing, Caithness (2010)

[10] http://www.legislation.gov.uk/uksi/1999/743/schedule/1/made

[11] http://www.hse.gov.uk/comah/sragtech/casestudyind.htm

Numerical problems

1. At a chemical factory, a quantity of 5 g of chlorine is accidentally leaked into a room of dimensions length 20 m, width 15 m, height 6.5 m. If the chlorine disperses uniformly, ascertain whether the TLV of 1 p.p.m. will be exceeded.

2. As the result of failure of a vessel containing chlorine, 70 persons are exposed to chlorine at a level of 10 p.p.m. for 5 hours. Predict whether there will be any fatalities and if so, how many.

3. The South Africa (1973) leakage of ammonia caused there to be a 7% fatality rate. What concentration of chlorine for the same time would, over a 30 minute period, have led to the same fatality rate.

4. What concentration of hydrogen fluoride would have caused the fatality rate that was observed in the South African ammonia accident?

5. Hydrogen fluoride is to be transported, and the climate along the route is such that the ambient temperature might be as high as 40°C. The substance is to be carried in a vessel containing the liquid in equilibrium with its own vapour pressure. What pressure must the vessel be able to withstand? The boiling point of HF is 19°C and the latent heat of vaporisation $31.3\,kJ\,mol^{-1}$.

6. The Antoine constants (see Chapter 6) for benzene are:

$$A = 6.90565$$
$$B = 1211.033$$
$$C = 220.790$$

There is accidental spillage of benzene in a refrigerated enclosure where the temperature is −7°C. If the liquid equilibrates with its vapour, what will be the final p.p.m. of benzene in the enclosure?

7. The probit equation for carbon monoxide is:

$$Y = -37.98 + 3.7\ln(Ct)$$

The gases from a factory boiler are heat exchanged with the incoming air in order to raise its temperature to about 25°C. At the exchanger there is leakage which is undetected for 30 minutes with the result that the 135 employees in the factory are breathing air contaminated with 0.25% (2500 p.p.m.) of carbon monoxide for that time. Estimate how many fatalities can be expected.

8. Acetone burns according to:

$$CH_3COCH_3 + 4O_2\ (+15.0N_2) \rightarrow 3CO_2 + 3H_2O\ (+15.0N_2)$$

The odour threshold is 450 p.p.m. Using principles discussed in Chapter 3, ascertain whether acetone at its flash point is above or below its odour threshold.

9. Styrene ($C_6H_5CH{=}CH_2$) is a liquid at room temperature and has a vapour pressure of approximately 10 mm Hg at 30°C. At a plant where styrene is in use and the ambient temperature is 30°C, by what factor will styrene vapour have to be diluted, by ventilation, below its equilibrium vapour pressure in order for the OEL of 100 p.p.m. not to be exceeded?

10. Vinyl chloride monomer is stored in a tank under its own vapour pressure at 15°C. The vapour with which the bulk liquid is in equilibrium occupies a space, above the liquid surface and enclosed by the tank's internal wall, of 100 litre. If during subsequent withdrawal of the VCM from the tank this quantity of vapour is accidentally allowed to escape and if, through dispersion, it dilutes by a factor of 10^3 on entering the nearby works area, will it be above or below the odour threshold for VCM given in the main text? The vapour pressure of VCM at the temperature concerned is 2.8 bar.

11. BCF reacts, in fire extinguishment, in a way which can be summarised as:

$$CBrClF_2 + 2`H_2' + 0.5O_2 \rightarrow HBr + HCl + 2HF + CO$$

where 'H_2' has been put in inverted commas since it is not molecular hydrogen but hydrogen atoms, previously formed from the fuel, that the BCF targets.

The substance is supplied to a fire and has an initial concentration, in the combustion zone, of 5% (molar or volume basis). If the fractional extent of reaction is 0.2, what will be the p.p.m. of each of the three hydrogen halides and of unreacted BCF in the post-combustion gas if this, on dispersion and cooling to ambient temperature, is diluted by a factor of 10^3?

12. Return to the calculation in the main text where there are 2805 p.p.m. of carbon monoxide in a room following a hydrocarbon fire. Using the probit equation for carbon monoxide:

$$Y = -37.98 + 3.7\ln(Ct)$$

calculate what period of exposure to the post-combustion gases would be necessary for there to be a fatality rate of 10%.

13. At the accident at Scotland, in 1987 observers reported a 'ring of fire' and, notwithstanding the fatal result, there was no explosion. This suggests that the vapour was present at or close to a proportion corresponding to its flash point. Assigning the hydrocarbon vapour the formula C_nH2_n estimate 'n'.

Endnotes

[30] This is noted in the Canvey report.
[31] The death toll was about 3000 and the number of persons affected non-fatally was of the order of 200000.
[32] 'Xylenes' means a mixture of ortho-, para- and meta-xylene in any proportions.
[33] The primary source drawn on in the writing of this section has been the comprehensive review 'Toxicity Assessment of Combustion Products' by D.A. Purser in the Second Edition of the SFPE Handbook.

Summary of US classification of hazardous substances

Class 1 Explosive materials

Capable of exploding without an oxidant, and distinguished from flammable. Another way of putting that is that they can detonate.

They include pentaerythritol tetranitrate (PETN) and other high explosives.

Class 2 Gases

Three divisions, each with its own hazard class label.

Examples of flammable gases are numerous, including all hydrocarbon gases and hydrogen.

The third and fourth signs in the adjacent column have the same meaning in terms of hazards and would apply for example to chlorine.

The right hand sign is for imported toxic gases.

Class 3 Flammable liquids

Includes fuels, solvents, and alcohols.

Class 4 Flammable Solids

Several divisions.

The first of the signs is for such substances as phosphorus. The second is for finely ground metals (which are a constituent of pyrotechnics!).

The right hand sign is for substances which release flammable gases on contact with water, e.g. sodium (there are many more examples).

Class 5   Oxidizing Substances and Organic Peroxides

Oxidising substances include nitric acid and hydrogen peroxide.Organic peroxides can detonate, as noted earalier in the text.

Classes 6 and 7 are concerned respectively with infectious substances and with radioactive substances. These are outside the scope of the text.

Class 8 Corrosives

These include the mineral acids.

Class 9 Miscellaneous dangerous goods

Includes for example solid CO_2.

11

SAFE DISPOSAL OF UNWANTED HYDROCARBON

II.I Flaring

11.1.1 Introduction

In the oil industry flaring is continuous. It sometimes happens at oil production facilities – on- or offshore – that the associated gas cannot be economically utilised, perhaps because it occurs in quantities which are very small relative to those of the oil or perhaps because it contains too much sulphur (is too 'sour') to be routinely usable.

In the chemical processing industries flaring is sometimes intermittent. In such industries, flaring might be carried out as necessary to break down some toxic or odorous waste product before discharge. Hydrocarbon destined for disposal at a flare is sometimes called 'off-gas'. We are concerned in this discussion with flaring in routine operation. Another facet of the topic is of course diversion of hydrocarbon to a flare as an emergency measure. A flare is a turbulent non-premixed flame which releases heat powerfully (often at a rate of several MW) with accompanying hazards to persons and property. Some of the principles discussed in Chapter 3 in the context of accidental fires therefore apply.

11.1.2 Hazards in flaring

Malfunctioning of a flare can have serious consequences. One form of malfunction, which is known to have caused flares to explode, is entry of air into the structure of pipes and valves that conveys the gas for burning. If it happens that air enters and mixes with the influx hydrocarbon and the proportions enter the flammable range, the flame at the tip of the flare will provide a ready ignition source. Also, fuel supply can be difficult to control and this can lead to extinction which, if purging does not follow, can lead to there being, some time later, a spontaneously explosive mixture within the pipe. This is illustrated in the calculation below.

At a flare there is unscheduled extinction of the flame through erratic supply of fuel and subsequent stoppage of fuel supply, but no purging. The result is that air diffuses into the flare pipe work, with a diffusion coefficient[34] (symbol D) of $2 \times 10^{-5} \text{m}^2\text{s}^{-1}$. The upper flammability limit of the gas is 14% by volume. The pipe shape and dimensions are such that a spontaneous explosion will take place when the concentration of fuel at an axial distance of 10 cm from the flare front is at the upper flammability limit. How long will it take for this to occur? If the lower flammability limit is 5%, what length of the pipe will, at this time, be occupied by a mixture in the flammable range?

Solution
The situation is that at time zero (extinction of the flare) there is diffusion of air into a stagnant fluid in the flare pipe work. Figure 11.1 shows the arrangement. The time required is that at which the percentage air in the pipe at the depth specified is $(100 - 14) = 86$.

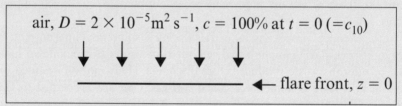

air, $D = 2 \times 10^{-5}\text{m}^2\,\text{s}^{-1}$, $c = 100\%$ at $t = 0\ (=c_{10})$

flare front, $z = 0$

Figure 11.1 *Schematic of air entry at the tip of an extinguished flare.*

This has been solved (originally by Boltzmann), to give [1]:

$$\frac{c(z,t)' - c_{10}}{c_{1\infty} - c_{10}} = \text{erf}\left\{z/\sqrt{(4Dt)}\right\}$$

where erf denotes the error function, obtainable from tables. Because the concentrations appear as a quotient, percentage units can be retained to give:

$$\frac{86-100}{0-100} = 0.14 = \text{erf}\left\{z/\sqrt{4Dt}\right\}$$

$$\Downarrow$$

$$z/\sqrt{4Dt} = 0.13 \Rightarrow t = 7396\,\text{s}\ (\approx 2\ \text{hours})$$

The lower flammability limit occurs when the concentration of air is

$$(100-5) = 95\%$$

$$\Downarrow$$

$$\mathrm{erf}\left\{z/\sqrt{4\times2\times10^{-5}\times7396}\right\} = 0.05$$

$$\Downarrow$$

$$z = 0.043\,\mathrm{m}\,(4.3\,\mathrm{cm})$$

Therefore, at this time the pipe work to the flare is occupied by a flammable mixture between the depths, measured from the flare tip, of 4.3 cm and 10 cm.

Other dangers with flares include flashback. The flare tip is, of course, acting as a burner, and any burner can only function within certain limits of fuel supply rate ('thermal delivery'). If the gas pressure drops so that the thermal delivery is outside the range within which the flare tip can anchor the flame there will be flashback. This can be overcome by incorporating a fuel gas to supplement that in the waste stream, thus ensuring that the total thermal delivery remains within the range for which the flare is designed. There are two other means of preventing flashback:

(i) By use of 'flame arresters'. These are simply solid surfaces which will absorb heat and, possibly, reactive chemical intermediates at incipient flashback, and prevent its development.
(ii) Use of 'molecular seals' to restrict air ingress in the event of flame extinction.

A further difficulty with flares is that they can become blocked through icing (having regard to the fact that steam is sometimes used as a purging agent) or by freezing of organic constituents including benzene and cyclohexane, which each have freezing points higher than those of water. Flame arresters can themselves provide a site for blockage.

11.2 Afterburning [2]

11.2.1 Introduction and basic principles

A gas having a heat value below about $4\,\mathrm{MJ\,m^{-3}}$ is too lean to sustain a flame on a burner. It sometimes happens in processing industries that an effluent gas which, for environmental reasons, cannot simply be discharged has too low a heat value to be burnt as it is. One option in such circumstances is for the gas to be mixed with enough of an additional fuel gas, usually natural gas, to enable it to sustain a flame. The principles of this are illustrated in the calculation below.

The effluent gas from a manufacturing process contains 2000 p.p.m. of benzene vapour, 700 p.p.m. of toluene vapour and 100 p.p.m. xylene vapour, balance non-combustibles. Will this gas require supplementing with a flammable gas (e.g., natural gas) for afterburning?

Heats of combustion/$kJ\,mol^{-1}$:

$$benzene = 3265$$
$$toluene = 3907$$
$$xylene = 4549$$

Solution

$1\,m^3$ of gas at 288 K, 1 atm. contains 42 mol.

amount of benzene = $2000 \times 10^{-6} \times 42\,mol = 0.084\,mol$, capable of releasing $0.084 \times 3265\,kJ = 0.27\,MJ$

amount of toluene = $700 \times 10^{-6} \times 42\,mol = 0.029\,mol$, capable of releasing $0.029 \times 3907\,kJ = 0.11\,MJ$

amount of xylene = $100 \times 10^{-6} \times 42\,mol = 0.0042\,mol$, capable of releasing $0.0042 \times 4549\,kJ = 0.02\,MJ$

$$Total = 0.40\,MJ$$

Not sufficient to sustain a flame on a burner.

Some blending with natural gas will be needed.

Of course, a gas burner is never usable for all fuel gases. A burner is designed and subsequently adjusted only for quite a narrow range of heat values, a point which was touched on in the previous section. One would not, therefore, attempt to burn an effluent gas of $5\,MJ\,m^{-3}$ on a burner designed and adjusted for natural gas. When afterburning is carried out there has to be combustion plant suitable for the purpose.

11.2.2 Catalytic afterburning

Smaller amounts of supplementary gas (or none at all) and lower temperatures are required if afterburning is carried out with the aid of a catalyst [2]. The catalyst surface area per unit amount of effluent gas flowed through is important in designing a catalytic afterburning process. Suitable catalysts contain such metals as platinum and palladium and much research and development in catalytic combustion (as in other branches of applied catalysis) is directed at maximising the life expectancy of the catalyst. Catalysts used in afterburning can, for example, be adversely affected by sulphur compounds in the gas which it receives. Similarly, surges in the pressure of gas passing through a catalytic afterburner can impair or even totally destroy the catalyst.

11.2.3 Heat recovery

When a hydrocarbon is flared, its (desired) destruction is fully accomplished though no return on the heat released is obtained. In afterburning – with or without a catalyst – the efflux gas can be heat exchanged with air in order to provide energy for some other plant process. Perhaps better still, the heat thus obtainable can be made available to the influx gas to the afterburner, thereby reducing the amount of supplementary fuel needed ('recuperative burning'). This is more fully explained in the context of the calculation below.

A waste stream from a hydrocarbon process has, on account of its own methane content (balance CO_2/N_2), a heat value of $1\,MJ\,m^{-3}$ and, in non-catalytic afterburning starting with cold influx gas, is supplemented by natural gas in order to raise this value to $5\,MJ\,m^{-3}$. How much natural gas per unit amount of the process gas will be required? The influx rate of enriched process gas is $15\,m^3\,min^{-1}$ (measured at 288 K, 1 bar) plus a stoichiometric amount of air. What will be the total rate of production of efflux gas?

A modification is proposed whereby efflux gas from the afterburner is heat-exchanged with the influx fuel gas with the result that the latter is raised in temperature by 600 K. The rate of influx of the process gas is unchanged. What will now be the proportion of natural gas then required for the afterburning? Take the heat capacity of the influx gas to be $1295\,J\,m^{-3}\,K^{-1}$.

Solution
The heat released by $1\,m^3$ of the process gas is 1 MJ. The heat released by $1\,m^3$ of the gas once blended with natural gas (heat value $37\,MJ\,m^{-3}$) must be equal to 5 MJ, hence, letting the proportion of natural gas $= X$:

$$X + 37(1 - X) = 5 \implies X = 0.89$$

i.e., 0.89 units (volume or molar) of process gas to 0.11 units of natural gas. Taking water to remain in the vapour phase, it follows from the stoichiometry of methane combustion:

$$CH_4 + 2O_2\ (+7.52\ N_2) \rightarrow CO_2 + 2H_2O\ (+7.52\ N_2)$$

that there will be no increase in numbers of moles due to the incineration.

Therefore the total efflux rate is:

$$(15\,m^3\,min^{-1})(1 + 9.52) = 158\,m^3\,min^{-1}$$

Without recuperation:

heat released on burning $1\,m^3 = [(0.89 \times 1) + (0.11 \times 37)]\,MJ = 5\,MJ$

With recuperation:

$$Y + 37(1 - Y) + h = 5\,MJ\,m^{-3}$$

where Y is the proportion of the process gas in the initial process gas/natural gas blend and h is the heat in MJ transferred from the efflux per m^3 of influx.

Now $h = 1295\,J\,m^{-3}\,K^{-1} \times 600\,K = 0.78\,MJ\,m^{-3}$

Solving for Y:

$$Y = 0.91$$

that is, the recuperation enables a mixture containing 91% of the lean gas to be burnt, whereas without recuperation only a mixture of 89% (balance natural gas in each case) could be burnt. The proportion of natural gas accordingly goes down from 0.11 to 0.09.

There are portable, small-scale incinerators available for destruction of small amounts of waste hydrocarbons, and these are used in places including Alaska [3]. There is no attempt with these to utilise the heat.

11.3 Use of adsorbent carbons

Adsorbent carbons feature in Chapter 5. Where there are small amounts of hydrocarbons or hydrocarbon derivatives in a gas stream and these have to be removed, an alternative to afterburning is adsorption on to an 'activated carbon'. Activated carbons are made from feedstocks such as coal, wood and peat and, after carbonisation, are treated either with steam or with a chemical agent such as phosphoric acid to give a final product with an internal surface area of the order of $1000\,m^2\,g^{-1}$. The basic principles of this are brought out in the context of the calculation below.

The effluent gas from an industrial process contains 30 p.p.m. of acetone vapour which is removed by passage through a column of activated carbon of internal surface area $500\,m^2\,g^{-1}$. $50\,m^3$ per minute (288 K, 1 atm. approx.) are passed through 500 kg of the carbon. If the area occupied by acetone molecule is $0.25\,nm^2$ and the acetone forms a monolayer on the internal pore structure of the carbon, how long will the charge of activated carbon last in continuous use before needing renewal?

Solution

$$\text{area occupied by 1 acetone molecule} = 2.5 \times 10^{-19} \, \text{m}^2$$

Amount of acetone which can be adsorbed by 500 kg

$$= \{(500 \times 10^3 \times 500)/[(2.5 \times 10^{-19}) \times (6.02 \times 10^{23})]\} \, \text{mol}$$

$$= 1661 \, \text{mol}$$

Rate of passage of gas through the column

$$= 50 \, \text{m}^3 \text{ per minute}$$

$$= 2100 \, \text{mol per minute}$$

$$\text{or } (2100 \times 30/10^6) \, \text{mol per minute of acetone}$$

$$= 0.063 \, \text{mol per minute of acetone.}$$

The column will therefore function for (1661/0.063) minutes or 439 hours.

Activated carbons can sometimes be made from starting materials collected as wastes, e.g., used tyres. Also 'olive stone' – the husk remaining from olives when the oil has been extracted – is a very suitable and quite inexpensive starting material for making adsorbent carbons. One UK patent appertains to the preparation of activated carbon from olive stone.

11.4 Venting

Venting, that is, discharge of hydrocarbon inventory into the atmosphere, is not so much destruction of large amounts as release of relatively small amounts. However, its coverage here is appropriate in that the result of the process is to transfer hydrocarbon material from a situation where its presence is potentially hazardous to one where it is not.

Hydrocarbon processing plant is commonly fitted with relief systems to prevent build-up of excessive pressures in the event of heating or of accidental oversupply of the influx material, as has been pointed out in previous chapters. Similarly, tanks used in storage and transportation are provided with pressure relief valves. Once pressure relief is in operation the hydrocarbon so discharged will, unless it is diverted in some way, simply enter the atmosphere. Lees *op. cit.* emphasises that this is straightforward and reliable, there being nothing beyond the pressure relief device itself that can malfunction.

Obviously certain requirements apply to the simple venting of hydrocarbon substances to the atmosphere, otherwise there will be a need for diversion of the hydrocarbon, e.g. to a flare. One requirement is that, once released, the hydrocarbon will be rapidly diluted below its lower flammability limit. Lees quotes a criterion for this as:

$$Re > 1.54 \times 10^4 \, \sigma_j/\sigma_a$$

where σ_j is the density of the hydrocarbon at the vent outlet and σ_a the density of air. It follows from this, surprisingly at first consideration, that the denser the vapour on exit the safer it is from the flammability angle. This is because the more dense the vapour relative to that of the air it is entering the greater the momentum with which the vapour contacts the air, and this promotes mixing and hence dilution. The Reynolds number is referred to the diameter of the vent and the external (air) temperature. One of the appended numerical problems appertains to this.

Once a hydrocarbon has been released at a vent there is the possibility, especially with a fairly high-molecular-weight example, that there will be mist formation. This is to be expected in circumstances where the temperature on exit is below the dew point, i.e. the temperature at which the pressure of the vapour in the atmosphere is equal to the saturated vapour pressure of the liquid[35].

11.5   Disposal methods in which the hydrocarbon is utilised

11.5.1   Introduction

The disposal methods considered so far, however effective they might be at eliminating hazards, result in total loss of any utilisation potential the hydrocarbon for disposal might have had. In this section, we consider two ways in which unwanted hydrocarbon is disposed of without total loss of its fuel value.

11.5.2   Blending with solid waste

This example, though a little mundane, is worth a brief mention. Where solid waste is used as a fuel, as it sometimes is in applications including steam raising, its calorific value can be enhanced by allowing it to absorb waste liquid hydrocarbon. One of the appended numerical problems is concerned with the improvement in such fuels obtainable by this means.

11.5.3   Gasification

Gasification of liquid fuels is an important area of fuel technology. In this text we are concerned only with gasification of liquid hydrocarbons which might otherwise have been incinerated, that is, low-value refinery products and unwanted by-products of cracking processes.

Residual hydrocarbon from a distillation process is typically [4] of composition carbon 89%, hydrogen 11%. If for the purposes of simple stoichiometric calculations the residue is taken to be a single hydrocarbon compound, this corresponds to an empirical formula of C_2H_3, therefore gasification with steam would proceed ideally according to:

$$‘C_2H_3’ + 2H_2O \rightarrow 2CO + 3.5H_2$$

yielding a product gas of composition:

$$CO: [(2/5.5) \times 100]\% = 36\%$$

$$H_2: [(3.5/5.5) \times 100]\% = 64\%$$

This somewhat oversimplifies the process in that steam gasification of liquid hydrocarbons, in the presence of a suitable catalyst, also produces significant amounts of carbon dioxide. There might also be small amounts of methane.

A whole genre of gasification technologies, called the 'Pacific Coast processes', is concerned with the conversion of low-value refinery products to quality fuel gas suitable for reticulation to homes and industrial premises. The simplest such process, known as the 'Jones process', consists of reacting the hydrocarbon with steam without a catalyst, whereupon such a gas was yielded. There is also a tendency for elemental carbon to form and deposit from the gas phase. Carbon so formed, known as 'carbon black', is a saleable substance and the Jones process can be controlled so as to produce this in high yield, the fuel gas being in these circumstances a by-product. Some of the Pacific Coast methods of gasification are however catalytic, including the Segas process [4] which utilises a bauxite catalyst. One of the appended numerical examples relates to steam reforming of waste aromatics.

11.6 Non-destructive disposal on land

This topic was touched on in Chapter 9, where it was stated that the oil pollution of the sea is to a large extent attributable to the drift of hydrocarbon previously disposed of on land into the sea. This accounts for much more marine oil pollution than the (happily rare) catastrophic leakages from ships such as happened off the Cornish coast in 1967.

Oil waste in small quantities can be buried in the ground. There are approved techniques for this, which include selection of a site where the soil is fairly absorbent and where the layered structure of the soil is such that it will provide filtration of suspended solids as the liquid waste descends. Most importantly, the site must be chosen so that the liquid is not transferred into rivers and streams. Solid hydrocarbon waste, including waxes produced in small quantities, can also be buried. Mention was made in Chapter 8 of the use of a 'pig' to scrape the inside surfaces of pipelines conveying crude oil from a platform to shore. The waxy deposit so removed is in no way saleable and is often simply buried. With this substance there is sometimes the difficulty that radioactive isotopes from the well have become concentrated in it, necessitating its disposal as a radioactive substance. Spent catalyst from hydrocarbon processes can be buried after encapsulation in concrete to prevent the leaching out of toxic ingredients.

Mention was made in an earlier chapter that small amounts of spilled hydrocarbon which make their way into drains are also a potential hazard, and it is appropriate to conclude this section with a case study for such [5]. It involved an aircraft company, where the inevitable spills of hydrocarbons were, with the help of flowing rainwater, entering the drains. Two measures were required to overcome the problem. At the outlet from the company's site separation of the oil and water was carried out as necessary. During rainy weather water was collected and diverted into the municipal drainage system in such a way as to avoid contact with oil so removed from the site.

11.7 Re-refining [3]

In the USA about 400 000 gallons of waste motor oil are processed by re-refining, and the practice occurs in a number of other countries including Australia and New Zealand. Germany is often said to be a leader in oil re-refining and accordingly a case study from that country will be outlined [6].

At a plant in Zeitz in Germany diesel fuels are produced from waste lubrication oil in a process which involves hydrogenation prior to refining. The diesel fuel obtained is desulphurised to a degree equivalent to that of a diesel fuel made from crude oil and also has a good cetane number. Performance of fuels so manufactured and their endorsement by car manufacturers is most unlikely ever to be a restriction in their proliferation. What might constitute such a restriction is a poor 'energy return on energy invested' (EROEI) on such fuels. EROEI was less important when oils were seen as being plentiful and wells were shallow. But it has become important in these days of consciousness of finiteness of reserves and deeper and deeper wells, as discussed in [7] amongst many other discourses on the topic. The author has no knowledge of the EROEI on the diesel fuel manufactured in the German case study being considered, but is making the point that the viability of processes of this genre will depend on the EROEI which in turn will require close and precise energy accounting during re-refining and the other processes which accompany it. Fuels from waste lubrication oil are not of course carbon neutral, so no offset of a poor EROEI by generation of carbon credits can be obtained. Such offset is possible with biodiesel made from waste cooking oil. That, perhaps, is where re-refining holds the most promise.

11.8 Steam raising

If a waste hydrocarbon is to be used in an internal combustion engine precise specifications are needed, as in the case study in the previous section. On the other hand in a steam engine the only purpose of the fuel is to raise the steam which is itself the working substance in the thermodynamic cycle according to which the engine works. Relatively poor fuels can therefore be used in steam applications which would not be suitable in internal combustion applications and this is obviously relevant to the use of waste hydrocarbon. The EROEI disadvantage applying to extensively processed waste fuels is eliminated if the fuel can be used as received. One of the numerical examples which follow appertains to this.

References

[1] Cussler E.L. *Diffusion: Mass Transfer in Fluid Systems* Cambridge University Press, London (1984)
[2] Wark K., Warner C.F. *Air Pollution*, Second Edition. Harper and Row, New York (1981)
[3] Jones J.C. *Thermal Processing of Waste*, 99 pp. Ventus Publishing, Fredricksberg (2010)
[4] *Gas Making and Natural Gas*, First Edition. British Petroleum, London (1972)
[5] Besselievre E.B., Schwartz M. *The Treatment of Industrial Waste*, Second Edition. McGraw-Hill, New York (1976)
[6] http://www.prokop-engineering.cz/Nemecko/Regenerace%20pouzitych%20ole-juEN.htm
[7] Jones J.C. 'Energy-Return-on-Energy-Invested in petroleum fuel production' *Hydrocarbon World* **7** (1) 10–11 (2012)

Numerical problems

1. An organic waste stream is conveyed to a flare, which has a nominal thermal delivery requirement of 500 kW and can operate safely at deliveries of up to 15% either side of the nominal value. The waste stream comprises vapour of heat value 100 MJ m^{-3}. The vapour flow can, if it drops, be supplemented by a flow of natural gas, heat value 37 MJ m^{-3}. The flow rate of the waste stream is set at a value which matches the nominal requirement of the flare burner. In the event that through accidental blockage the supply rate of the waste stream drops by half, at what minimum rate must natural gas be supplied in order that there will not be flashback?

2. A particular refinery flare burns 50 tonne of waste hydrocarbon gas per day. As a result of an energy audit at the refinery, it is recommended that a quarter of the heat from the flare should be diverted to steam raising for refinery use. Making a reasonable estimate of your own for the calorific value of the fuel received by the flare, calculate how much steam at 100°C could be raised per day starting with liquid water previously raised by heat exchange to 100°C. The latent heat of vaporisation is 40.6 kJ mol^{-1}. *From the MSc (Safety and Reliability) Examination, University of Aberdeen, 2001.*

3. One of the calculations in of Chapter 4 was concerned with storage of *n*-decane ($C_{10}H_{22}$, molar weight 142 g) and a safety pressure relief valve set to open at a pressure of 5 bar. Assigning the vent diameter in the *n*-decane storage vessel a value of a quarter of an inch (6.35×10^{-3} m), calculate at what minimum linear speed the vapour must exit in order for that there shall be sufficiently rapid dilution to below the lower flammability limit. Use a value of 6×10^{-6} kg m^{-1}s^{-1} for the viscosity of the vapour.

4. The effluent gas from a hydrocarbon processing plant contains itself negligible combustibles, but has to be afterburnt in order to break down certain harmful constituents before discharge into the atmosphere. It is desired to do this with propane on a burner designed for 8 MJ m^{-3} gas. How much propane must be blended with unit amount (molar basis) of the effluent gas?

5. The effluent gas from a manufacturing process contains 18% molar basis of carbon monoxide, balance inerts. It is desired to afterburn this on a burner adjusted for gas of heat value 10 MJ m^{-3} (measured at 1 atm., 15°C). In what proportions will the effluent gas have to be mixed with methane in order for such afterburning to be effected? Heat of combustion of carbon monoxide = 282 kJ mol^{-1}. Heat of combustion of methane = 889 kJ mol^{-1}. *From the MSc (Safety and Reliability) Examination, University of Aberdeen, 1999.*

6. Effluent gas from an industrial process contains 3% butane (C_4H_{10}) 2% propane (C_3H_8) and 3% ethylene (C_2H_4), balance non-combustibles. Find the calorific value of this gas in MJ m^{-3} at 288 K, 1 atm., and make a judgement as to whether it can be 'afterburnt' as it is or whether it will need blending with natural gas.

Heat of reaction/kJ mol^{-1}

butane = 2874

propane = 2217

ethylene = 1409

7. Return to the calculation in section 11.2.3 of the main text, where after-burning of an effluent stream with and without recuperation is under consideration. If the afterburning is carried out, at the rate given, for 4 hours a day 5 days a week, what is the annual saving of gas afforded by the recuperation. Express your answer in m^3 measured at 15°C, 1 bar.

8. The effluent gas from an industrial process contains 35 p.p.m. of methanol. $50\,m^3$ per minute of the effluent gas is passed at room temperature and pressure through an adsorption column containing a quantity of 0.3 tonne of activated carbon of internal surface area $1000\,m^2 g^{-1}$. If the area occupied by one methanol molecule is $0.17\,nm^2$ estimate the time for which the adsorption column will function before needing replacement. Express your answer in days and state any assumptions made in your calculation. *From the MSc (Safety and Reliability) Examination, University of Aberdeen, 2001.*

9. Pelletised household waste of $17\,MJ\,kg^{-1}$ heat value is to be upgraded to $21\,MJ\,kg^{-1}$ by absorption of waste oil. Making a reasonable estimate of your own for the heat value of the oil waste, determine in what proportions the pellets and the oil will need to be blended. *From the MSc (Safety and Reliability) Examination, University of Aberdeen, 2001.*

10. A waste hydrocarbon gas of carbon to hydrogen ratio (by weight) 6.5:1 is available as a gasification feedstock. It is reacted with steam and, so as to enhance its heat value, is subsequently methanated, that is, its carbon dioxide content is partially converted to methane by reaction with hydrogen over a catalyst. If the gas required as the final product is to be used on a burner adjusted for $12.5\,MJ\,m^{-3}$, what extent of methanation will be required?

Heats of combustion

hydrogen = $285\,kJ\,mol^{-1}$

methane = $889\,kJ\,mol^{-1}$

11. Return to the calculation in section 11.2.3. in which recuperation is used to reduce the amount of natural gas needed in an afterburning application. The 2001 price of natural gas is around $US2.00 per million BTU. If the process is round-the-clock, what would be the cost saving per annum brought about by the recuperation?

12. What rate of supply of waste lubricating oil, calorific value about $40\,MJ\,kg^{-1}$, would be necessary to sustain round-the-clock a stationary steam engine working at 100 horse power with 30% efficiency? Express your answer in tonnes per year.

13. BTX (benzene, toluene, xylenes) feature in the previous chapter. The respective formulae are C_6H_6, C_7H_8 and C_8H_{10}, empirical formulae CH, $CH_{1.14}$ and $CH_{1.25}$ so $CH_{1.1}$ would be a reasonable formula for engineering calculations. Write down the chemical equation for steam reforming of BTX on this basis and determine the composition of the resulting gas.

Endnotes

[34] The value for methane-air [1].
[35] For water vapour, these circumstances correspond to 100% humidity.

12

MEANS OF OBTAINING HYDROCARBONS OTHER THAN FROM CRUDE OIL AND RELATED SAFETY ISSUES

12.1 Introduction

Mention has already been made in this book of certain means of hydrocarbon production other than from crude oil, notably from shale. As we saw earlier, shale predates crude oil usage and there has been major use of shale products throughout the last century. Interest in the use of shale products continues to the present day [1]. In this chapter we outline some of these alternative means of obtaining hydrocarbons and draw attention to related safety issues.

12.2 Oil from shale

12.2.1 Background on shale oil

Shale consists of a small proportion of organic material called kerogen in combination with rock. The rock containing kerogen has to be excavated and crushed in readiness for retorting: heating to decompose the kerogen to give a liquid which is an approximate equivalent of crude oil. This can be distilled into fractions which can be used interchangeably with the respective petroleum fractions. Shale products tend to be higher in sulphur and nitrogen than their counterparts from crude oil. One must, however, avoid the view that shale products are necessarily less satisfactory than their petroleum analogues: standards of fuel processing can be high or low with any starting material, and liquid fuels of very high quality have been produced from shale. Reference has been made more than once already in this book of the serious shortage of oil reserves in Japan. A consequence of this was the use of shale-derived fuels by the Japanese navy during World War II.

Mention has been made of the need to excavate the rock containing kerogen, and this necessitates use of high explosives and all that that entails in

Table 12.1 *Outline of methods of above-ground retorting of shale, chiefly taken from [2]*

Method	Details	Comments
'Scottish shale retort'	Externally heated retort. By-product gas from the retorting process, possibly supplemented with producer gas, used to heat the retort externally	Contrasts with the more recent method outlined in the row below, where heating by the product gas is internal
Gas combustion retort	Vertical cylindrical vessel. Downward movement under gravity of crushed shale. Supply of heat by combustion of part of the by-product gases and the residual carbon in the shale. Hydrocarbon vapour so released rises and is cooled at the top of the vessel, where it is collected after condensation. Spent shale exiting the vessel at the base used to heat 'recycle gas', present as a diluent in the retorting part of the process	Several technologies of this genre developed between roughly 1945 and 1970 by the US Bureau of Mines. Requires no cooling water (see calculation below on temperature control)
Union Oil Company retort	Similar in principle to the Bureau of Mines retort, but shale passed vertically *up* the vessel by means of a 'rock pump'	Requires no cooling water
The Oil Shale Corporation ('TOSCO') method	Crushed shale retorted by contact with previously heated ceramic spheres in a rotary kiln	Successful development work not immediately followed by commercial implementation [3]

terms of hazards. There has even been interest, and indeed planning activity, in the use of nuclear processes to break up huge amounts of shale all at once.

Of more relevance to this text is the retorting process and safety features built into retorting plant. Temperatures reached in such a process are typically in the range 425–475°C, whereupon the kerogen decomposes to oil and by-product gas, leaving a solid residue which, of course, has to be suitably disposed of.

12.2.2 Retorting processes

Established processes for above-ground retorting of shale are summarised in Table 12.1. The retorting of shale *in situ* will be considered separately.

It is obviously economically beneficial to have a shale retort close to the excavation site, and as shale often occurs in dry areas this means that cooling water might not be available. The gas combustion retort was therefore developed in such a way that the retorting temperature is controlled by influx of recycle gas. The calculation below illustrates the principles.

In the processing of shale in a 'gas combustion' type of retorting vessel, per tonne of raw shale processed $7.2\,m^3$ of by-product gas are burnt in the combustion zone of the vessel. The temperature in the combustion zone of the retort must not exceed 450°C. In the limit where heat losses through the walls are negligible, how much of the recycle gas diluent, if supplied to the combustion zone at 145°C, is required per tonne of raw shale processed if the necessary temperature control is to be achieved. Approximate the composition of the by-product gas to ethane, C_2H_6.

Solution
Per tonne of shale processed, $7.2\,m^3$ of by-product gas are burnt to provide heat. From the stoichiometry:

$$C_2H_6 + 3.5O_2 + 13.2N_2 \rightarrow 2CO_2 + 3H_2O + 13.2N_2$$

this gives rise to:

$$7.2 \times 18.2\,m^3 = 131\,m^3$$

of post-combustion gas, which if there were no dilution would have a temperature of the order of 'adiabatic' values, i.e., approximately 2000°C. Now the specific heat of the product gas is the same as that of the recycle gas, the two having the same composition. Calling this $C\,J\,m^{-3}K^{-1}$, heat balance gives:

$$VC(450 - 145) = 131C\,(2000 - 450)$$

where V is the required volume of recycle gas, giving:

$$V = 666\,m^3$$

This is the amount of recycle gas required per tonne of raw shale processed.

In situ retorting is attractive for several reasons, not least that the useless part of the shale – the rock residue – remains in its original place and does not have to be disposed of. *In situ* retorting practices require the introduction of wells into a shale deposit with a high degree of intercommunication between them. Shale in one or more of the wells can then be ignited, using compressed air as oxidant. The combustion gases so released are hot enough to decompose kerogen which they subsequently contact, so that the combustion products resurfacing carry with them the gas and vapour decomposition products.

Figure 12.1 shows the site of exploratory work on shale oil production in Morocco.

Figure 12.1 *Exploratory work on shale oil production. Reproduced courtesy of Shell International.*

12.3 Hydrocarbons from tar sands

Tar sands have been a primary source of liquid hydrocarbon fuels in the west of Canada for very many years; the term 'Athabasca sands' is a synonym for 'tar sands'. They consist of porous rocks with hydrocarbon in the pores. Where recovery of the organic component is 'above ground', hot water treatment of the excavated rock suffices. However, as with shale, there are a number of in situ methods of removing separating the hydrocarbon from the inorganic component. Table 12.2 summarises some of these.

Note that whereas hot water treatment to separate the organic material is (unlike shale retorting) simply a physical process, the *in situ* methods in the table above may also involve some pyrolysis.

Table 12.2 *Methods for* in situ *separation of tar sands from porous rocks. Information taken from [4]*

Method	Description	Comments
Solvent or diluent injection	The geological formation first of all fractured by means of explosives, and injection of an organic substance to dissolve the tar sands and convey then to the surface	Petroleum distillate in the diesel boiling range a suitable solvent
Steam injection combined with emulsification	Entry of steam at the injection well and collection of emulsified hydrocarbon and the production well	Possible to incorporate an additive such as pyridine to promote emulsification
'Stimulation by combustion'	Ignition close to the surface of a deep well, penetration of the post-combustion gas and heating of the tar sands which rise to the surface and exit. The process operates continuously for several months	Injection well and production well one and the same in this configuration. Temperature control necessary to eliminate carbon particle formation
'Forward combustion'	'Pneumatic fracturing' of the formation, initiation of combustion at the injection well and collection of tar vapour at the production well	Injected air provides both an oxidant for combustion and, by reason of its pressure, a means of fracturing the formation in the required direction. Oxygen consumption incomplete, i.e., the combustion performed in excess air

12.4 Hydrocarbons from coal

We saw in Chapter 5 that hydrocarbons can be made from coal via synthesis gas. In this section we are primarily concerned with production of hydrocarbons from coal by liquefaction. This involves high-pressure treatment of a coal either with hydrogen or with a hydrogen-donating solvent, possibly in the presence of a catalyst. The hydrogen-to-carbon ratio of the coal substance is consequently raised and liquid products form, together with by-product gas. Alternatively, there may be no addition of hydrogen from the liquefying reagent, simply breakdown of the polymeric structure of the coal.

Pressures involved are high [5], 200 bar for processes using hydrogen gas and possibly this high for processes using a donor solvent. The latter uses lower temperatures, about 260°C compared with about 450°C when elemental hydrogen is the hydrogenating agent. Safety requirements for high-pressure processes apply and there is the additional factor of the incompatibility of hydrogen gas with certain alloys including any containing nickel.

Hydrocarbons are also obtainable from coal by pyrolysis. In coal combustion the liquid pyrolysis products – tars and oils – burn and contribute to the total heat release. When coal is carbonised these constitute the liquid product. In terms of petroleum substitutes, there has been considerable research activity into flash pyrolysis of coals. This involves heating coal, of particle size of the order a few millimetres, in a fluidised bed with a non-oxidising gas mixture as the atmosphere. The rapid heating of the coal particles leads to release of condensable pyrolysis products possibly suitable for further processing into substitutes for conventional liquid fuels.

12.5 Tight gas, CBM and hydraulic fracture

Tight gas, a.k.a. shale gas, is very topical at present, President Obama having expressed the view, presumably on the basis of information from his advisors, that there is sufficient of it to keep the US supplied with gas for 'nearly 100 years'. 'Tight' means from a rock formation having low porosity. Porosity has units of darcy and dimensions of area : 1 darcy is $1\,\mu m^2$. Conventional natural gas occurs in rock formations of millidarcy or tens (possibly hundreds) thereof whereas tight gas is occurs in formations of permeability down to 0.001 millidarcy.

Hydraulic fracture is frequently required to access tight gas [6]. Hydraulic fracture has been used in oil and gas production since about 1940, and most fracture fluids are water based although because of surface tension effects a hydrocarbon based fracturing fluid might be more effective where the formation permeability is low. Another reason for preference of an organic base fluid is that where clay is present water might cause it to swell. An aqueous or non-aqueous fracturing fluid will use suspended material called a proppant which will deposit in a fracture created by the fluid and prevent it from closing under the weight of the formation when fluid passage ceases. The permeability of the proppant once in place will such that oil passage through it is straightforward.

There is tight gas production on a large scale in the US at places including Haynesville, which extends into three states Texas, Louisiana and Arkansas. There is hydraulic fracture with sand as proppant and production is 22 million cubic metres of gas per day [6]. The only current scene of tight gas exploration in the UK is at the Lancashire coast near Blackpool where, again, there has been hydraulic fracture. Water contamination and geological instability are seen as the factors requiring hazard evaluation in hydraulic fracture to produce tight gas.

It was mentioned in Chapter 9 that much was known about the hazards of methane long before the proliferation of natural gas, because of experience in coal mining. There has for decades been an interest in methane from coal mines as an alternative to, or 'top-up' for, natural gas, notably in the Rocky Mountain states of the US. Internationally there has been major expansion in this activity during the decade between the first and second editions of this book. Coal-bed methane (CBM) is a major source of fuel in countries including Queensland Australia, where it is converted to LNG. It is intended that this will be delivered by tanker along the east coast.

The term 'tight gas' takes in CBM, and hydraulic fracture is used to access it. Even so some CBM reserves are of good permeability, for example that at Barrhead, Alberta [7]. Well completion for CBM is as for natural gas reserves. A point made in [6] is that hydraulic fracture does not open up small pores in the reservoir: it provides space into which gas contained in such pores can expand. In both utility and safety terms there is one major difference between tight gas in a rock structure (shale gas, which term does not include gas from coal seams) and CBM: the latter contains no condensate.

12.6 Concluding remarks

In meeting the energy requirements of the current century the technologies discussed in this brief chapter have their parts to play. Perhaps what is required is a deeper appreciation that these technologies are not simply to be drawn on as a contingency measure when there is some threat to oil supply, or an intensifying awareness that such supply is not inexhaustible, but that they have a role in a stable regime for meeting world hydrocarbon needs. Sometimes R&D done in an emergency scenario has not been subsequently applied to advantage once the emergency was over. As an example of this the author will invoke a personal recollection.

In the mid 1980s it was his privilege to be an associate of G.E. Baragwanath, the eminent Australian coal chemist. Baragwanath was by that time well past retirement age but still taking a great deal of interest in his profession. At the outbreak of World War II it was feared that, because of closure of shipping routes in the Pacific, crude oil imports would cease, leaving Australia critically short of liquid fuels. Consequently Baragwanath, then in his thirties, was sent to the Fuel Research Station in Greenwich, England, to conduct research into the production of liquid fuels from certain Australian coals. The work was very successful, but once the war was over a decision was made by the author-ities not to release the detailed findings. It was not until well into the 1970s that they finally entered the public domain as a set of bound volumes. It is not for the present author to criticise this course of events, for which there were no doubt very good reasons. It is simply cited as an example of R&D on fuel production carried out, so to speak, under pressure, and not implemented once the pressure was off.

References

[1] Jones J.C. 'Once and future fuels' *Chemistry in Britain* **35** (10) 20 (1999)

[2] Dinneen G.U. in Yen T.F., Chilingarian G.V. *Oil Shale*, First Edition. Elsevier, Amsterdam (1976)

[3] Russell P.L. *History of Western Shale Oil*, First Edition. Applied Science Publishers, Barking, UK (1980)

[4] Burger J.G. in Chilingarian G.V., Yen T.F. *Bitumens, Ashphalts and Tar Sands*, First Edition. Elsevier, Amsterdam (1978)

[5] Goodger E.M. *Alternative Fuels: Chemical Energy Resources*, First Edition. Macmillan, London (1980)

[6] Jones J.C. *Dictionary of Oil and Gas Production*, Whittles Publishing, Caithness (2012)

Numerical example

1. Return to the problem in the main text in which a shale retort is heated internally by means of the by-product gas diluted with recycle gas. If an equivalent amount of the gas was first allowed to exit the retort vessel and then mixed with air at 25°C for re-admittance to the retort and combustion, what proportions of gas and air would be required if the operating temperature were to be 450°C?

Calorific value of the by-product gas = 60 MJ m^{-3} (1 bar, 288 K). Specific heats: air 1000 J kg^{-1}K^{-1}; nitrogen 1040 J kg^{-1}K^{-1}; water vapour 1900 J kg^{-1}K^{-1}; carbon dioxide 850 J kg^{-1}K^{-1}.

2. It is believed that the UK has ≈104 trillion cubic feet of CBM. To how many barrels of oil is this thermally equivalent?

Appendix

THE CANVEY AND RIJNMOND STUDIES

Two Canvey reports were put out, the first in 1978, the second in 1981. The study comprises a very complex, in-depth treatment of many interacting occurrences, according to the most suitable analyses and equations known at the time.

The still extensive use of the Canvey reports reflects the fact that features they contain have a long shelf life. The realities of storing and transporting hydrocarbons are very much as they were in 1978, and further back than that. Although there have certainly been advances in hydrocarbon combustion chemistry since then, these have on the whole been only marginally relevant to storage practices. Adiabatic flame temperatures, reactivities of different sorts of hydrocarbon to oxidation, flame spread, minimum ignition energies and so on were all well understood before the Canvey reports. Advances in chemical processing since then have been largely in instrumentation and control and have accompanied advances in computing. However, in storage and transportation situations this usually means that the same measurements that were made 20 years ago are still made but with different sorts of sensing and recording devices.

Lees *op. cit.* describes the study as 'the most comprehensive hazard assessment of non-nuclear installations in the UK'. In this appendix there will also be reference to the Rijnmond report, which in some respects is comparable to the Canvey studies and approximately contemporary with them.

Background to the study

In this section, we consider first why the study was commissioned. Canvey Island is in the Thames estuary, east of London, and the scene of several chemical facilities of various sorts. At the time of commissioning of the study these were as described in the box below. At the time of the commissioning of the study, then, there were two refineries, and by 1972–73 planning permission

---

On Canvey Island
- A British Gas methane terminal used to receive shipments of LNG. The terminal also used for refrigerated butane storage and for the commissioning of new LNG-transporting ships by treatment to lower the temperature of the tanks
- A Texaco storage facility for liquid hydrocarbons, e.g., kerosine, diesel
- A storage facility (London and Coastal Wharves Ltd.) for chemicals, some flammable and some toxic
- A facility for transferring explosives between ships

On nearby mainland, close enough for hazards to be 'interactive' with those on the island
- A Mobil refinery. Crude oil brought by sea to storage tanks and drawn off from these for refining. Products transported out by sea, road, rail and pipeline
- A Shell refinery. A larger stored hydrocarbon inventory at any one time than at the Mobil refinery. Also a refrigerated ammonia facility *and* stored hydrogen fluoride (HF)
- A terminal for LPG, operated by Calor Gas. LPG sent from here to consumers by road
- An ammonium nitrate plant (Fison's)

---

for a third (actually on Canvey Island) had been granted. This refinery was under construction when there was a proposal to reconsider the planning permission, so construction was halted. The study was commissioned at this point, and obviously a key consideration was interaction between hazards from the many installations in the area, and the increased risk, due to the new refinery, of a major 'interactive' accident. The Coryton refinery, which features in Chapter 4, is on Canvey Island.

Some points from the First Canvey Report (1978)

This report considered the conditions without the projected third refinery. Hazardous events identified are given in the box below. These hazardous events may be either initiating or consequential, e.g., failure of an LPG vessel may be 'spontaneous' or the result of impact by a 'missile' from a nearby explosion.

> • Crude oil or refined product leakage not contained by a 'bund' (dike)
> • LNG release from the terminal on Canvey island or from a vessel carrying LNG
> • LPG release at either of the existing refineries
> • Ammonium nitrate explosion
> • Ammonia leak
> • Leakage of the HF

Various plausible mechanisms for each of these were suggested, having regard to the particular conditions at the site. Frequencies were assigned to them from records and informed judgement, for example:

failure of an LPG vessel at the Shell refinery 1.4×10^{-4} per year

LPG pipe failure summed across all the installations

4.5×10^{-4} per year.

Some of the other data of this sort used in the First Canvey Report are summarised in Table A1. Table A2 summarises some of the engineering principles invoked in the First Report.

Recommendations of the Second Canvey Report

Ammonia storage at the Shell facility had ceased between the First and Second Reports. Some risks were believed to have been overestimated in the First Report. Improved mathematical formulae for gas dispersion became available between the First and Second Reports. The improved treatments gave smaller travel distance of clouds of leaked hydrocarbons, with consequently reduced risks.

Concluding remarks on the Canvey study

In the process safety/loss prevention literature one frequently encounters reference to the Canvey study, hence its very proper mention, at a number of places, in the main text of this book. This appendix provides a reader with sufficient background knowledge to appreciate, and benefit from, such references to the study.

There was a rather remarkable spin-off from the failure of the second refinery at Canvey to materialise. Roads to the ultimately non-existent refinery on the estuary site had river dredgings as part of their base, and through non-usage the road construction as far as it had gone became a varied and disorderly landscape incorporating sand and silt which in the event became over the decades a habitat for rare invertebrates and is now a site of some interest for that reason. It is called Canvey Wick.

Table A1 *Selected frequency and probability data used in the First Canvey Report*

Activity or operation	Faults or failures	Frequency or probability
Total refinery operation and running	A major fire	0.1 per year
	Probability of development of a major fire into an explosion with significant overpressure	0.5
	Probability of 'missile' generation resulting from such an explosion	0.1
		i.e., frequency of a missile-generating explosion $= 0.1 \times 0.5 \times 0.1$ $= 5 \times 10^{-3}$ one every 200 years
Above-ground storage of LNG	Tank failure and leakage	2×10^{-4} per year
	'Rollover' (instabilities through density gradients)	10^{-5} to 10^{-4} per year
	Excess pressure through overfilling	10^{-5} to 10^{-4} per year
Storage of HF, LPG or NH_3 in a pressure vessel	Failure of the vessel and leakage of contents	10^{-5} to 10^{-4} per year
Transport of hydrocarbon stock by road	Accident involving spillage	1.6×10^{-8} per km travelled

The Rijnmond Report (brief)

This appertains to a vast complex of refineries and plant on the Dutch coast, as mentioned in Chapter 4. The origin of the Rijnmond Report was that a local authority report on hazards to residents of the region, published in 1976, was strongly criticised by the industry. Accordingly a commission, in which the local authority, the industrial firms at Rijnmond and the national government were represented was set up to conduct a pilot study of the hazards posed by six of the activities at Rijnmond (Lees, *op. cit.*). The Rijnmond Report was published in 1982, one year after the Second Canvey Report. From the selection of activities which were studied in the Report it

Table A2 *Scientific and engineering principles used to model particular occurrences at Canvey Island*

Occurrence	Principles invoked and conclusions reached
Flow of leaked oil from a 'catastrophically failed' vessel towards a populated area	Equations for flow of liquid 'slumped' from a failed upright cylindrical cylinder: radius of the flooded region (increasing linearly with time) and velocity of the flowing liquid
Exposure of an above ground LNG storage tank to heat flux from an LNG pool fire nearby	Heat flux of $140 \, \text{kW m}^{-2}$ from the fire. $30 \, \text{MW}$ received by the tank exterior
	Heat transferred from tank wall to surroundings both by radiation and convection. Radiation greater
	An eventually steady tank wall temperature of about $650 \, \text{K}$, attained 42 minutes after commencement of the pool fire
Ammonium nitrate explosion	$1.5 \, \text{MJ}$ heat released per kg ammonium nitrate explosively decomposed
	Overpressures at particular distances from the explosion predicted by comparison with data for TNT
Hydrogen fluoride leakage	HF less dense than air, but often highly associated $(H_n F_n)$ making it effectively denser than air. Dissociation on dilution endothermic
Ammonia leakage	Ammonia gas itself is significantly lighter than air. However, clouds of leaked ammonia will contain liquid droplets which, if sufficiently numerous, will cause the cloud to behave as if it were heavier than air. Conditions in such a cloud are far from 'phase equilibrium'
	Summary of the prediction of fluid modelling for a leaked cloud: (i) An initially tall, cylindrically symmetrical leak becomes a very squat cylinder by slumping within time of the order of minutes, during which the cloud drifts about 2–3 km (ii) Further dispersion leads to attainment of sub-lethal concentrations about 5 km beyond the cessation of slumping

Table A3

Hazard	Estimated frequencies of leakage	Predicted fatalities amongst the public
Acrylonitrile storage	Major rupture of a tank estimated as having a frequency of 4.3×10^{-5} per year. Other means of leakage, including that due to overfilling and that from hoses and lines, also assigned frequencies	Frequency of fatalities amongst the public due to all of the identified means of leakage 7.6×10^{-6} per year
Ammonia storage	Catastrophic failure of a spherical vessel full of ammonia assigned a frequency of 2.3×10^{-7} per year. Leakages during transfer and leakages arising from cracks in the vessel also considered quantitatively	Frequency of fatalities amongst the public due to all of the identified means of leakage 197×10^{-6} per year
Chlorine leakage	Catastrophic burst of a vessel full of chlorine assigned a frequency of 0.93×10^{-7} per year. Catastrophic burst of a half-full vessel assigned a frequency of 0.74×10^{-6} per year. Leakages from connectors at various orientations also assigned frequencies	Frequency of fatalities amongst the public due to all of the identified means of leakage 3.57×10^{-3} per year
LNG storage	Catastrophic failure of a tank assigned a frequency of 0.8×10^{-6} per year. Fractures of lines and loss during pumping also considered	Frequency of fatalities amongst the public due to all of the identified means of leakage 0.68×10^{-10} per year
Propylene storage	Failure of a full sphere assigned a frequency of 0.028×10^{-6} per year. Connectors and liquid lines also assigned frequencies of leakage	Frequency of fatalities amongst the public due to all of the identified means of leakage 3.66×10^{-5} per year
Hydrodesulphurisation	H_2S leakage from various stages of the process assigned frequencies	No fatalities to the public predicted

will be appreciated that there is considerable similarity to the Canvey Report. These were acrylonitrile storage, ammonia storage, chlorine storage, LNG storage, propylene storage and hydrodesulphurisation. Table A3 gives selected information from the report in relation to each of these.

Numerical problems

1. At Canvey Island at the time of the reports, the Occidental plant contained two containers for propane storage at ordinary temperatures (say 25°C). Calculate the vapour pressure of the propane at such temperatures. Latent heat of vaporisation of propane at 25°C = 18.5 kJ mol^{-1}. Normal boiling point of propane = –42°C (231 K). *From the MSc (Safety and Reliability) Examination, University of Aberdeen, 1999.*

2. Reference is made in one of the tables above to the fact that ammonia, though less dense than air, acts like a 'dense gas' on leakage, largely because of its tendency to form droplets. Repeat question 6 in Chapter 2 with ammonia rather than propane as the gas, i.e., a quantity of leaked ammonia, initial radius of the leak 15 m, initial height 4 m. On the basis of the van Ulden treatment calculate the radius after: (a) 10 s, (b) 3 minutes. Assume that droplets account for 50% of the total leaked quantity of ammonia.

Density of liquid ammonia at the leakage temperature = 645 kg m^{-3}. Density of gaseous ammonia at the leakage temperature = 0.69 kg m^{-3}. Density of air at the leakage temperature = 1.17 kg m^{-3}. NH_3 = 17 g mol^{-1}.

Solutions to numerical examples

Probit tables and temperature/e.m.f. tables for type K thermocouples are attached at the end of the solutions.

Chapter I

1. 1099 trillion BTU $= 1099 \times 10^{12} \times 252 \times 4.2\,J = 1.2 \times 10^{18}\,J$

28 million ton ≈ 28 million tonne, 2.8×10^{10}kg capable of releasing when burnt $\approx 2.8 \times 10^{10} \times 44 \times 10^{6}\,J = 1.2 \times 10^{18}\,J$

Agreement appears to be exact.

2. Volume of the oil in SI units $= 26.6 \times 10^{9}$ barrels $\times 0.159$ m^3 barrel^{-1}
$= 4.2 \times 10^{9}$ m^3
From web sources, 1 cubic mile $= 4.16818183 \times 10^{9}$ m^{3}, the 2011 production is therefore, to one place of decimals, exactly a cubic mile.

3.   One million BTU $= 1.055 \times 10^{9}$ J

Now 889 kJ mol^{-1} becomes 889×40 kJ m$^{-3} = 3.6 \times 10^{7}$ J m^{-3}

One million BTU are therefore released on burning
$1.055 \times 10^{9}/(3.6 \times 10^{7})$ m^3

$$= 30\,m^3$$

4. 1 cubic mile $= 4.16818183 \times 10^9$ m^3 so the quantity of gas corresponds to:
$(10^{14}$ ft$^3 \times 0.028$ m^3 ft$^{-3}/4.16818183 \times 10^9$ m^3 cubic mile$^{-1}) = 670$ cubic miles.

The calorific value of the gas is about 37 MJ m^{-3}
Heat releasable from the quantity of gas =
37×10^6 J m$^{-3} \times 10^{14}$ ft$^3 \times 0.028$ m^3 ft$^{-3} = 10^{20}$ J
1 cubic mile of oil $= 4.17 \times 10^9$ m^3
Setting the density at 900 kg m^{-3} and the calorific value at 42 MJ kg^{-1}
Heat releasable $= 4.17 \times 10^9$ m$^3 \times 900$ kg m$^{-3} \times 42 \times 10^6$ J kg^{-1}
$= 1.6 \times 10^{20}$ J

5.   1150 BTU per cubic foot $= 43.3$ MJ m^{-3}
Now 1 cubic metre of gas at 1 bar, 15°C contains 40 moles.
Let the proportion of ethane (fractional basis) in the gas be x:
$40[1560x + 889(1-x)] \times 10^{-3} = 43.3 \rightarrow x = 0.29$

Chapter 2

1. Using the gas laws:

$$PV = nRT \Rightarrow 7.5 \, \text{m}^3\text{s}^{-1} \equiv 313 \text{ mol s}^{-1} \equiv 5.0 \, \text{kg s}^{-1}$$

$$Q = C_d AP\sqrt{(M\gamma/RT)[2/(\gamma+1)]^{(\gamma+1)/(\gamma-1)}}$$

(equation 2.1 in main text)
$\Downarrow$

$$A = 3 \times 10^{-4} \, \text{m}^2 \, (3 \, \text{cm}^2)$$

2. $$Q = C_d AP(t)\sqrt{(M\gamma/RT)[2/(\gamma+1)]^{(\gamma+1)(\gamma-1)}}$$

$$P(t) = P_{\text{initial}}\exp\{-(Kt/MV)RT\} \text{ (equation 2.2 in main text)}$$

$$K = C_d A\sqrt{(M\gamma/RT)[2/(\gamma+1)]^{(\gamma+1)/(\gamma-1)}}$$

$$= 8.62 \times 10^{-8} \text{ SI units}$$

Substituting:

At $t = 0$, $Q = 0.086$ kg s^{-1}

After 15 minutes, $P = 9.26 \times 10^5$ Pa, $Q = 0.080$ kg s^{-1}

3. Bernoulli's equation: $z_1 g + v_1^2/2 + P_1/\sigma = z_2 g + v_2^2/2 + P_2/\sigma$ (equation 2.4 in main text)

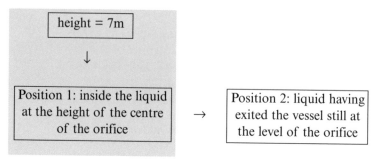

Since the vessel diameter is very much larger than the leak diameter, v_1 negligible in comparison with v_2. Also, $z_1 = z_2$ therefore:

$$0.5(v_2^2 - v_1^2) = (1/\sigma)(P_1 - P_2) \Rightarrow v_2 = \sqrt{2(1/\sigma)(P_1 - P_2)}$$

At position 2 the pressure is atmospheric. At position 1 the pressure exceeds atmospheric by $\sigma g H$, where H(m) is the depth below the vessel surface of the leak, therefore:

$$v_2 = \sqrt{2gH} = 12 \text{ m s}^{-1}$$

4. Fitting to: $d = d(0)e^{-ku} \Rightarrow \ln d = \ln d(0) - ku$ (equation 2.6 in main text)
Inserting the data points:

$$5.011 = \ln d(0) - 3.9k$$

$$4.700 = \ln d(0) - 4.8k$$

$$\ln d(0) = 6.360,\ d(0) = 578 \text{ m},\ k = 0.346$$

$$d = 578\, e^{-0.346u}\,\text{m}$$

In still air, drift distance needed for dilution to sub-flammable = 578 m

At a wind speed of 3 m s^{-1}, distance = 205 m

5. Fitting to $d = d(0)e^{-ku}$ gives $k = 0.463$, hence:

$$d/m = 9.1\, e^{-0.463u} \text{ with } u \text{ in m s}^{-1}$$

6. Using the equation from the main text:

$$R^2 - R_o^2 = 2c_E t \sqrt{(\sigma_o - \sigma_a)gV_o / \pi\sigma_o}$$

for the ratio of densities we can substitute the molar weights to give:

$$(\sigma_o - \sigma_a)/\sigma_o = (58 - 28.8)/58 = 0.50$$

Assigning the initial height the symbol H_o, other symbols as defined in the main text:

$$(V_o / \pi) = R_o^2 H_o = 900 \text{ m}^3$$

$$c_E = 1$$

$$\Downarrow$$

$$R^2(10\text{ s}) = 225 + 2 \times 10 \left\{ \sqrt{0.5 \times 900 \times 9.81} \right\} \Rightarrow R(10\text{ s}) = 39 \text{ m}$$

$$R^2(180\text{ s}) = 225 + 2 \times 180 \left\{ \sqrt{0.5 \times 900 \times 9.81} \right\}$$

$$= 24144 \text{ m}^2 \Rightarrow R(180\text{ s}) = 155 \text{ m}$$

7. $\qquad \log_{10}h = 1.48 + 0.656 \log_{10}d - 0.122 (\log_{10}d)^2$

Putting $h = 100$ m $(\log h = 2)$ gives:
$$0.79 = \log_{10}d - 0.186 (\log_{10}d)^2$$

Solving by trial and error gives:
$$d = 9.3 \text{ km}$$

Chapter 3

1. Calorific value of gasoline about 45 MJ kg^{-1}.

$$2.257 \text{ MJ released by } (2.257/45)\text{ kg} = 0.050 \text{ kg}$$

$$2.257 \text{ MJ released by } (2.257/25) \text{ kg coal} = 0.090 \text{ kg}$$

2. $\qquad C_8H_{10} + 10.5O_2 (+ 39.5N_2) \rightarrow 8CO_2 + 5H_2O (+ 39.5N_2)$

Proportion of m-Xylene at stoichiometric $= 1/[1 + 10.5 + 39.5]$
$$= 0.0196$$

pressure (total pressure 1 bar) corresponding to this $= 1960$ Pa pressure at half-stoichiometric = 980 Pa

Since the vapour pressure at 28°C is higher than this at 1316 Pa, the compound is above its flash point at this temperature.

(the literature value of the flash point is 25°C, so the prediction of the calculation is upheld)

3. Clausius–Clapeyron equation:

$$\frac{d(\ln P)}{d(1/T)} = -\frac{\Delta H_{vap}}{R}$$

(symbols as defined in the main text). Now for meta-Xylene, the normal boiling point is 412 K and the heat of vaporisation is 43 kJ mol^{-1}.

$$\int_{1\,atm}^{P} d(\ln P) = -(\Delta H_{vap}/R) \int_{412\,K}^{288\,K} d(1/T)$$

$$\ln(P/1\,atm) = -(\Delta H_{vap}/R) \{(1/288) - (1/412)\}$$

from which:

$$P = 4.49 \times 10^{-3} \text{ bar (449 Pa)}$$

Now the proportion of m-Xylene at 1.5 times stoichiometric is:

$$1.5/\{1 + 10.5 + 39.5\} = 0.0295$$

For a total pressure of 1 bar, the pressure of m-Xylene has to be 0.0295 bar. Returning to the integrated form of the Clausius–Clapeyron equation:

$$\ln(0.0295) = -(\Delta H_{vap}/R) \{(1/T) - (1/412)\}$$

$$\text{temperature} = 322 \text{ K } (49°C)$$

4. $$C_8H_8 + 10O_2 (+ 37.6N_2) \rightarrow 8CO_2 + 4H_2O (+ 37.6N_2)$$

Total units at stoichiometric = (1 + 10 + 37.6) = 48.6

Proportion styrene vapour at half-stoichiometric (the proportion required for a flash) = 0.0103. Pressure of styrene = 0.0103 × 10^5 Pa = 1030 Pa. Actual pressure given as 658 Pa. The liquid is therefore below its flash point.

5. $$C_6H_{12} + 9O_2 (+ 33.8N_2) \rightarrow 6CO_2 + 6H_2O (+ 33.8N_2)$$

Temperature rise = 3658000/{(6 × 43) + (6 × 57) + (33.8 × 32)} K
= 2175°C, final temperature 2200°C

6.
$$\text{Total area} = 2\pi \times 0.075 \times 14\,\text{m}^2 = 6.6\,\text{m}^2$$
$$\text{Total rate of heat release} = 0.52 \times 55.6 \times 10^6\,\text{W}$$
$$= 2.89 \times 10^7\,\text{W}$$

$$\text{Rate of radiative heat transfer} = 0.2 \times 2.89 \times 10^7\,\text{W} = 5.782\,\text{MW}$$
$$\equiv 5782\;\text{kW}$$

$$\text{flux} = 5782/6.6\,\text{kW m}^{-2} = 876\,\text{kW m}^{-2}$$

a black body releases energy according to:

$$\text{flux} = \sigma T^4 = 876000\,\text{W m}^{-2} \Rightarrow T = 1980\,\text{K}\ (1707^\circ\text{C})$$

7.
$$\text{Methane about } 55\,\text{MJ kg}^{-1}$$
$$\text{Crude oil about } 43\,\text{MJ kg}^{-1}$$

$$\text{mean } 49\,\text{MJ kg}^{-1}$$

$$\text{rate of heat release} = 5 \times 49\,\text{MW} = 245\,\text{MW}$$

8. First calculate the pool fire height from the equation in the main text:

$$H/D = 42\left\{ m'\middle/\left[\sigma_a \sqrt{gD} \right] \right\}^{0.61} = 2.8 \Rightarrow H = 14\,\text{m}$$

$$\text{mass loss rate from the pool} = 0.1 \times \pi \times 6.25\,\text{kg s}^{-1}$$
$$= 2.0\,\text{kg s}^{-1}$$

$$\text{rate of radiation from the flame}$$
$$= 0.25 \times 2.0 \times 44 \times 10^6\,\text{W} = 22 \times 10^6\,\text{W}$$

$$\text{radiative flux from the flame}$$
$$= 22 \times 10^6/\{2\pi \times 2.5 \times 14\}\,\text{W m}^{-2} = 100\,\text{kW m}^{-2}$$
$$q/A = 5.7 \times 10^{-8}\{T^4 - T_o^4\} \Rightarrow T = 1152\,\text{K}$$

9. Let $\lambda = 0.2$, i.e., 20% of the heat transferred by radiation. In Figure 3.2, the total area of the isosceles triangle therefore represents 20% of the total heat released by the propane for the duration of the fireball, i.e., the area is:

$$0.2 \times 50 \times 10^6\,\text{J kg}^{-1} \times 45 \times 10^3\,\text{kg} = 4.5 \times 10^{11}\,\text{J}$$

Recalling the definition of the dimensionless time t':

$$t' = g^{0.5}t/V^{1/6}$$
$$V = (45 \times 10^3/0.044)/(P/RT) = 24\,489\,\text{m}^3$$

$$t' = 0 \text{ corresponds to } t = 0$$

$$t' = 6 \text{ corresponds to } t = 10.3\,\text{s}$$

$$t' = 12 \text{ corresponds to } t = 20.6\,\text{s}$$

$$4.5 \times 10^{11}\,\text{J} = 0.5 \times 20.6 \times Q_{max}$$

where Q_{max} is the maximum radiation rate

$$\Downarrow$$

$$Q_{max} = 4.4 \times 10^{10}\,\text{W}$$

Diameter given by equation 3.8:

$$D(\text{m}) = 5.25[M(\text{kg})]^{0.314} \Rightarrow D = 152\,\text{m}$$

$$4\pi \times (76^2)\,\text{m}^2 = 72\,593\,\text{m}^2$$

Flux $= (4.4 \times 10^{10}/72\,593)\,\text{W m}^{-2} = 606\,\text{kW m}^{-2}$, corresponding to a temperature of:

$$\{606000/(5.7 \times 10^{-8})\}^{1/4}\,\text{K} = 1805\,\text{K}\ (1532°\text{C})$$

10. The relevant equation is 3.2 in the main text:

$$E = \sigma(T^4 - T_o^4)$$

where E = radiative flux (W m^{-2})
σ = Stefan's constant $= 5.7 \times 10^{-8}\,\text{W m}^{-2}\text{K}^{-4}$
T = temperature of the black body in Kelvin
T_o = temperature of the surroundings in Kelvin

Let $T_o = 300\,\text{K}$ and substitute to give: $E = 0.9\,\text{MW m}^{-2}$

11. Using the 'isosceles triangle' model:

Total area of the triangle $= 0.2 \times 5 \times 50 \times 10^6\,\text{J} = 5 \times 10^7\,\text{J}$
vapour volume leaked $= (5000/44) \times 8.314 \times 288/(1 \times 10^5)\,\text{m}^3 = 2.7\,\text{m}^3$

maximum flux at $t' = g^{0.5}t/V^{1/6} = 6$, from which $t = 2.3\,\text{s}$

diameter $= 5.25[M(\text{kg})]^{0.314} = 8.7\,\text{m}$

duration of fireball $= 4.6\,\text{s}$

$\frac{1}{2}\{\text{maximum rate of release of energy} \times 4.6\} = 5 \times 10^7\,\text{J}$

maximum rate of release of energy $= 2 \times 10^7\,\text{W}$

maximum flux $= 2 \times 10^7/\{4\pi\,(8.7/2)^2\} = 84\,\text{kW m}^{-2}$

Corresponding temperature $= 1100\,\text{K}$

12. Take the middle of the range of TNT equivalences:

72.5 tonne of TNT would provide 304500 MJ of blast energy

63 tonne of isobutane would release 3150000 MJ of combustion energy

$$yield = 10\%$$

13. Overpressure $= 2 \times 10^4$ Pa $\Rightarrow Y = -15.6 \times 1.93$ {ln20000} = 3.51, equivalent to 7%, therefore 10–11 persons are expected to suffer eardrum rupture.

14. The calculation is the same as previously in terms of the rate of radiation from the fireball, which is:

$$1.25 \times 10^{10} \text{ W}$$

The diameter is now:

$$7.71(5437)^{1/3} \text{ m} = 136 \text{ m}$$

the flux is therefore:

$$1.25 \times 10^{10}/(4\pi \times 68^2) \text{ W m}^{-2} = 215 \text{ kW m}^{-2}$$
$$\text{temperature} = \{215000/5.7 \times 10^{-8}\}^{1/4} = 1393 \text{ K}$$

This is about as far on the low side of the measured value as the previously calculated value was on the high side.

15. $C_4H_9OH + 6O_2 + 22.6N_2 \rightarrow 4CO_2 + 5H_2O + 22.6N_2$
the flash point corresponds to the temperature at which the *n*-butyl alcohol is present in a quantity of half-stoichiometric.
This gives:

$$[0.5/(1 + 5.5 + 20.7)] \times 10^5 \text{ Pa} = 1838 \text{Pa}$$

Since the vapour pressure at 30°C is actually 1316 Pa the substance will be below its flash point at that temperature.

16.   Total rate of heat release $= 55 \text{ MJ kg}^{-1} \times 2 \text{ kg s}^{-1} = 110 \text{ MW}$
              radiation accounts for about 20% of this, or 22 MW

Now the diameter is about $1/100 \times$ the length, i.e., about 0.25 m. The curved surface area is then:

$$2\pi \times 0.125 \times 25 \text{ m}^2 = 20 \text{ m}^2$$

$$\text{flux} = (22 \times 10^6/20) \text{ W m}^{-2} = 1100 \text{ kW m}^{-2}$$

$$1100000 = 5.7 \times 10^{-8}\{T^4 - 300^4\} \Rightarrow T = 2096 \text{ K } (1823°C)$$

17. The outside surface of the vessel, being described as quite tarnished, will have a high absorptivity and will thus be treated as a black surface (absorptivity 1.0). As far as this is in error it is on the safe side. Since the petroleum fraction starts to boil at about 60°C it is clearly a light fraction, and a density of about 850 kg m^{-3} is a reasonable estimate. The quantity of the material is therefore:

$$(4/3)\pi \times 1^3 \times 850 \, \text{kg} \times 0.8 \, \text{kg} = 2850 \, \text{kg}$$

Heat required to raise this quantity of the hydrocarbon from 15 to 60°C

$$= 2850 \, \text{kg} \times 2000 \, \text{J} \, \text{kg}^{-1} \text{K}^{-1} \times 45 \, \text{K} = 2.6 \times 10^8 \, \text{J}$$

$$\text{rate of supply of heat} = 1400 \times 4 \times \pi \times 1^2 \, \text{W} = 17600 \, \text{W}$$

$$\text{Time taken for the vessel contents to reach } 60°C = 2.6 \times 10^8 / 17600 \, \text{s}$$
$$= 4.1 \, \text{hour}$$

The absorptivity must be reduced so that this amount of heat is received in a time of 10 hours or longer, say 12 hours to build in a safety margin.

Letting the absorptivity = 4.1/12 = 0.34:

$$\text{rate of supply of heat} = 0.34 \times 1400 \times 4 \times \pi \times 1^2 \text{W} = 6012 \, \text{W}$$

$$\text{time taken for the vessel contents to reach } 60°C = 2.6 \times 10^8 / 6012 \, \text{s}$$
$$= 12 \, \text{hours.}$$

The absorptivity therefore needs to be reduced to a value not exceeding 0.35.

$$\text{Thermal e.m.f. corresponding to } 60°C = 2.43 \, \text{mV}$$

It should be noted that in the above calculation the total outside surface area of the vessel was put into the calculations, which is equivalent to approximating the vessel, for heat transfer purposes, to a flat plate or disc of equivalent surface area to the outside surface of the vessel. Otherwise the frontal area, πr^2, might have been used. This would however give a result erring on the hazardous side as it would not take account of radiation incident upon the parts of the tank not directly exposed to solar radiation, receiving it in reflected and therefore attenuated form. The approximation used in the above model answer errs on the safe side, perhaps too much so. More detailed knowledge of the circumstances, including the proximity of the vessel to other large objects and the nature of their surfaces, would enable some figure between 1.0 (frontal area only) and 4.0 (entire surface) as the coefficient of πr^2 to be estimated.

Note that for a cylindrical horizontal vessel the frontal area would be that of a rectangle of length that of the cylinder and width $2r$. In this event the total area is:

$$2\pi r(r + L)$$

and the frontal area $2rL$. The factor by which the two differ, 4.0 for a sphere, is therefore for the horizontal cylinder:

$$2\pi r(r + L)/(2rL) = \pi[(r/L) + 1]$$

18. 4×10^{10} J = 40000 MJ requiring 40000/4.2 kg of TNT, or 9500 kg, approx. 9.5 tonne, i.e., the 'TNT equivalence' is 9.5 tonne.

19. Rate of burning of a pool fire = 0.1 kg m^{-2} s^{-1}

For fuel of 44 MJ kg^{-1}, rate of heat release =

0.1 kg m^{-2} s^{-1} $\times$ 44 $\times$ 10^6 J kg^{-1} = 4.4 MJ m^{-2} s^{-1}

So a 1MW fire need have a pool area only of about 0.25 m^2, equivalent to a circle of about 30 cm radius.

Note on the above: an 'enclosure' in the sense used in the question need not be tightly enclosed. It might even consist of a space covered with a roof but without walls, provided that there was enough within the enclosure to absorb radiation from the pre-flashover fire and to restrict escape of hot product gases. Note also that at hydrocarbon plant fires flashover is often precluded by the fact that the initiating event itself has a heat-release rate in excess of 1MW, so the fire enters the post-flashover regime immediately.

20. As instructed in the question, we treat the structure as a black body radiating at the combustion temperature T, whereupon:

$$q = \sigma A(T^4 - T_o{}^4) \text{ W}$$

where q = rate of heat transfer from structure to surroundings (W), σ = Stefan's constant = 5.7×10^{-8} W m^{-2}K^{-4}, A = area of the radiating surface, T = combustion temperature (K), T_o = surrounding temperature (K).

Flashover requires about 1 MW, therefore:

$$q = 5.7 \times 10^{-8} \times 2\pi \times 0.5^2 \times L \ \{1073^4 - 300^4\} = 10^6 \text{ W}$$

where L is the height of the structure $\Rightarrow L = 8.5$m

$$V = \pi \times 0.5^2 \times 8.5 \text{ m}^3 = 6.7 \text{ m}^3$$

$$PV = nRT \rightarrow n = 26862 \text{ mol capable of releasing:}$$

26862 mol $\times$ 889 kJ mol^{-1} $\times$ 10^{-3} MJ = 23880 MJ

This would sustain a 0.1 MW flame for 66 hours.

21. Quantity of energy released $= 70\,000 \times 10^3$ kg $\times 42 \times 10^6$ J kg^{-1} $= 3 \times 10^{15}$ J

Blast energy $= 0.05 \times 3 \times 10^{15}$ J $= 1.5 \times 10^{14}$ J

TNT equivalence $= [1.5 \times 10^{14}$ J/$(4.2 \times 10^6$ J kg$^{-1}) \times 10^{-3}$ tonne $= 36$ kilotonne.

The following point of interest can be added in relation to the Avonmouth terminal. About 20 years later than the accident under discussion oil was discovered under the picturesque Dorset coast, including the town of Poole which is a holiday town and a prestigious place to reside. The field became known as Wytch Farm and was found to be the largest onshore field in Europe. Vertical drilling was precluded by the setting, and horizontal drilling was developed. When oil was eventually produced at Wytch farm it went to Avonmouth. Advances in horizontal drilling made at Wytch farm were later very effectively applied at offshore fields.

22.   Blast energy released by 15 kilotonnes of TNT

$= 15000000$ kg $\times 4.2 \times 10^6$ J kg^{-1} $= 6.3 \times 10^{13}$ J

Total energy to be released by the natural gas from this amount of blast

energy $= (6.3 \times 10^{13}/0.05)$ J $= 1.3 \times 10^{15}$ J

Putting the calorific value at 37 MJ m^{-3}, amount required

$= (1.3 \times 10^{15}$ J/$37 \times 10^6)$ m^3

$= 35$ million m^3 (a sphere of 0.4 km diameter).

23. Heat released $= 10^6$ bbl $\times 0.159$ m^3 bbl^{-1} $\times 900$ kg m$^{-3} \times 42 \times 10^6$ J kg^{-1}

$= 6 \times 10^{15}$ J

Blast energy $= 5\%$ of this $= 3 \times 10^{14}$ J

TNT equivalence $= 3 \times 10^{14}$ J/$(4.2 \times 10^6$ J kg$^{-1}) \times 10^{-6}$ kilotonne

$= 70$ kilotonne.

24. The flash point occurs when the vapour pressure of a fuel is at its lower flammability limit, and this usually approximates to half-stoichiometric at a total pressure of 1 bar. So in the case of 1,8 cineole the vapour pressure at the flash point would be:

$[0.5 \times 1/(1 + 14 + 52.6)]$ bar $= 0.0074$ bar

Utilising the Clausius-Clapeyron equation:

$\ln(0.0074$ bar/1 bar$) = [-44.2 \times 10^3$ J mol$^{-1}/8.314$ J K^{-1}mol$^{-1}]$ $\{1/T_f - 1/T_b\}$

where T_b is the normal boiling point and T_f the temperature at which the equilibrium vapour pressure would be 0.0074 bar.

$$T_f = 44.5°C$$

The measured value (closed cup) is 49°C

For terpinen-4-ol

$$\ln(0.0074 \text{ bar}/1\text{bar}) = [-51.8 \times 10^3 \text{ J mol}^{-1}/8.314 \text{ J K}^{-1}\text{mol}^{-1}] \{1/T_f - 1/485\}$$

giving a value of 78°C for the flash point

The measured value (closed cup) is 83°C

Chapter 4

1. Clausius–Clapeyron equation:

$$\frac{d(\ln P)}{d(1/T)} = -\frac{\Delta H_{vap}}{R}$$

(symbols as defined in the main text). Now for diethyl ether the normal boiling point is 308 K and the heat of vaporisation is 27 kJ mol^{-1}. Substituting:

$$\int_{1\,bar}^{4\,bar} d(\ln P) = -(\Delta H_{vap}/R) \int_{308\,K}^{T} d(1/T)$$

where T is the temperature at which the vapour pressure is 4 bar. Integrating:

$$\ln(4) = -(\Delta H_{vap}/R) \{(1/T) - (1/308)\}$$

from which:

$$T = 355 \text{ K } (82°C)$$

From tables, the e.m.f. at which the valve must open is 3.35 mV

2. Clausius–Clapeyron equation, with symbols as defined in the main text:

$$\frac{d(\ln P)}{d(1/T)} = -\frac{\Delta H_{vap}}{R}$$

Now for benzene, the normal boiling point is 78°C and the heat of vaporisation is 34 kJ mol^{-1}, therefore:

$$\int_{1\,bar}^{7\,bar} d(\ln P) = -(\Delta H_{vap}/R) \int_{354\,K}^{T'K} d(1/T)$$

where T' is the temperature at which the valve will open. Integrating:

$$\ln(7) = -(\Delta H_{vap} / R)\{(1/T') - (1/351)\} \Rightarrow T' = 421 \text{ K } (148°C)$$

3. Clausius–Clapeyron equation:

$$\ln(P/1 \text{ atm}) = -(\Delta H_{vap}/R) \{(1/298) - (1/260)\}$$

$$\Downarrow$$

$$P = 3.4 \text{ atm}$$

$$\ln(8) = -(\Delta H_{vap}/R) \{(1/T) - (1/260)\}$$

$$T = 331 \text{ K } (58°\text{C})$$

4. Total blast energy $= 30000 \times 4.2 \text{ MJ} = 1.26 \times 10^{11} \text{ J}$. This will account for about 5% of the total combustion energy, hence this must have been:

$$1.26 \times 10^{11}/0.05 \text{ J} = 2.52 \times 10^{12} \text{ J}$$

The vapour must have had a calorific value of about 45 MJ kg^{-1}, hence the quantity reacted must have been:

$$2.52 \times 10^{12}/(45 \times 10^6) \text{ kg} = 56000 \text{ kg, 56 tonne}$$

(This question would have been equally suitable for Chapter 3, but is included here because of its refinery context.)

5. The heat received by the tank, and subsequently transferred through to the hydrocarbon, in 1 minute is:

$$0.8 \times 1 \times 60 \times 1400 \times [4\pi \times 0.25^2] \text{ J} = 52779 \text{ J}$$

hydrocarbon evaporated in 8 hours

$$= (52779/500000) \times 8 \times 60 \text{ kg} = 51 \text{ kg}$$

If the outside surface were made reflective clearly the loss would be a tenth of this, i.e., 5.1 kg. So the benefit is a reduction in emission of 45.9 kg per 8-hour shift. See previous note re use of areas, total and projected, in such calculations.

6. The flash point will be reached when the tank has absorbed sufficient heat for the equilibrium pressure of m-Xylene to have reached the value corresponding to the lower flammability limit.

m-Xylene burns according to:

$$C_8H_{10} + 10.5O_2 \, (+ \, 39.5N_2) \rightarrow 8CO_2 + 5H_2O \, (+ \, 39.5N_2)$$

The lower flammability limit is taken to represent half-stoichiometric, i.e., a vapour pressure of the hydrocarbon of:

$$[0.5/(1 + 10.5 + 39.5)] \times 10^5\,\mathrm{N\,m^{-2}} = 980\,\mathrm{N\,m^{-2}}$$

Using the Clausius–Clapeyron equation, with symbols as previously defined:

$$\frac{d(\ln P)}{d(1/T)} = -\frac{\Delta H_{\text{vap}}}{R}$$

Now for m-Xylene, the normal boiling point is 412 K and the heat of vaporisation is 43 kJ mol^{-1}, therefore:

$$\int_{980\,\text{Pa}}^{100000\,\text{Pa}} d(\ln P) = -(\Delta H_{\text{vap}}/R) \int_{T_{\text{f}}}^{412\,\text{K}} d(1/T)$$

where T_{f} denotes the flash point. Solving:

$$T_{\text{f}} = 301\,\text{K } (28°\text{C})$$

quantity of m-Xylene in the sphere $= 0.67 \times [(4/3)\pi \times (0.5)^3]\,\mathrm{m^3} \times 860\,\mathrm{kg\,m^{-3}}$

$$= 302\,\text{kg} \equiv 2844\,\text{mol}$$

Heat Q required to raise the liquid from its initial temperature to 28°C given by:

$$Q = 2844\,\text{mol} \times 180\,\mathrm{J\,mol^{-1}K^{-1}} \times (28{-}15)\,\text{K} = 6.65 \times 10^6\,\text{J}$$

rate of supply of heat to the sphere $= 4\pi \times (0.5)^2 \times 700\,\text{W} = 2199\,\text{W}$

(See comment from previous question on areas.)

time required for the heating $= 6.65 \times 10^6/2199\,\text{s} = 3023\,\text{s}$ (50 minutes)

7. We first have to analyse the situation according to the First Law of Thermodynamics. The system is 'closed' in the thermodynamic sense, and in steady operation the liquid maintains a constant temperature therefore $\Delta U = 0$. This means:

$$Q = -W$$

and the rate of heat transfer to the outside is numerically equal to the power. Accordingly we set up the heat balance equation for the vessel:

$$\text{heat received from stirring}$$
$$= \text{heat transferred by convection} + \text{heat transferred by radiation}$$

Substituting:

$$500\,\mathrm{W\,m^{-3}} \times (4000/900)\,\mathrm{m^3} = 2222\,\text{W}$$
$$= \{10 \times 15 \times (T_{\text{v}} - 303) + [15 \times 5.7 \times 10^{-8}(T_{\text{v}}^4 - 293^4)]\}$$

where T_v is the tank temperature, and we have taken the emissivity to be unity. Rearranging:

$$14.8 = [(T_v - 303) + 5.7 \times 10^{-9}(T_v^4 - 293^4)] = f(T_v)$$

and this transcendental equation has to be solved by trial and error, to give:

$$T_v = 308.3 \, \text{K} \, (35.3°\text{C})$$

There is no inconsistency in specifying different temperatures to the 'surrounding air' and to the 'surroundings', as the former is transparent to thermal radiation therefore the 'surroundings' are the closest solid surface and this can have a temperature quite different from that of the air.

8. The refrigerating effect – the heat removed per kg refrigerant circulated – is $135240 \, \text{J} \, \text{kg}^{-1}$ as before.

required rate of heat removal from the brine
$$= (150/60) \, \text{kg} \, \text{s}^{-1} \times 2730 \, \text{J} \, \text{kg}^{-1}\text{K}^{-1} \times 13 \, \text{K} = 88725 \, \text{W}$$

rate of circulation of refrigerant $= (88725/135240) = 0.7 \, \text{kg} \, \text{s}^{-1}$

It should be noted that the unit weight of the brine is not equivalent, as a coolant, to the unit weight of water because of the different specific heats.

9. The heat released per kg benzene reacted is 1.8 MJ, and we are told in the question to assume that the refrigerator acts as an ideal reversed Carnot cycle, and that there is a temperature step of 3°C between the refrigerant and the brine at the stage where they exchange heat.

rate of heat release by the reaction $= 1.8 \, \text{MJ} \, \text{kg}^{-1} \times (10/60) \, \text{kg} \, \text{s}^{-1}$
$$= 300 \, \text{kW}$$

This is the rate of uptake of heat by the brine, therefore, assigning the reticulation rate the symbol m':

$$300\,000 \, \text{W} = m' \times 3200 \times (27 - (-3)) \Rightarrow m' = 3.1 \, \text{kg} \, \text{s}^{-1}$$

This is the rate at which the refrigerant must remove heat. Hence, for the refrigerator, retaining the symbols used in the main text:

$$T_h = 27°\text{C} \, (300 \, \text{K})$$

T_c (having regard to the temperature step) $= -6°\text{C} \, (267 \, \text{K})$

From tables, e.g., Perry *op. cit.*:

saturation pressure of R-11 at $267 \, \text{K} = 31.2 \, \text{kPa} =$ evaporator pressure

saturation pressure of R-11 at $300 \, \text{K} = 113.4 \, \text{kPa}$
$$= \text{pressure at the compressor}$$

refrigerating effect $= T_c(\text{Kelvin})\{s_g - s_f\} \, \text{J} \, \text{kg}^{-1}$

where s_g and s_f are respectively the specific entropies of refrigerant vapour and liquid at T_h. From tables:

$$s_g = 812.4 \, \text{J} \, \text{kg}^{-1} \text{K}^{-1}$$

$$s_f = 210.2 \, \text{J} \, \text{kg}^{-1} \text{K}^{-1}$$

⇓

refrigerating effect $= 160\,787 \, \text{J} \, \text{kg}^{-1}$

required rate of circulation $= 300\,000 \, \text{J} \, \text{s}^{-1} / 160\,787 \, \text{J} \, \text{kg}^{-1} = 1.9 \, \text{kg} \, \text{s}^{-1}$

10. The outside surface of the vessel, being described as quite tarnished, will have a high absorptivity and will thus be treated as a black surface (absorptivity 1.0). As far as this is in error it is on the safe side. Since the petroleum fraction starts to boil at about 60°C it is clearly a light fraction, and a density of about $850 \, \text{kg} \, \text{m}^{-3}$ is a reasonable estimate. The quantity of the material is therefore:

$$(4/3)\pi \times 1^3 \times 850 \, \text{kg} \, \text{m}^{-3} \times 0.8 = 2850 \text{kg}$$

Heat required to raise this quantity of the hydrocarbon from 15 to 60°C
$$= 2850 \, \text{kg} \times 2000 \, \text{J} \, \text{kg}^{-1} \text{K}^{-1} \times 45 \, \text{K} = 2.6 \times 10^8 \text{J}$$

rate of supply of heat $= 1400 \times 4 \times \pi \times 1^2 \, \text{W} = 17600 \text{W}$

Time taken for the vessel contents to reach 60°C $= 2.6 \times 10^8 / 17600 \text{s}$
$$= 4.1 \text{ hour}$$

The absorptivity must be reduced so that this amount of heat is received in a time of 10 hours or longer, say 12 hours to build in a safety margin.

Letting the absorptivity $= 4.1/12 = 0.34$

rate of supply of heat $= 0.34 \times 1400 \times 4 \times \pi \times 1^2 \, \text{W} = 5982 \text{W}$

Time taken for the vessel contents to reach 60°C $= 2.6 \times 10^8 / 5982 \text{s}$
$$= 12 \text{ hours.}$$

The absorptivity therefore needs to be reduced to a value
not exceeding 0.34.

Thermal e.m.f. corresponding to 60°C $= 2.43 \text{mV}$

(See note at the solution to problem 5 for this chapter regard to surface areas.)

11.
$$\delta = \frac{r_o^2 Q E \sigma A \exp(-E/RT_o)}{k R T_o^2}$$

Inserting the numbers given, with $\delta_{crit} = 3.32$ for a sphere, gives:

$$r = 8.6 \times 10^5\,m \;(\approx 3.5 \times 10^{18} \text{ tonne for a full container})$$

This staggeringly high value is a consequence of the very high activation energy. It is of course inconceivable that the substance would be stored in quantities of this magnitude. The reader should note that the figures given in the question relate to an actual, not a 'hypothetical', cetane enhancer.

12. Returning to:

$$\delta_{crit} = \frac{r_o^2 Q E \sigma A \exp(-E/RT_o)}{kRT_o^2} = 3.32$$

100 tonne $\equiv 100\,000$ kg occupying a volume of $(100000/1380\,m^3) = 72\,m^3$

a sphere having this volume has radius 2.59 m, hence in the FK
equation $r_o = 2.59\,m$

Rearranging the FK equation:

$$T_o^2 \exp(23815\,/\,T_o) = 9.8 \times 10^{25}$$

giving the trial-and-error solution $T_o = 502.5\,K$ (229.5°C)

13.

$$\sqrt{\delta_{crit}} = \sqrt{\delta_{crit}^*} \times \left[1 + \frac{(\theta_i - 3)^2 b(j+1)}{30k_r^{2/3}(1+3b^{2/3})}\right]$$

$$b = \frac{(\text{heat capacity} \times \text{density})_{\text{reacting medium}}}{(\text{heat capacity} \times \text{density})_{\text{hot spot}}} = \frac{900 \times 1860}{600 \times 7500} = 0.372$$

$$\varepsilon = RT_i/E = (8.314 \times 973\,/\,143000) = 0.098$$

$$k_r = \frac{\text{thermal conductivity of the hot spot material}}{\text{thermal conductivity of the reacting material}} = (25\,/\,0.027) = 926$$

$$\theta_i = E\,/\,RT_i^2\{T_i - T_A\} = (125000\,/\,8.314 \times 1473^2)\{1473 - 273\} = 8.32$$

substituting:

$$\sqrt{\delta_{crit}} = \sqrt{\delta_{crit}^*} \times \left[1 + \frac{(\theta_i - 3)^2 b(j+1)}{30k_r^{2/3}(1+3b^{2/3})}\right]$$

$$\Downarrow$$

$$\sqrt{\delta_{\text{crit}}} = \sqrt{\delta^*_{\text{crit}}} \times 1.004$$

hence δ^*_{crit} is a more than adequate approximation to δ_{crit}.

$$\sqrt{\delta^*_{\text{crit}}} = 0.4\sqrt{b^2 + 0.25\,j(j+1)(b+0.1b^3)}\ [\theta_i + 2.25(j-1)]^2[1+0.5\varepsilon\theta_i]$$

$$\Downarrow$$

$$\delta_{\text{crit}} = 2787 = \frac{QA\sigma Er^2 \exp(-E/RT_i)}{RT_i^2 k}$$

$$\Downarrow$$

$$r = 9 \times 10^{-7}\text{m}\ (0.9\,\mu\text{m})$$

For the reactive hot spot we use:

$$\delta_{\text{crit}} = \frac{r_0^2 QE\sigma A \exp(-E/RT_0)}{kRT_0^2} = 25$$

$$\Downarrow$$

$$T_0^2 \exp(E/RT_0) = 7.1 \times 10^{12}$$

trial-and-error solution yields $T_i = 946.5$ K

So a reactive hot spot the same size as the inert one but over 525 K cooler will have the same effect.

14. A value of 740 kg m^{-3} will be used for the density of the gasoline. By far the major component of the energy will be pressure energy, the kinetic and potential energy changes on pumping being small. This can be expressed:

(a) Pressure energy $= P/\rho$ where P is the pressure and r is the density. Putting our value for the pressure and 740 kg m^{-3} for the density:

Pressure energy $= 2 \times 10^5$ N m^{-2}/740 kg m$^{-3} = 270$ m^{2}s^{-2} ($^\circ$J kg^{-1})

Fuel to be transferred in 1 hour $= 6000$ m^3 or 4 440 000 kg

Rate of supply of mechanical energy by the pump
$= 270$ J kg$^{-1} \times 4$ 440 000 kg/(3600 s) W $= 330$ kW

(b) Speed of flow $= (100/60)$ m^3 s$^{-1}/(\pi \times 0.25^2)$ m s^{-1}

$= 9$ m s^{-1} to the nearest whole number.

$$Re = ud/\upsilon \text{ symbols as defined in the text}$$
$$\Downarrow$$
$$Re = 9 \times (0.5/10^{-6}) = 5 \times 10^5$$

(c)
$$Nu = 0.023Re0.8Pr0.3$$
$$\Downarrow$$
$$Nu = 1663 h = 470 \text{ W m}^{-2}\text{K}^{-1}$$

Rate of heat transfer $= (100/60) \text{ m}^3\text{s}^{-1} \times 740 \text{ kg m}^{-3} \times 2000 \text{ J kg}^{-1}\text{K}^{-1} \times 10 \text{ K}$
$$= 25 \text{ MW}$$

Outer radius of the pipe = 0.25 m
$$25 \times 10^6 \text{ W} = 2\pi \times 0.25 \times X \times 470 \times \{[(60 + 45)/2] - 10\}$$
where X(m) is the pipe length
$$\Downarrow$$
$$X = 800 \text{ m}$$

The problem began as one in fluid flow and became one in heat transfer. This is an example of how unit ops. can be interactive.

15. Saturated vapour pressure of water at 40°C
$$= 7375 \text{ Pa} = \text{compressor pressure}$$
Saturated vapour pressure of water at 5°C = 871.9 Pa = evaporator pressure
$$\text{COP} = 288/(313 - 278) = 8 \text{ to the nearest whole number}$$
specific entropy of liquid water at 40°C = 572 J kg^{-1}K^{-1}
specific entropy of water vapour at 40°C = 8256 J kg^{-1}K^{-1}
Refrigerating effect = 278K ×(8256 − 572) J kg^{-1}K^{-1} = 2136 kJ kg^{-1}

The higher refrigerating effect than, for example, that of R12 calculated in the main text is due to the larger entropy of vaporisation. At the molecular level this can be identified with hydrogen bonding in water, there being no such bonding in R12.

16. Amount of heat released
$$= 0.6 \times 10^6 \text{ bbl} \times 0.159 \text{ m}^3\text{bbl}^{-1} \times 900 \text{ kg m}^{-3} \times 44 \text{ MJ kg}^{-1}$$
$$= 3.8 \times 10^{15} \text{ J}$$
Blast energy $= 0.05 \times 3.8 \times 10^{15} \text{ J} = 1.9 \times 10^{14} \text{ J}$
TNT equivalence $= 1.9 \times 10^{14} \text{ J}/(4.4 \times 10^6 \text{ J kg}^{-1}) = 42 \text{ kilotonne}$
High, but 'only chemical'. Nuclear explosions are thousands of kilotonnes or higher.

17. Volume of the cylinder = $\pi \times (0.115)^2 \times 1.46$ m^3 = 0.061 m^3

Moles of hydrogen in the cylinder as supplied

= 200×10^5 N m^{-2} × 0.061 m^3/(8.314 × 298) mol = 492 mol or 1 kg

Fire load = $(1 \times 2.205$ lb/0.45 ft$^2)$ × $(143/17)$ lb ft^{-2} = 41 lb ft^{-2}

18.   40 000 tonne × 10^3 kg tonne^{-1}/550 kg m^{-3}
= 0.9 × (4/3)πr^3 where r is the radius
⇓
r = 27 m, diameter = 54 m
Minimum separation = 0.25 × (54 + 54) m = 27 m

19.   2m = 0.25 × 2d where d is the tank diameter
↓
d = 4 m, radius (r) 2 m

weight of LPG = $(4/3)\pi r^3$ m^3 × 550 kg m^{-3} = 18433 kg (≈ 18.5 tonne)
This would be the actual capacity, nameplate capacity 0.8 to 0.9 of this.

20.   Fractionation tower, descriptor 29
Low-pressure storage tank, descriptor   34
60 m
Hydrocarbon compressor, descriptor   19
Heat exchanger 'above ignition temperature', descriptor 24
7.5 m
Pipe way, descriptor 31
Reactor 'below ignition temperature', descriptor 23
5m

21.   Flare and fractionation tower
Minimum distance 60m.

Fire hydrant and reactor below ignition temperature
Minimum distance 15 m.

Process pump handling LPG and flare
Minimum distance 60m.

Hydrocarbon compressor and main plant roads
Minimum distance 60 m.

Equipment handling non-flammables and open flame equipment
Minimum distance 5 m

Open flame equipment and property boundary
na, outside the scope of the standard.

Chapter 5

1. (a) We calculate the vapour pressure from the Clausius–Clapeyron equation using symbols as previously defined:

$$\frac{d(\ln P)}{d(1/T)} = -\frac{\Delta H_{vap}}{R}$$

For cyclohexane, the normal boiling point is 81°C and the heat of vaporisation is 33 kJ mol⁻¹, therefore:

$$\int_{1\,atm}^{P} d(\ln P) = -(\Delta H_{vap} / R) \int_{354\,K}^{428\,K} d(1/T)$$

$$\ln(P/1\,atm) = -(\Delta H_{vap}/R) \{(1/428) - (1/354)\} \Rightarrow P = 6.9\,bar$$

(b) Pressure of air = $(8 - 6.9)\,bar = 1.1\,bar$

(c) Heat released = $10^{-5} \times 35000 \times 4$ MJ = 1.4 MJ

amount vaporised = $(1.4 \times 10^{6}\,J)/(33 \times 10^{3}\,J\,mol^{-1}/0.084\,kg\,mol^{-1})$
$$= 3.6\,kg$$

(d) $PV = nRT$

$P = nRT/V = \{[(3600/84) \times 8.314 \times 428]/10\} \times 10^{-5}\,bar = 0.2\,bar$

total pressure becomes 8.2 bar

2. The significance of the stirring is that the oil has effectively infinite thermal conductivity therefore, at any one time, its temperature profile is flat. Heat balance on the spherical vessel:

heat transferred from the vessel surface by convection
= heat lost by the vessel contents

$$-hA(T(t) - T_o) = c\rho V dT(t)/dt$$

(symbols as used in the main text)

Reversing the contents of the bracket on the left-hand side so as to eliminate the minus sign, and preparing to integrate:

$$\frac{\mathrm{d}T(t)}{\mathrm{d}t} = [hA / c\rho V](T_{\mathrm{o}} - T(t))$$

$$\int_{T_{\mathrm{i}}}^{T} \mathrm{d}T(t) / (T_{\mathrm{o}} - T(t)) = [hA / c\rho V] \int_{0}^{t} \mathrm{d}t$$

where T_{i} is the initial temperature of the oil. Integrating:

$$-\ln \frac{(T_{\mathrm{o}} - T(t))}{(T_{\mathrm{o}} - T_{\mathrm{i}})} = [hA / c\rho V]t$$

for a sphere: $A/V = 3/r = 1.5\,\mathrm{m}^{-1}$

$[hA/c\rho V] = 1.3 \times 10^{-4}\mathrm{s}^{-1}$

put $T(t) = 358\,\mathrm{K}$

$T_{\mathrm{o}} = 298\,\mathrm{K}$

$T_{\mathrm{i}} = 473\,\mathrm{K}$

$\Downarrow$

$t = 8234\,\mathrm{s}$ (2.3 hours)

3. Total volume of the room $= 3500\,\mathrm{m}^3$. Total number of moles of gas (i.e., air) in it initially $= PV/RT$

$$146\,172\,\mathrm{mol} = 1 \times 10^5 \times 3500/(8.314 \times 288)\ \mathrm{mol}$$

$1\,\mathrm{p.p.m.}$ of this $= 0.15\,\mathrm{mol}$

leakage rate of $5 \times 10^{-6}\,\mathrm{m}^3\,\mathrm{s}^{-1} = 2.1 \times 10^{-4}\,\mathrm{mol}\,\mathrm{s}^{-1}$

time for $0.15\,\mathrm{mol}$ to leak $= 0.15/(2.1 \times 10^{-4})\,\mathrm{s}$
$= 714\,\mathrm{s}$ (approx. 12 minutes)

4. The composition of the reactor contents in steady operation is clearly, in molar terms:

$C_2H_6 = 0.175 \quad Cl_2 = 0.175 \quad C_2H_5Cl = 0.325 \quad HCl = 0.325$

all figures on a fractional basis. So the heat capacity per mol of reactant supplied is:

$$(0.175 \times 53) + (0.175 \times 34) + (0.325 \times 63) + (0.325 \times 29) = 45\,\mathrm{J}\,\mathrm{K}^{-1}$$

the heat released per mol of ethane supplied is:

$$0.65 \times 119\,\mathrm{kJ} = 77\,\mathrm{kJ}$$

if the temperature rise is to be kept to 500 K, the heat retained is not to exceed:

$$45 \times 500 \text{ J} = 22.5 \text{ kJ}$$

the heat to be removed by the water is therefore:

$$(77 - 22.5) \text{ kJ} = 54.5 \text{ kJ}$$

the amount of water required is then:

$$54500 / [4180 \times (80 - 25)] \text{ kg} = 0.24 \text{ kg}$$

per kg of ethane admitted the quantity of water required is then:

$$0.24 \times (1000/30) \text{ kg} = 8 \text{ kg}$$

5. Per mol of ethane admitted, 0.65 mol of HCl produced, requiring neutralisation by:

$$HCl \rightarrow H^+ + Cl^-, \quad \text{followed by:}$$
$$CaCO_3 + 2H^+ \rightarrow Ca^{2+} + CO_2 + H_2O$$
$$0.65 \text{ mol HCl} \rightarrow 0.65 \text{ mol } H^+ \text{ requiring } 0.325 \text{ mol CaCO}_3$$
$$\text{for neutralisation} \equiv 0.0325 \text{ kg}$$

per kg of ethane admitted to the continuous reactor, lime requirement given by:

$$0.0325 \times (1000/30) \times 1.2 \text{ kg} = 1.3 \text{ kg}$$

6. Calling the amount of reactant initially admitted z mol, the stoichiometry is:

$$
\begin{array}{ccccc}
C_{10}H_8 & + & 2H_2 & \rightarrow & C_{10}H_{12} \\
0.333z(1-\alpha) & & 0.667z(1-\alpha) & & 0.333\alpha z
\end{array}
$$

initially	after reacting
$0.333z$ mol $C_{10}H_8$	$0.333z(1 - \alpha)$ mol $C_{10}H_8$
$0.667z$ mol H_2	$0.667z(1 - \alpha)$ mol H_2
no $C_{10}H_{12}$	$0.333\alpha z$ mol $C_{10}H_{12}$
total z mol	total $z(1 - 0.667\alpha)$ mol

Applying the ideal gas equation to the initial and final states (assigned subscripts 1 and 2 respectively):

$$P_1 V_1 = n_1 R T_1 \quad \text{and} \quad P_2 V_2 = n_2 R T_2$$
$$V_1 = V_2$$
$$n_1 = z$$

$$n_2 = z(1-0.667\alpha)$$

$$\Downarrow$$

$$P_2 = P_1 \times (1-0.667\alpha)(T_2/T_1) = 9.35 \text{ bar}$$

i.e., a drop in the pressure.

7. 1 tonne of aniline is produced from $1 \times (123/93)$ tonne of nitrobenzene, 1.32 tonne, 1.08×10^4 mol. Hence the heat released is:

$$2.08 \times 10^4 \times 540000 \text{ J} = 5.8 \text{ GJ}$$

8.   Heat requiring removal $= 0.85 \times 1.55 \times 10^6 \times 5000 \text{ J} = 6.6 \times 10^9 \text{ J}$

       rate of removal $= [(6.6 \times 10^9)/(6 \times 3600)] \times 10^{-3} \text{kW} = 305 \text{ kW}$

       flow rate of water $= \{305000/[4180 \times (55-1)]\} \text{ kg s}^{-1} = 1.4 \text{ kg s}^{-1}$

9. Let the rate of heat transfer $= q\text{W}$:

$$q = 7.5 \text{ kg s}^{-1} \times 2000 \text{ J kg}^{-1}\text{K}^{-1} \times 160 \text{ K} = 2.4 \times 10^6 \text{ W}$$

$$\text{also, } q = UA\Delta T$$

where U = heat transfer coefficient, A = area = LA^*, where L is the length and A^* the specific area, ΔT = temperature difference between the ends of the exchanger

Temperature profile of the heat exchanger in steady operation:

oil entry at 250°C   *oil exit at 90°C*
water exit at 75°C   *water entry at 20°C*
$\downarrow$ $\downarrow$
axial distance along exchanger, total L *m*

arithmetic mean temperature difference between the ends of the heat exchanger $= \{(250-75)+(90-20)]/2\}°C = 122.5°C$

     log mean temperature difference between the ends of the exchanger

$$= \left\{ \left[(250-75)-(90-20) \right]/\ln(175/70) \right\}°C = 114.6°C$$

and the log mean temperature difference is a better value to use with the heat balance equation for the heat exchanger, whereupon:

$$L = \left[2.4 \times 10^6 / (65 \times 70 \times 114.6) \right] m = 4.6\,m$$

(using the arithmetic mean temperature difference gives 4.3 m)
building in a safety margin of about 10%, an exchanger should be installed which is at least 5 m long

10. Estimating the density of the crude oil as $900\,kg\,m^{-3}$, weight of 25 000 barrels

$$= 25\,000\;bbl \times 0.159\;m^3 bbl^{-1} \times 900\;kg\;m^{-3} = 3.58 \times 10^6\;kg$$
Number of moles of 'CH_2' $= 2.6 \times 10^8$
Number of moles of hydrogen $= 2.6 \times 10^8 \times 0.0625 = 1.6 \times 10^7$
Volume of hydrogen
$= 1.6 \times 10^7\;mol/40\;mol\;m^{-3} = 0.4$ million m^3

11. Sulphur requiring removal per day
$= (10^5\,bbl \times 0.159\;m^3 bbl^{-1} \times 800\,kg\;m^{-3} \times 0.005 \times 0.98/0.032\;kg\,mol^{-1})\,mol$
$= 2 \times 10^6\,mol$

It is by the reaction $\phi SH + H_2 \rightarrow \phi H + H_2 S$

where ϕ denotes the organic structure to which the sulphur is bonded, which is analogous to simple mercaptans (thiols).

Moles hydrogen required $= 2 \times 10^6$ or $(2 \times 10^6/40)\;m^3 = 50\,000\,m^3$
Amount of methane needed if the hydrogen generation is by steam reforming of methane about 16 000 cubic metres, quite a small amount.

Conditions in a hydrodesulphurisation unit: about 400°C, up to 130 bar pressure, catalyst.

12. Heat released in the reaction of one tonne $= 4 \times 10^9$ J
Number of kg of monomer required to remove this heat by condensation
$= [(4 \times 10^9\;J)/(480 \times 10^3)] \times 10^{-3}$ tonne $= 8.3$ tonne

So eight recirculations to the nearest whole number.

13. Heat released in poylmerisation of 1 tonne $= 500$ MJ and this quantity of heat requires removal.

$$44\;kJ\;mol^{-1} = 423\;kJ\;kg^{-1}\;\text{or}\;0.423\;MJ\;kg^{-1}$$

Amount requiring recirculation = (500/0.423) kg = 1182 kg, 1.2 tonne approx.

So only just over one recirculation of the entire amount.

14. Recalling,

$$3C_nH_{2n} + nH_2 \rightarrow nC_3H_8$$

$$P_1V_1 = n_1RT_1$$
$$P_2V_2 = n_2RT_2$$

For isothermal conditions is a rigid reactor:
$$P_2/P_1 = n_2/n_1 = n/(3 + n) = 0.87 \text{ for } n = 20$$

15. Heat of reaction = $[-239 - (-75)]$kJ mol^{-1} = -164 kJ mol^{-1}
Molar mass of methanol = 0.032 kg
Heat released on producing 1 kg = $(164/0.032) \times 10^{-3}$ MJ = 5.1 MJ

Note: The difficulty with implementation of the process has been low selectivity to methanol. Too much of the methane goes to oxides of carbon!

16. Heat of the reaction as written = -26 kJ per mol benzene reacted.
0.078 kg of benzene gives on reaction 0.157 kg of bromobenzene.

1 kg of bromobenzene is obtained from $(0.078/0.157)$ kg benzene = 0.5 kg benzene or 6.4 mol.

Heat accompanying formation of 1 kg of bromobenzene = $6.4 \times 26 \times 10^{-3}$ MJ
$$= 0.17 \text{ MJ}$$

17. $CH(CH_3)_3 + 4.5O_2 (+ 16.9 N_2) \rightarrow 2CO_2 + 5H_2O (+ 16.9 N_2)$

Proportion of isobutane at half-stoichiometric = $0.5/(1 + 4.5 + 16.9)$ = 0.022
So pressure = 0.022 bar, 2200 Pa

18. From the stoichiometry, 0.091 kg of the scavenger will remove 0.017 kg of the hydrogen sulphide.
Ratio = 5.4
So totally to desulphurise a gas of 1 p.p.m. weight basis H_2S would require 5 to 6 p.p.m. of the scavenger. This figure is about typical for amine scavengers.

19. per m^3 of the gas, 100 p.p.m. represents $40 \times 100 \times 10^{-6}$ mol H$_2$S requiring:
$$[(40 \times 100 \times 10^{-6}) \text{ m}^{-3} \times 0.12 \times 10^{-18} \text{m}^2 \times 6 \times 10^{23}] \text{ m}^2$$
$$= 288 \text{ m}^2 \text{ of adsorbent area m}^{-3} \text{ of gas}$$
Quantity of gas so desulphurised by 1 tonne of the carbon =
$$1000 \text{ m}^2\text{g}^{-1} \times 10^9 \text{ g}/(288 \text{ m}^{-1}) = 3.5 \text{ billion m}^3$$

Perspective on the figure: A natural gas storage facility in the southern North Sea, one of the largest of its kind, can hold 4.6 billion cubic metres. It is operated by Eni.
No information on the H$_2$S content, which will vary.
An activated carbon used for this purpose might contain a metal catalyst.

20. $2\text{NH}^i\text{Pr}_2 + \text{H}_2\text{S} \rightarrow [(\text{NH2}^i\text{Pr}_2) +]2\text{S}^{2-}$

Molar mass of the amine = 0.101 kg capable of reacting with 0.017 mole of hydrogen sulphide. So weight ratio required = 5.9.

Chapter 6

1. $$e = \frac{PD}{2f - 0.2P} \Rightarrow e = 10 \text{ mm}$$

A domed end is much more efficient and economical that a flat one, requiring a less thick section of metal.

2. The conditions are 8 bar absolute pressure, 331 K (58°C). The design stress of stainless steel [1] is 160 MPa. The pressure experienced by the wall is 7 bar, 0.7 MPa.

$$e = \frac{PD}{4f - 1.2P} = 5.5 \text{ mm}$$

Further parts:
(a) If the weld is fully radiographed it can be taken to be as strong as the virgin plate, hence the answer is still 5.5 mm.
(b) If a J factor of 0.85 applies:

$$e = \frac{PD}{4Jf - 1.2P} = 6.4 \text{ mm}$$

3. Equation 6.3, with symbols as defined in the main text:

$$\text{schedule number} = \frac{1000 P}{S}$$

Now at 250°C stainless steels have a design stress of about 120 MPa [1], hence a suitable schedule number would be:

$$1000 \times 1.2 / 120 = 10$$

A delivery rate of 1450 US gallon min^{-1} at 3 m s^{-1} (9.8 ft s^{-1}) flow speed is equivalent to one of 150 US gallon min^{-1} at 1 ft s^{-1} flow speed. From tables [3] we see that schedule 10 piping of nominal outer diameter 8 inch has a delivery rate of 169.8 US gallon min^{-1} at 1 ft s^{-1}.

4. Using symbols as in the main text:

$$e = \frac{PD}{2f - P}$$

Now the pressure difference P is 33 bar, 3.3 MPa. From [1], design stress of 304 stainless steel at 305°C is 105 MPa. Substituting, with $D = 5$ m, gives:

$$e = 0.08 \text{ m (8 cm)}$$

5. The fluid density will be about 800 kg m^{-3}. The delivery rate of 5.5 m^3 min^{-1} is therefore equivalent to 4400 kg min^{-1} = 582120 lb hour^{-1}. The density in Imperial units is 48 lb ft^{-3}. Substituting:

$$\text{most economic diameter (inch)} = 0.098 m'^{0.45} / (\sigma^{0.31}) = 11.6 \text{ inch}$$

6.
$$V_{\text{mix}} = [(0.61 \times 5.86 \times 10^{-5}) + (0.39 \times 1.81 \times 10^{-5})] \text{ m}^3 \text{mol}^{-1}$$
$$= 4.28 \times 10^{-5} \text{ m}^3 \text{mol}^{-1}$$

Now one mol of the mixture (i.e., an Avogadro number of molecules) weighs:

$$(0.61 \times 0.046) + (0.39 \times 0.018) \text{ kg} = 0.0351 \text{ kg}$$
$$\text{density} = 820 \text{ kg m}^{-3}$$

This is in effect the same procedure as the first of the two related calculations in the main text and therefore gives the same result. It is imprecise in that it uses 'mole-fraction averaged' volumes instead of partial molar volumes. It is, however, less crude than the argument:

$$\text{density} = (0.39 \times 996) + (0.61 \times 785) = 867 \text{ kg m}^{-3}$$

In this case densities were 'mole-fraction averaged', and this result is actually about as far above the correct value as the Amagat value is below it.

7. The Reynolds number Re is given by:

$$Re = 0.05 \times 0.04 \times 700/(3 \times 10^{-4}) = 4667$$

hence the convection coefficient is given by:

$$h = 0.05(4667)^{0.8} \, Wm^{-2}K^{-1} = 43 \, Wm^{-2}K^{-1}$$

In order to set up the heat balance equation for the liquid, we need the mass flow rate, which is clearly:

$$0.04 \, m \, s^{-1} \times \pi(0.025)^2 \, m^2 \times 700 \, kg \, m^{-3} = 0.055 \, kg \, s^{-1}$$

$$\text{rate of heat transfer to the surroundings}$$
$$= \text{rate of heat loss by the liquid}$$
$$h \times 2\pi rL(T_{liquid} - T_{wall}) = m'c\Delta T$$

where $r = $ tube radius $= 0.025$ m; $L = $ tube length, required; $T_{liquid} = $ liquid temperature, taken as the mean of the entry and exit temperatures i.e. 42.5°C; $T_{wall} = $ wall temperature $= 30$°C; $m' = $ mass flow rate of liquid $= 0.055$ kg s^{-1}; $c = $ specific heat of the liquid $= 2000$ J kg^{-1}s^{-1}; $\Delta T = $ temperature drop of the liquid $= 9$°C.

Substituting and solving for L (and rounding up to the nearest half-metre):

$$L = 12 \, m$$

A tube of this length or longer will afford the necessary cooling.

8. The rate of heat loss is given by:

$$1135 \, J \, kg^{-1}K^{-1} \times 0.01 \, kg \, s^{-1} \times 18 \, K = 204.3 \, W$$

whis is about 18% lower than the value calculated previously, which is more accurate.

9. $P^* = 7$ bar $= 5320$ mm Hg
Substituting into the Antoine expression, with the constants given:

$$T = 160°C,$$

This is 12°C higher than the value calculated from the the Clausius–Clapeyron equation. The Antoine expression is more accurate.

10. At half-stoichiometric, total pressure 1 bar, the pressure of ethanol vapour is:

$$0.5/\{1+3+11.3\} \text{ bar} = 0.0326 \text{ bar} = 24.84 \text{ mm Hg}$$

Inserting in the Antoine equation (equation 6.6 in the main text):

$$\log P^* = A - \frac{B}{T+C}$$

$$\Downarrow$$

$$T = 11°C$$

11. From the Antoine equation, the vapour pressure of methanol at 25°C is 126 mm Hg, 16579 Pa. In a total pressure of 1 bar, the proportion of methanol is then:

$$(16579/10^5) \times 100\% = 16.6\%$$

Dilution to below 6.7% requires (16.6/6.7) vessel volumes of gas = 2.5 vessel volumes, $2.5 \times 5 \times 40$ mol or equivalently:

$$2.5 \times 5 \times 40 \times 0.038 \text{ kg} = 19 \text{ kg}$$

12. (i) Clausius–Clapeyron equation:

$$\frac{d(\ln P)}{d(1/T)} = -\frac{\Delta H_{vap}}{R}$$

(symbols as defined in the main text). Now for butane the normal boiling point is 272.5 K and the heat of vaporisation is 22 kJ mol^{-1}. Substituting:

$$\int_{1 \text{ bar}}^{P} d(\ln P) = -\frac{\Delta H_{vap}}{R} \int_{272.5 \text{ K}}^{308 \text{ K}} d(1/T)$$

where P is the vapour pressure at 308 K. Integrating:

$$\ln(P/1 \text{ bar}) = -(\Delta H_{vap}/R)\{(1/308)-(1/272.5)\}$$

from which:

$$P = 3.06 \text{ bar} \ (3.06 \times 10^5 \text{ Nm}^{-2}, \ 0.306 \text{ MPa})$$

(ii) The pressure differential across the vessel wall = (0.306 − 0.1) MPa
$$= 0.206 \text{ MPa}$$

Substituting, and noting that $J = 1$ if the seam is fully radiographed and that $N\,mm^{-2} \equiv MPa$

$$e = 8\,mm$$

(iii) For $J = 0.85$ $e = 9\,mm$

13. $\qquad\qquad P_1V = n_1RT \qquad P_2V = n_2RT$

where 1 denotes gasoline and 2 denotes air.

$$n_1/(n_1 + n_2) = 1/[(1 + (P_2/P_1)] = 0.07$$
$$\downarrow$$

$$P_2/P_1 = 13.3$$
$$P_1 + P_2 = 1\ bar$$
$$14.3P_1 = 1\ bar \quad P_1 = 0.070\ bar$$

From tables, vapour pressure of iso-octane at 25°C is 0.065 bar, very close to the above result.

The RVP would be much higher than this, indicating that in general if there is phase equilibrium in such a vessel the vapour will be too rich to ignite.

14. $\qquad\qquad P_1V = n_1RT \qquad P_2V = n_2RT$

where 1 denotes diesel and 2 denotes air.

$$n_1/(n_1 + n_2) = 1/[(1 + (P_2/P_1)] = 0.006$$
$$\downarrow$$

$$P_2/P_1 = 167$$
$$P_1 + P_2 = 1\ bar$$
$$168P_1 = 1\ bar$$
$$P_1 = 10^5/168 \quad Pa = 595\ Pa$$

The RVP of diesels is around 0.2 p.s.i. or 1350 Pa, over twice the above value.

Chapter 7

1. Using symbols as defined in the main text:

$$v_2 = \sqrt{\left[(2/\sigma)(p_1 - p_2)\right]/\left[1 - (D_2/D_1)^4\right]}$$

Inserting $(p_1 - p_2) = 80000\ N\ m^{-2}$, $\sigma = 780\ kg\,m^{-3}$ and $D_2/D_1 = 0.25$ gives:

$$v_2 = 14.5\ m\,s^{-1}$$

the flow rate is clearly:

$$m' = 5.7\ kg\,s^{-1}$$

2. Clearly a case where in equation 7.1 (below), $n = 1$.

$$Q = 1.84[B - (0.1 \times nD)]D^{1.5}$$

$$\Downarrow$$

$$Q = 1.84(B - 0.1D)D^{1.5}$$

At the maximum safe flow rate, $Q = 10\,\text{kg s}^{-1}/800\,\text{kg m}^{-3} = 0.0125\,\text{m}^3\,\text{s}^{-1}$.

Putting $B = 0.25\,\text{m}$ gives:

$$0.0068 = 0.25D^{1.5} - 0.1D^{2.5}$$

$$\Downarrow \qquad \begin{array}{l} \textit{trial and error} \\ \textit{solution} \end{array}$$

$$D = 0.093\,\text{m} \ (9.3\,\text{cm})$$

3. Substituting into equation 6.6, the vapour pressure of methanol at 40°C is 264.6 mm Hg $\equiv$ 0.348 bar, $3.48 \times 10^4\,\text{N m}^{-2}$.

(a) If the open-ended manometer were applied, with methanol vapour only in the space above the liquid, the pressure of the atmosphere would exceed that of the vapour by:

$$(10^5 - 3.48 \times 10^4)\ \text{N m}^{-2} = 6.52 \times 10^4\,\text{N m}^{-2}$$

Expressed as a height of mercury, this is:

$$[6.52 \times 10^4/(13590 \times 9.81)]\ \text{m} = 0.49\,\text{m}$$

The difference in heights will be 0.49 m, and the mercury surface on the side connected to the vessel will be above that open to the atmosphere.

(b) If there is also air at 1 bar, to a good approximation[*] the two pressures simply add to give a total of $1.348 \times 10^5\,\text{N m}^{-2}$, or a height of mercury of:

$$[1.348 \times 10^5/(13590 \times 9.81)]\ \text{m} = 1.01\,\text{m}$$

This then will be the difference in heights, and this time the mercury surface on the side open to the atmosphere will be above that on the side connected to the vessel.

---

[*] In fact the vapour pressure of the methanol will in principle depend on the pressure of the air, as the nitrogen and oxygen molecules comprising the air will stimulate evaporation by bombardment of the methanol liquid surface. But if the air is at atmospheric pressure this effect will be very small and can be neglected.

4. (a) Errors are assessed as follows.

2.2°C due to intrinsic calibration uncertainty
$2 \times (13/200)$°C $= 0.13$°C due to the extension cable
say 1°C due to the cold-junction compensation

⇓

$$\text{error} = \sqrt{(2.2^2 + 0.13^2 + 1^2)}\text{°C} = 2.4\text{°C}$$

When the reading was 85.5°C the true temperature was in the range 83.1–87.9°C, so the conclusion is that the liquid was below 90°C. When the reading was 87.7°C the true temperature was between 85.3 and 90.1°C.

(b) Errors are assessed as follows.

4.5°C due to intrinsic calibration uncertainty.
other errors are as before

$$\text{error} = \sqrt{(4.5^2 + 0.13^2 + 1^2)}\text{°C} = 4.6\text{°C}$$

When the reading was 85.5°C the true temperature was in the range 80.9 to 90.1°C. When the reading was 87.7°C the true temperature was between 83.1 and 92.3°C.

Either from substitution into the polynomial or from tables:

$$R(90\text{°C}) = 134.71 \text{ ohm}$$

Differentiating the polynomial:

$$dR/d\theta = AR_o + 2BR_o\theta$$

Substituting A, B $\theta = +90$°C, gives:

$$dR/d\theta = 0.38 \text{ ohm °C}^{-1}$$

Hence for a resistance uncertainty of 0.38 ohm the temperature uncertainty is given by:

$$\delta\theta = \delta R/(dR/d\theta) = (0.38/0.4) \text{ °C} = 1\text{°C} \text{ approx.}$$

The RTD affords much better accuracy.

5. (a) No effect at all. Thermocouples of the same type displaying the colour codes of different standards authorities are interchangeable.

(b) The temperature reading will be in error by twice the temperature difference between the terminal head of the thermocouple and the temperature of the recorder terminals, in this case (2×22)°C $= 44$°C.

6. Since one of the thermal e.m.f.s is +0.718 mV with respect to an arbitrary zero and the other is −0.701 mV with respect to the same arbitrary zero, the e.m.f. in the thermocouple circuit is clearly:

$$[0.718 - (-0.701)]\,\text{mV} = 1.419\,\text{mV}$$

Using the Seebeck coefficient, the e.m.f. is:

$$36 \times 39.5 \times 10^{-3}\,\text{mV} = 1.422\,\text{mV}$$

The first calculation is more accurate, as the Seebeck coefficient is not, as has been assumed in the second calculation, constant across the temperature range. If it were the e.m.f.s at +18°C and −18°C would be equal and opposite, which they are not, quite. The difference is, however, insignificant in practical terms.

7. $R(\theta) = R_o\{1 + A\theta + B\theta^2 + C(\theta - 100)\,\theta^3\}$, symbols as defined in the main text, values of A, B and C also given there. Putting $\theta = -18°C$:

$$R(-18°C) = 92.98\,\Omega$$

It should be noted that:
(i) The resistance has to be below R_o, in this case 100 Ω, if the temperature is below 0°C.

(ii) The labour involved in applying the above equation can be saved if tables of temperature against resistance, analogous to temperature e.m.f. tables for thermocouples, are used. Such tables are available in [7] (amongst other sources), and actually give the resistance for a 100 Ω RTD at −18°C as 92.95 Ω. Note the minute difference.

8. From application of the equation in the main text or from tables, the resistance of an RTD100 detector at −88°C is 65.1 ohms.

$$39.5\,\mu\text{V}\,°\text{C}^{-1} \times -88°\text{C} = -3.476\,\text{mV}$$

The tables give −3.492 mV. Close.

Probably better NOT to use MIMS. Little seems to be known about the behaviour of sheath materials at such temperatures. Recommend bare-wire with a twin-bore ceramic support.

9. Heat balance at a thermocouple tip:

heat transferred to the tip = heat transferred from the
by convection   tip by radiation

$$hA\,(T_g - T_t) = \varepsilon\sigma A(T_t^4 - T_w^4)$$

h = convection coefficient ($W\,m^{-2}K^{-1}$), A = tip area (m^2), T_t = tip temperature (K), T_g = gas temperature (K), T_w = wall temperature (K), ε = emissivity, σ = Stefan's constant = $5.7 \times 10^{-8}\,W\,m^{-2}K^{-4}$.

Inserting numbers gives h = 127 W $m^{-2}K^{-1}$

Obtainable in a gas with good stirring. The meaning is that when the t.c. reads 300°C the gas is at 301°C.

10. Heat to be taken up = $61s \times 77\,000$ W = 4.7 MJ

4.7×10^6 J = 6 kg × 1000 J $kg^{-1}°C^{-1}$ × (700 – 25)°C + m kg × 571 000 J kg^{-1}

where m is the mass in kg of solid CO_2 and allowance has been made for the fact that gas once produced from the solid will rapidly attain 700°C.

Solving, m = 1.1 kg or 18% of the total.

The message of the calculation is that the sublimation effect is very small. The heat of sublimation of carbon dioxide on a weight basis is only about a quarter of the heat of vaporisation of water at its boiling point and this is the origin of the unimportance of the sublimation.

11. From steam tables, temperature of saturated steam at 6 bar = 159°C
$$\delta\theta = \delta R/(dR/d\theta)$$
Now $dR/d\theta = AR_o + 2BR_o\theta$ giving a value of
$$= 0.372 \text{ ohm } °C^{-1} \text{ at } \theta = 159°C$$
$$\delta\theta = (0.4/0.372)°C = 1.1°C$$

Chapter 8

1.

Occurrence	Probability
Both valves fail to close	0.0025
A closes, B fails to close	0.0475
A fails to close, B closes	0.0475
Both valves close	0.9025
	Total 1.0

2. Densities of crudes vary in the approximate range 800 to 990 $kg\,m^{-3}$. We assign a value of 900 kg m^{-3} to the density, whereupon 100 $m^3 \equiv 90\,000$ kg. The calorific value will be ≈44 MJ kg^{-1}, therefore the heat released is ≈4 TJ.

3. For the course of events whereby there is leakage and ignition, the frequency is:

$$0.25 \times 5 \times 10^{-3} \text{ per year}$$

The frequency with which this occurs and is accompanied by failure of one or both ESD valves is:

$$(1 - 0.98^2) \times 0.25 \times 5 \times 10^{-3} \text{ per year}$$

which is the frequency with which there will be a jet fire long enough to impinge upon the truss. However, it also has to be in the right direction to impinge upon the truss, probability of this $\approx 1/6$, hence the frequency whereby there is:

leakage
ignition
failure of one or both ESD valves
resulting flame in the direction of the truss is:

$$(1/6) \times (1 - 0.98^2) \times 0.25 \times 5 \times 10^{-3} \text{ per year} = 8 \times 10^{-6} \text{ per year}$$

If five deaths result, then using equation 8.1 with symbols as in the main text:

$$N\phi(N) = 4 \times 10^{-5} = 10^{-4.4}$$

One death every 25000 years, towards the high end of the 'acceptable risk' area.

4.   Frequency of leakage with ignition $= 1 \times 10^{-3} \times 0.35$ per year

$$= 3.5 \times 10^{-4} \text{ per year}$$

this will result in (5×0.1) deaths $= 0.5$ deaths

$$N\phi(N) = 0.5 \times 3.5 \times 10^{-4} \text{ per year} = 1.75 \times 10^{-4} = 10^{-3.80}$$

This is well into the values of n for which ALARP applies. If, for example, the susceptible pipe work could, at reasonable cost, be relocated in a part of the platform with fewer occupants, that would be required.

5. The cumulative frequencies are:

$$v_1 = 0.1 \times 0.02 = 2 \times 10^{-3} \text{ per year}$$

$$v_2 = 2 \times 10^{-3} \times 0.81 = 1.6 \times 10^{-3} \text{ per year}$$

$$v_3 = 1.6 \times 10^{-3} \times 0.166 = 2.7 \times 10^{-4} \text{ per year}$$

$$v_4 = 2.7 \times 10^{-4} \times 0.98 = 2.6 \times 10^{-4} \text{ per year}$$

$$v_5 = 0.81 \times 2.6 \times 10^{-4} = 2.1 \times 10^{-4} \text{ per year}$$

Meaning a course of events whereby:

there is leak at the gas export riser
the leaking hydrocarbon ignites without overpressure
both emergency shutdown valves work thereby limiting the leak
the resulting jet fire points vertically downwards
the fire water pump works
both deluge sets work

has a frequency of 2.1×10^{-4} per year (approx. once every 5000 years).

Deaths due to the initiating event $0.2 \times 0.25 = 0.05$

6. After 1 hour:

$$\frac{T(t) - T_i}{T_F - T_i} = 1 - \exp(h^2 \alpha t / k^2) \mathrm{erfc}\left[h\sqrt{(\alpha t)/k} \right]$$

$$= 1 - (1.016 \times 0.881) = 0.105$$

$$T_t = 81°C$$

7. Putting $k_m = 5 \, m^2 \, g^{-1}$ and $m = 0.1 \, g \, m^{-3}$ gives:
$k = 0.5 \, m^{-1}$, an illuminated sign visible at $8/0.5 = 16 \, m$

8. Smoke requires a burning rate of about $5 \, kg \, s^{-1}$. A pool fire burns at about $0.1 \, kg \, m^{-2} s^{-1}$. So the maximum area for there to be no smoke is $50 \, m^2$.
N.B.: This calculation assumes that the pool fire is the only combustion taking place at that part of the platform. If there is also a jet fire, it is the sum of the two burning rates that has to be compared with $5 \, kg \, s^{-1}$ to ascertain whether smoke will occur.

9. For $I = 16554 \, W \, m^{-2}$ and $t = 7 \, s$:

$$Y = 5.16 \equiv 56\%$$

Cooling reduces the flux from the surface to $24246 \, W \, m^{-2}$, that incident upon persons to half of this, or $12123 \, W \, m^{-2}$. Substituting into the probit equation, with $t = 7 \, s$:

$$Y = 3.91 \equiv 14\%$$

10. $(100 \times 10^{-3} g/96 \, g \, mol^{-1})/(1000 \, g/18 \, g \, mol^{-1}) \times 10^6 \, p.p.m. = 19 \, p.p.m.$ molar basis of sulphate, same for the sulphide.

11. There will be a blast, that is an overpressure, if the fireball has a diameter equal to or greater than the module volume. This gives a value for M of:

$$(1000/5.25)1^{/0.314} \text{ kg} = 1.8 \times 10^7 \text{ kg}$$

Volume of a spherical container holding this amount $= 1.8 \times 10^7 \text{ kg}/950 \text{ kg m}^{-3}$
$= 19000 \text{ m}^3 = (4/3)\pi r^3$ where r is the radius.

$\downarrow$

$r = 16$ m rounding down to the nearest whole number.

12. One death every 150 years $= 6.7 \times 10^{-3}$ per year

$$1 \times 6.7 \times 10^{-3} = 10^{-n}$$

$$n = 2.2$$

One death every 1200 years $= 8.3 \times 10^{-4}$ per year

$$n = 3.1$$

One death every half a million years $= 2 \times 10^{-6}$ per year

$$n = 5.7$$

Chapter 9

1. Total amount leaked $= 0.59 \times 120 \text{ m}^3 = 70.8 \text{ m}^3$.
This must be 5% (or less) of the volume across which it is dispersed in order for the lower flammability limit not to be exceeded. Hence the volume is:

$$70.8/0.05 = 1416 \text{ m}^3$$
(approx. an 11 m cube)

The fact that methane is much less dense than air means that natural mixing will not provide a uniform concentration of methane in the air unless mixing is aided, e.g., if the enclosure contains a fan.

2. Heat supplied $= 700 \times 400 \times 3600 \text{ J}$

amount vaporised $= (700 \times 400 \times 3600) / 535\,000 \text{ kg}$
$= 1884 \text{ kg (approaching 2 tonne!)}$

3. Rate of heat loss by sea water = rate of heat uptake by LNG

$$40 \times 4200 \times 12 = m \times 512\,000$$

where m is the rate of evaporation (kg s^{-1}) assuming 100% efficiency.

$\Downarrow$

$$m = 3.9 \text{ kg s}^{-1}$$

actual rate of evaporation $= 3.2 \text{ kg s}^{-1}$, 276 tonne day^{-1}

4. First, the temperature at which toluene has a vapour pressure of 6 bar has to be calculated:

Clausius–Clapeyron equation:

$$\ln(6 \text{ atm}/1 \text{ atm}) = -(\Delta H_{vap}/R) \{(1/T) - (1/384)\} \Rightarrow T = 460 \text{ K}$$

$$\text{let the mass of toluene} = m$$

$$\text{heat required to heat the toluene to } 460 \text{ K from } 288 \text{ K}$$

$$= 2000 \, m\{460 - 288\} \text{J} = 3.44 \times 10^5 m \text{ J}$$

Energy supplied in 30 minutes by the pool fire $= 30 \times 10^6 \times 1800$ J
Equating:

$$m = 30 \times 10^6 \times 1800/\{3.44 \times 10^5\} \text{ kg} = 1.6 \times 10^5 \text{kg (160 tonne)}$$

N.B.: Our having disregarded the thermal resistance of the tank itself involves an error on the side of safety. Note also the comment in the Appendix on the Canvey and Rijnmond studies that a full vessel is in some respects less of a hazard than a half-full one.

5. $\qquad T(t) = 650 \text{ K}, \ T_i = 298 \text{ K} \ \Rightarrow \ T(t) - T_i = 352 \text{ K}$
Rearranging the equation for the semi-infinite solid given in the question, and substituting,

$$t = 321 \text{ s (about 5 minutes)}$$

Now other things being equal the time depends on $1/q^2$, hence:

$$t_1 q_1^2 = t_2 q_2^2$$

$$\Downarrow$$

$$\{q(\text{received})/q(\text{total})\} = \{5/42\}^{0.5}$$

$$q(\text{received}) = 48 \text{ kW m}^{-2} \text{ (about a third of the total)}$$

6. Frequency of derailment, leakage of contents and jet fire burning with the first car $= 0.2 \times 0.25 \times 5 \times 10^{-3}$ per year $= 2.5 \times 10^{-4}$ per year.

Frequency with which this will occur and the paired car also derail $= 0.6 \times 2.5 \times 10^{-4}$ per year $= 1.5 \times 10^{-4}$ per year.

Frequency with which this will occur and the paired car derail without leakage of its contents $= (1 - 0.2) \times 1.5 \times 10^{-4}$ per year $= 1.2 \times 10^{-4}$ per year.

The jet fire can be vertically up, or horizontal N, S, E or W. It cannot be vertically down in these circumstances. One of these five directions will be taken to be that of the second tank car, hence the frequency of a scenario whereby:

one tank car derails

the contents leak

there is a jet fire

the neighbouring tank car derails without leaking its contents

the jet fire from the first derailed car torches the second is:

$$1.2 \times 10^{-4} \times (1/5) \text{ per year} = 2.4 \times 10^{-5} \text{ per year}$$

We assume that if the second tank car is torched eventual leakage of the contents to form a fireball is inevitable.

7. We equate the ratio of sizes to that of quantities, and assume that the process of ammonium nitrate explosion began when the respective ships started to become heated by previously relatively mild burning. Letting the amount of ammonium nitrate on the second ship be M ton:

$$(1.25/7.2) = (2300/M)^2 \Rightarrow M = 5500 \text{ ton approx.}$$

i.e., about twice as much as was on the first ship. The author has been unable to verify this answer; the amount of ammonium nitrate on the second ship appears not to be known.

8. The following thermal influences operate:

(i) Radiation of heat from the cover of the tank to the surroundings:

(rate q_1)

(ii) Transfer of heat from the atmosphere to the tank cover by convection:

(rate q_2, convection coefficient $h = 20 \text{ W m}^{-2}\text{K}^{-1}$)

The following approximations and assumptions are made:

• Since the vessel lid is heavily insulated underneath it will be assumed that conduction through the lid is negligible.
• For heat balance purposes we treat the sky as a black body 10 K below the temperature of the earth's surface to which the heat balance relates.
• The lid is 'non-gray', being highly reflective towards most of what it receives from the sun but having a high emissivity when itself radiating at long wavelength.

In formulating an expression for q_1, we have to take the view factor into account, viz.:

$$q_1 = A\sigma\varepsilon F_{\text{ground-sky}}(T_{\text{cover}}^4 - T_{\text{sky}}^4)$$

where A is the area (m^2) of the radiating source, σ is the Stefan–Boltzmann constant, ε is the emissivity and $F_{\text{ground-sky}}$ the appropriate view factor. We are told that the cover is 'black', therefore:

$$\varepsilon = 1$$

Now to obtain a value for the view factor, we treat the tank cover and the sky as parallel surfaces of unequal area. For 2 surfaces there have to be $2^2 = 4$ view factors, easily characterised as follows:

$$F_{\text{ground-ground}} = 0$$

since the ground, being flat, does not 'see itself', likewise the sky which is being treated as a flat emissive surface, hence:

$$F_{\text{sky-sky}} = 0$$

Now $F_{\text{sky-ground}}$ can also be taken to be zero; the sky is so expansive in comparison with the cover of the tank that any radiation from it has an infinitesimal likelihood of striking the cover, therefore, from:

$$\sum F = 1.0 \implies F_{\text{ground-sky}} = 1$$

So the equation for radiation becomes:

$$q_1 = A\sigma(T^4_{\text{cover}} - T^4_{\text{sky}})$$

Now under steady conditions:

$$q_1 = q_2$$

We therefore have two equations to combine and solve. These are:

$$q_1 = A\sigma(T^4_{\text{cover}} - T^4_{\text{sky}})$$

and

$$q_2 = hA(T_{\text{ambient}} - T_{\text{cover}})$$

Combining these, having regard to the fact that conditions are steady, and inserting the given value for h ($20\,\text{W}\,\text{m}^{-2}\text{K}^{-1}$), the value for the Stefan–Boltzmann constant ($5.7 \times 10^{-8}\,\text{W}\,\text{m}^{-2}\text{K}^{-1}$), a value of 313 K for T_{ambient} and a value of 303 K for the sky temperature gives:

$$2.85 \times 10^{-9}\,T^4_{\text{cover}} + T_{\text{cover}} = 337.02$$

The above equation has to be solved by trial and error, yielding a solution of 310.5 K (37.5°C). The hydrocarbon would indeed boil if brought into contact with the surface.

9. We use the same correlation that was used for ammonium nitrate in the main text to give:

$$\text{time for explosion} = 21 \times (2.76/0.91)^2 \text{ days} = 193 \text{ days}$$

Reference [13] gives a value for the time to ignition, calculated by a more precise method, of the 2.76 m radius cylinder of 161 days. Though order-of-magnitude agreement is there, it should be noted with caution that the very simple treatment above errs on the hazardous side.

10. On the basis of the 'non-gray body model', the following thermal influences operate:

(a) heat transfer q_1 (W) from the outside surface of the cover to the sky by radiation.

(b) heat transfer from the q_2 (W) from the air to the outside of the cover by convection.

In considering (a) it has to be remembered that since there are two 'surfaces' – the cover and the sky – there are $2^2 = 4$ view factors. This follows from basic radiation heat transfer principles. It is shown in the main text as well as in the solution to question 8, and will therefore not be repeated here, that in an application such as this the equation is simply:

$$q_1 = A\sigma\varepsilon(T_{cover}^4 - T_{sky}^4)$$

where A is the cover area, ε the emissivity of the cover material and σ the Stefan–Boltzmann constant $(5.7 \times 10^{-8}\,W\,m^{-2}K^{-4})$; the subscripts on the temperatures are taken to be self-explanatory. If conditions approximate to being steady, the heat balance equation is then:

$$A\sigma\varepsilon\,(T_{cover}^4 - T_{sky}^4) = hA(T_o - T_{cover})$$

$$\Downarrow$$

$$\sigma\varepsilon\,(T_{cover}^4 - T_{sky}^4) = h(T_o - T_{cover})$$

The air is at 0°C (273 K). The ground temperature will be close to that of the air, so we take the sky to be at 258 K. Substituting, recalling that $\varepsilon = 0.9$, gives:

$$5.13 \times 10^{-9} T_{cover}^4 + T_{cover} = 295.73$$

This has to be solved by trial and error, to give:

$$T_{cover} = 268.9\,K\ (-4.1°C)$$

11. The heat balance equation becomes:

$$5.13 \times 10^{-9} T_{cover}^4 + T_{cover} = 361.24$$

yielding the trial-and-error solution:

$$T_{cover} = 312.4 \, K \, (39.4°C)$$

12.
$$\delta_{crit} = \frac{r_0^2 Q E \sigma A \exp(-E/RT_0)}{kRT_0^2} = 2.57$$

$$\text{for } T_0 = (T_0)_{crit} = 318 \, K$$

Substituting:
$$r_0 = 165 \, m \Rightarrow \text{cube side} = 330 \, m$$

13. Rate of heat loss by the water = Rate of heat uptake by LNG
$$40 \times 4200 \times 35 = m \times 512\,000$$
where m is the rate of evaporation (kg s^{-1}) assuming 100% efficiency.
$$\Downarrow$$
$$m = 11.5 \, kg \, s^{-1}$$

Actual rate of evaporation = 9.2 kg s^{-1}, 800 tonne day^{-1}

In designing the SCV a heat transfer coefficient and an area appropriate to the LMTD and the heat exchange rate would have to be factored in and would be different from those for the conventional exchanger.

14. 1150 BTU per cubic foot = 43.3 MJ m^{-3}
Now 1 cubic metre of gas at 1 bar, 15°C contains 40 moles.
Let the proportion of ethane (fractional basis) in the gas be x:
$$40[1560x + 889(1-x)] \times 10^{-3} = 43.3 \rightarrow x = 0.29$$

15. From steam tables:
enthalpy of 1 kg of saturated steam at 1 bar, dryness fraction unity
$$= 2675.8 \, kJ$$
enthalpy of 1 kg of water vapour at 30°C = 2555.7 kJ
enthalpy of 1 kg of liquid water at 30°C = 125.7 kJ
enthalpy of the steam after turbine exit
$$= [(0.68 \times 2555.7) + (0.32 \times 125.7)] \, kJ$$
$$= 1778.1 \, kJ$$
work done per kg steam = (2675.8 – 1778.1) kJ = 897.7 kJ
enthalpy of 1 kg of liquid water at 25°C = 104.8 kJ
enthalpy required to raise 1 kg of the steam = (2675.8 – 104.8) kJ = 2571 kJ
efficiency = work out/heat in = (897.7/2571) × 100% = 35%

$$100\,000 \text{ horse power} = 74\,570 \text{ kW}$$

rate of steam supply = $74\,570$ kJ s^{-1}/897.7 kJ kg^{-1}) = 83 kg s^{-1} of steam or approximately 7000 tonne for 24 hours of operation.

16. The hole area is:

$$\pi \times 0.009^2 \text{ m}^2 = 2.5 \times 10^{-4} \text{ m}^2$$

The molar mass of propane is 0.044 kg mol^{-1}.

Putting $P = 9 \times 10^5$ N m^{-2}, R = 8.314 J K^{-1}mol^{-1} and T = 298K gives on substitution:

$$Q = 0.6 \text{ kg s}^{-1}$$

Chapter 10

1. We saw in the main text that 1 m^3 of gas at room temperature and atmospheric pressure contains about 40 mol. Hence the amount of air in the room is:

$$20 \times 15 \times 6.5 \times 40 \text{ mol} = 78000 \text{ mol}$$

$$\text{moles of chlorine leaked} = 5/71 = 0.070$$

$$\text{p.p.m. chlorine} = (0.07/78000) \times 10^6 = 0.9$$

The TLV is therefore approached but not exceeded.

2. The probit equation for chlorine deaths (equation 10.1) is:

$$Y = -8.29 + 0.92 \ln\{C^2 t_i\}$$

Putting $C = 10$ p.p.m. and $t_i = 300$ minutes gives:

$$Y = 1.2$$

This indicates a nil death rate.

3. Using equation 10.1:

$$Y = -8.29 + 0.92 \ln\{C^2 t_i\} = 3.52 \text{ (corresponding to 7\%)}$$

$$\text{Putting } t_i = 30 \text{ minutes, } C = 112 \text{ p.p.m.}$$

4. $Y = -26.36 + 2.854 \ln\{C \times 30\} = 3.52 \Rightarrow C = 1175$ p.p.m. (0.1%)

5. Clausius–Clapeyron equation:

$$\frac{\mathrm{d}(\ln P)}{\mathrm{d}(1/T)} = -\frac{\Delta H_{vap}}{R}$$

(symbols as defined in the main text). Now for HF the normal boiling point is 292 K and the heat of vaporisation is 31.3 kJ mol^{-1}. Substituting:

$$\int_{1bar}^{P} d(\ln P) = -(\Delta H_{vap} / R) \int_{292K}^{313K} d(1/T)$$

where P is the vapour pressure at 313 K. Integrating:

$$\ln(P) = -(\Delta H_{vap}/R)\{(1/313) - (1/292)\}$$

from which:

$$P = 2.4 \text{ bar}$$

6. Using the Antoine equation:

$$\log P^* = A - \frac{B}{T + C} \Rightarrow P^* = 17.4 \text{ mm Hg } (0.023 \text{ bar})$$

$$P_{total} = n_{total}RT/V$$

$$P_{benzene} = n_{benzene}RT/V$$

Therefore in a total pressure of 1 bar:

$$n_{benzene}/n_{total} = 0.023 \equiv 23\,000 \text{ p.p.m. } (2.3\%)$$

7. Substituting into the probit equation for CO gives:

$$Y = 3.55 \equiv 7\% \text{ approx., so 10 deaths}$$

8. $\quad CH_3COCH_3 + 4O_2 (+ 15.0N_2) \rightarrow 3CO_2 + 3H_2O (+ 15.0N_2)$

The flash point occurs at half-stoichiometric, hence the proportion is:

$$[0.5/(1 + 4 + 15)] = 0.025 = 25000 \text{ p.p.m.}$$

There is no ignition hazard at the odour threshold.

9. $\qquad\qquad\qquad\qquad P_{total} = n_{total}RT/V$

$$P_{styrene} = n_{styrene}RT/V$$

$$(n_{styrene}/n_{total}) = (P_{styrene}/P_{total}) = (10/760)$$

$$= 0.013 \equiv 13\,000 \text{ p.p.m.}$$

The factor by which the vapour must be kept below its equilibrium vapour pressure by ventilation is therefore:

$$(13\,000/100) = 130$$

10. Applying the ideal gas equation, the number of moles of VCM in the space inside the tank is:

$$(2.8 \times 10^5 \times 0.1)/(8.314 \times 288) = 11.7\,\text{mol}$$

This previously occupies a space of $0.1\,\text{m}^3$, so after dispersion it occupies a space of $100\,\text{m}^3$, containing ≈ 4000 moles of gas.

p.p.m. of VCM $= (11.7/4000) \times 10^6 = 2925$, above the odour threshold.

11. $5\% \equiv 50\,000\,\text{p.p.m.}$ For fractional extent of reaction 0.2
$$\Downarrow \textit{on reaction}$$

$$10\,000\,\text{p.p.m. HBr}$$
$$10\,000\,\text{p.p.m. HCl}$$
$$20\,000\,\text{p.p.m. HF}$$
$$40\,000\,\text{p.p.m. BCF}$$

When diluted these concentrations become:

$$10\,\text{p.p.m. HBr}$$
$$10\,\text{p.p.m. HCl}$$
$$20\,\text{p.p.m. HF}$$
$$40\,\text{p.p.m. BCF}$$

12. 10% corresponds to a probit of 3.72, therefore

$$Y = 3.72 = -37.98 + 3.7\ln(Ct)$$

Putting $C = 2805$ p.p.m. gives $t = 28$ minutes

13. $C_nH_{2n} + 1.5nO_2\,(+\,5.6nN_2) \rightarrow nCO_2 + nH_2O + (+\,5.6nN_2)$
$$\text{Proportion at half-stoichiometric} = 0.5/(1 + 7.1n) = 0.05$$
$$9 = 7.1n \quad n = 1.3 \quad \text{giving } C_{1.3}H_{2.6}$$

It is at first consideration surprising that a vapour at its LFL should cause a fatal injury, less so when it is known that when a closed cup flash point apparatus is in inexperienced hands the cup and lid can both be 'lifted'. The author in his laboratory instructing duties has seen this many times!

Chapter II

1. Setting of the waste stream flow rate $= 0.5\,\text{MW}/(100\,\text{MJ m}^{-3})$

$$= 5 \times 10^{-3}\,\text{m}^3\,\text{s}^{-1}$$

In the event that this drops by half, the thermal delivery becomes 0.25 MW, and must be brought at least up to 0.425 MW by the natural gas, therefore, letting x be the natural gas flow rate in $m^3\ s^{-1}$:

$$[(2.5 \times 10^{-3}) \times 100] + 37x = 0.425 \Rightarrow x = 4.7 \times 10^{-3}\ m^3\ s^{-1}$$

This flow rate of gas will maintain the thermal delivery at the minimum value for safe operation.

N.B.: This approach considers only the thermal delivery. In substitution of one gas or gas mixture for another on a flare or burner there is sometimes the additional issue of matching flame speeds in order for the substitution to work. In this particular example, that might mean 'pepping up' the relatively unreactive natural gas with some hydrogen to increase its flame speed.

2. Assign the hydrocarbon a heat value of $45\ MJ\ kg^{-1}$:

$$\text{heat diverted per day to steam raising}$$

$$= 12.5 \times 10^3 \times 45 \times 10^6\ J = 5.6 \times 10^{11}\ J$$

$$\text{amount of steam generated per day}$$

$$= (5.6 \times 10^{11}/40.6 \times 10^3)\ mol \equiv 249\ tonne$$

3. Returning to the definition of Reynolds number:

$$Re = ud\sigma_j/\mu$$

where u = speed, d = diameter and μ = dynamic viscosity of the hydrocarbon. The ratio of densities of the hydrocarbon vapour and air can be taken to be the ratio of their molar weights, scaled to allow for the pressure difference, viz.:

$$\sigma_j/\sigma_a = 5 \times 142/28.8 = 25 \Rightarrow (Re)_{minimum} = 3.8 \times 10^5$$

The dynamic viscosity of n-decane vapour at the reference temperature is $6 \times 10^{-6}\ kg\,m^{-1}s^{-1}$. The density is easily estimated by approximating it to an ideal gas at 5 bar, 528 k giving a value of $16\ kg\,m^{-3}$, from which:

$$u_{minimum} = 22\ m\,s^{-1}$$

4.   Heat of combustion of pure propane $= 2217 \times 42 \times 10^{-3}\ MJ\ m^{-3}$
$$= 93\ MJ\ m^{-3}$$
Dilution with the effluent process gas therefore has to be
by a factor $(93/8) = 11.6$

1 molar unit of propane must be blended with 10.6 molar
units of the process gas

5. The heat value of the effluent gas is 0.18×282 kJ mol^{-1}. The heat released by 42 mol of the blended gas must be equal to 10 MJ (10 000 kJ), hence:

$$42\{x[0.18 \times 282] + 889(1 - x)\} = 10000$$

where x is the proportion of the original effluent in the blend.
Solving:

$$x = 0.78$$

Hence the gases must be blended in the proportions 0.78 effluent to 0.22 methane or equivalently:

$$3.5 \text{ effluent to 1 methane}$$

6. Number of moles in 1 m^3 of the gas at 1 atm.,

$$288 \text{ K} = (1 \times 10^5 \times 1)/(8.314 \times 288) = 42$$

Heat released on burning 1 m^3 of the effluent gas

$$= \{[0.03 \times 2874] + [0.02 \times 2217] + [0.03 \times 1409]\} \; 42 \times 10^{-3} \text{ MJ}$$

$$= 7.3 \text{ MJ m}^{-3}$$

Will burn without adding natural gas

7. Without recuperation, burning for one minute requires:

$$15 \times (0.11/0.89) \text{ m}^3 \text{ of natural gas} = 1.854 \text{ m}^3$$

With recuperation, burning for one minute requires:

$$15 \times (0.09/0.91) \text{ m}^3 \text{ of natural gas} = 1.484 \text{ m}^3$$

$$\text{saving in 1 minute} = 0.370 \text{ m}^3$$

annual saving at the rate of burning given in the question

$$= 23100 \text{ m}^3 \text{ approx}$$

8. 50 m^3 per minute of the effluent gas $\approx$2000 mol per minute. Quantity of methanol passed into the column per minute $= 2000 \times (35/10^6)$ mol $= 0.07$ mol.

Now 0.07 mol $\equiv 4.21 \times 10^{22}$ molecules, occupying, if a monolayer:

$$4.21 \times 10^{22} \times 0.17 \times 10^{-18} \text{ m}^2 = 7164 \text{ m}^2 \text{ or } (7164/1000) = 7.2 \text{ g carbon}$$

Life expectancy of the column is therefore $(300000/7.2)$ minutes $= 41876$ minutes $= 29$ days.

9. Let there be x g of the pellets and y g of the oil per kg of the final product.

$$\{10^{-3}x \times 17\} + \{10^{-3}y \times 44\} = 21$$

$$x + y = 1000$$

$$17x + 44y = 21000$$

$$17x + 44(1000 - x) = 21000 \implies x = 852 \, \text{g} \; y = 148 \, \text{g}$$

10. The empirical formula is $CH_{1.85}$, and if the steam gasification is carried out so as to produce a preponderance of carbon dioxide to the exclusion of the monoxide the chemical equation is:

$$CH_{1.85} + 2H_2O \rightarrow 2.925H_2 + CO_2$$

A cubic metre of this gas at room temperature and pressure has a heat value of:

$$(2.925/3.925) \times 40 \times 0.285 \, \text{MJ} = 8.5 \, \text{MJ}$$

Methanation proceeds according to:

$$CO_2 + 4H_2 \rightarrow CH_4 + 2H_2O \, \text{(liq.)}$$

That is, one molar unit of carbon dioxide is replaced by one of methane, the product water going to liquid. Hence, if the number of moles of methane required is x:

$$[(2.925 / 3.925) \times 40 \times 0.285] + [(x / 3.925) \times 40 \times 0.889] \, \text{MJ} = 12.5 \, \text{MJ}$$

$$\Downarrow$$

$$x = 0.44$$

That is, methanation is required to the extent of 44%, and the overall chemical balance is:

$$CH_{1.85} + 1.12H_2O \rightarrow 1.165H_2 + 0.56CO_2 + 0.44CH_4$$

11. Without recuperation:

  Influx gas consists of 0.89 parts waste gas + 0.11 parts natural gas
With recuperation:

  Influx gas consists of 0.91 parts waste gas + 0.09 parts natural gas

  Now $15 \, \text{m}^3$ of the waste gas are incinerated per minute, requiring:
Without recuperation:

  $(15 \times 0.11/0.89) \, \text{m}^3$ of natural gas $= 1.854 \, \text{m}^3$ of natural gas
With recuperation:

  $(15 \times 0.09/0.91) \, \text{m}^3$ of natural gas $= 1.484 \, \text{m}^3$ of natural gas

$$\text{saving} = (1.854 - 1.484) \, \text{m}^3 \, \text{per minute} = 0.370 \, \text{m}^3 \, \text{per minute}$$
$$\equiv 1.95 \times 10^5 \, \text{m}^3 \, \text{per year}$$
$$\text{energy saving} = 7.2 \times 10^6 \, \text{MJ}$$

$US2.00 per million BTU $\equiv$ $US2.00 per 1058 MJ
= $US0.00189 per MJ

cost saving = $US 13.6 k

This has to be compared with the cost of the recuperating plant in order to assess over how many years this would pay for itself.

12. 100 h.p. = 134 kW requiring (134/0.3) = 450 kW of heat.

Required rate of supply

= 450 kJ s^{-1}/(40 000 kJ kg^{-1}) × 3600 × 24 × 365 × 10^{-3} tonne per year

= 350 tonne per year.

The annual rate of production of waste oil in the UK is almost exactly two orders of magnitude higher than the above, so 100 such engines could be powered!

13. $$CH_{1.1} + H_2O \rightarrow CO + 1.55H_2$$
Composition of the gas CO (1/2.55) × 100% = 39%, H_2 61%

The above is less suitable for use as synthesis gas than that prepared from 'CH$_2$'. It can however be adjusted in composition by the well known water gas shift reaction:

$$CO + H_2O \rightarrow CO_2 + H_2$$

Chapter 12

1. Heat released by the gas on combustion
$$= 7.2 \times 60 \times 10^6 \text{ J} = 4.3 \times 10^8 \text{ J}$$

If the temperature is to be limited to 450°C, the temperature rise has to be limited to 425°C, whereupon:

$$450 = 4.3 \times 10^8/C$$

where C (J K^{-1}) is the total heat capacity of the post-reaction mixture. This gives:

$$C = 9.6 \times 10^5 \text{ J K}^{-1}$$

Now 7.2 m^3 of the gas leads to 131 m^3 of the combustion products, 5240 mol. Specific heat of the products given by:

$$\{[(2/18.2) \times 850] + [(3/18.2) \times 1900] + [(13.2/18.2) \times 1040]\} \text{ J kg}^{-1}\text{K}^{-1}$$
$$= 1160 \text{ J kg}^{-1}\text{K}^{-1}$$

One molar unit of the combustion products has weight:

$$\{[(2/18.2)\times 44]+[(3/18.2)\times 18]+[(13.2/18.2)\times 28]\}\,g = 28.1\,g$$

hence the specific heat in molar terms becomes:

$$= 1160\,J\,kg^{-1}\,K^{-1}\times 0.0281\,kg\,mol^{-1} = 32.6\,J\,mol^{-1}\,K^{-1}$$

$$\text{heat capacity of the products} = 32.6\,J\,mol^{-1}K^{-1}\times 5240\,mol$$

$$= 1.7\times 10^{5}\,J\,K^{-1}$$

$$\text{extra heat capacity required} = (9.6 - 1.7)\times 10^{5}\,J\,K^{-1}$$

$$= 7.9\times 10^{5}\,J\,K^{-1}, \text{ obtainable from:}$$

$$(7.9\times 10^{5}\,JK^{-1}/28.8\,J\,mol^{-1}K^{-1})\text{ mol of air, }2.74\times 10^{4}\,mol, 685\,m^{3}$$

This is the amount of excess air required. That required for combustion is:

$$7.2\times (3.5 + 13.2)\,m^{3} = 120\,m^{3}$$

$$\text{total air requirement} = 805\,m^{3}$$

2. 104 trillion cubic feet $= 1.04\times 10^{14}\times 0.028\,m^{3}$ capable of releasing when burnt:

$$1.04\times 10^{14}\times 0.028\,m^{3}\times 37\times 10^{6}\,J = 10^{20}\,J$$

Now a barrel of oil releases about 6 GJ when burnt, so this amount of heat could be released by:

$$10^{20}/(6\times 10^{9})\,\text{barrels} = 18\,\text{billion barrels.}$$

Appendix (Canvey)

1. Using the Clausius–Clapeyron equation:

$$d\ln P / dT = \Delta H / RT^{2}$$

$$\Downarrow$$

$$\int_{P^{0}}^{P} d\ln P = (\Delta H / R)\int_{T_{b}}^{T} 1/T^{2}\ dT$$

where P^{0} denotes atmospheric pressure, T_{b} the normal boiling point. Integrating,

$$\ln(P / P^\circ) = -(\Delta H / R)\{(1/T) - (1/T_b)\} = 2.17$$
$$P = P^\circ e^{2.17} = 9 \, \text{bar}$$

2. $NH_3 = 17 \, \text{g mol}^{-1}$

Unit mass of ammonia gas is, in the conditions of the question, accompanied by unit mass of ammonia liquid of negligible volume, hence the density is twice that of the gas only, i.e.:

$$\sigma_o = 1.38 \, \text{kg m}^{-3}$$

Retaining the symbols and subscripts used previously:

$$(\sigma_o - \sigma_a) / \sigma_o = (1.38 - 1.17) / 1.38 = 0.15$$

Using the equation:

$$R^2 - R_o^2 = 2c_E t \sqrt{(\sigma_o - \sigma_a) g V_o / \pi \sigma_o}$$

$$(V_o / \pi) = R_o^2 H_o = 900 \, \text{m}^3 \text{as before}$$

$$c_E = 1 \text{ as before}$$

$$\Downarrow$$

$$R^2(10 \, \text{s}) = 225 + 2 \times 10 \left\{ \sqrt{0.15 \times 900 \times 9.81} \right\} \Rightarrow R(10 \, \text{s}) = 31 \, \text{m}$$

$$R^2(180 \, \text{s}) = 225 + 2 \times 180 \left\{ \sqrt{[0.15 \times 900 \times 9.81]} \right\}$$

$$R(180 \, \text{s}) = 115 \, \text{m}$$

Transformation of percentages to probits

%	0	1	2	3	4	5	6	7	8	9
0	–	2.67	2.95	3.12	3.25	3.36	3.45	3.52	3.59	3.66
10	3.72	3.77	3.82	3.87	3.92	3.96	4.01	4.05	4.08	4.12
20	4.16	4.19	4.23	4.26	4.29	4.33	4.36	4.39	4.42	4.45
30	4.48	4.50	4.53	4.56	4.59	4.61	4.64	4.67	4.69	4.72
40	4.75	4.77	4.80	4.82	4.85	4.87	4.90	4.92	4.95	4.97
50	5.00	5.03	5.05	5.08	5.10	5.13	5.15	5.18	5.20	5.23
60	5.25	5.28	5.31	5.33	5.36	5.39	5.41	5.44	5.47	5.50
70	5.52	5.55	5.58	5.61	5.64	5.67	5.71	5.74	5.77	5.81
80	5.84	5.88	5.92	5.95	5.99	6.04	6.08	6.13	6.18	6.23
90	6.28	6.34	6.41	6.48	6.55	6.64	6.75	6.88	7.05	7.33
–	0.0	0.1	0.2	0.3	0.4	0.5	0.6	0.7	0.8	0.9
99	7.33	7.37	7.41	7.46	7.51	7.58	7.65	7.75	7.88	8.09

Reproduced from: Finney D.J. Probit Analysis Cambridge University Press (1971) with the permission of the publisher.

Type K thermocouple table *Nickel–chromium/nickel–aluminium, electromotive force as a function of temperature*

	E/µV										
T/°C	0	1	2	3	4	5	6	7	8	9	T/°C
0	0	39	79	119	158	198	238	277	317	357	0
10	397	437	477	517	557	597	637	677	718	758	10
20	798	838	879	919	960	1000	1041	1081	1122	1163	20
30	1203	1244	1285	1326	1366	1407	1448	1489	1530	1571	30
40	1612	1653	1694	1735	1776	1817	1858	1899	1941	1982	40
50	2023	2064	2106	2147	2188	2230	2271	2312	2354	2395	50
60	2436	2478	2519	2561	2602	2644	2685	2727	2768	2810	60
70	2851	2893	2934	2976	3017	3059	3100	3142	3184	3225	70
80	3267	3308	3350	3391	3433	3474	3516	3557	3599	3640	80
90	3682	3723	3765	3806	3848	3889	3931	3972	4013	4055	90
100	4096	4138	4179	4220	4262	4303	4344	4385	4427	4468	100
110	4509	4550	4591	4633	4674	4715	4756	4797	4838	4879	110
120	4920	4961	5002	5043	5084	5124	5165	5206	5247	5288	120
130	5328	5369	5410	5450	5491	5532	5572	5613	5653	5694	130
140	5735	5775	5815	5856	5896	5937	5977	6017	6058	6098	140
150	6138	6179	6219	6259	6299	6340	6380	6420	6460	6500	150
160	6540	6580	6620	6660	6701	6741	6781	6821	6861	6901	160
170	6941	6981	7021	7060	7100	7140	7180	7220	7260	7300	170
180	7340	7380	7420	7460	7500	7540	7579	7619	7659	7699	180
190	7739	7779	7819	7859	7899	7939	7979	8019	8059	8099	190
200	8138	8178	8218	8258	8298	8338	8378	8418	8458	8499	200
210	8539	8579	8619	8659	8699	8739	8779	8819	8860	8900	210
220	8940	8980	9020	9061	9101	9141	9181	9222	9262	9302	220
230	9343	9383	9423	9464	9504	9545	9585	9626	9666	9707	230
240	9747	9788	9828	9869	9909	9950	9991	10031	10072	10113	240
250	10153	10194	10235	10276	10316	10357	10398	10439	10480	10520	250
260	10561	10602	10643	10684	10725	10766	10807	10848	10889	10930	260
270	10971	11012	11053	11094	11135	11176	11217	11259	11300	11341	270
280	11382	11423	11465	11506	11547	11588	11630	11671	11712	11753	280
290	11795	11836	11877	11919	11960	12001	12043	12084	12126	12167	290
300	12209	12250	12291	12333	12374	12416	12457	12499	12540	12582	300
310	12624	12665	12707	12748	12790	12831	12873	12915	12956	12998	310
320	13040	13081	13123	13165	13206	13248	13290	13331	13373	13415	320
330	13457	13498	13540	13582	13624	13665	13707	13749	13791	13833	330
340	13874	13916	13958	14000	14042	14084	14126	14167	14209	14251	340
350	14293	14335	14377	14419	14461	14503	14545	14587	14629	14671	350
360	14713	14755	14797	14839	14881	14923	14965	15007	15049	15091	360
370	15133	15175	15217	15259	15301	15343	15385	15427	15469	15511	370
380	15554	15596	15638	15680	15722	15764	15806	15849	15891	15933	380
390	15975	16017	16059	16102	16144	16186	16228	16270	16313	16355	390
400	16397	16439	16482	16524	16566	16608	16651	16693	16735	16778	400
410	16820	16862	16904	16947	16989	17031	17074	17116	17158	17201	410
420	17243	17285	17328	17370	17413	17455	17497	17540	17582	17624	420
430	17667	17709	17752	17794	17837	17879	17921	17964	18006	18049	430
440	18091	18134	18176	18218	18261	18303	18346	18388	18431	18473	440

Type K thermocouple table   (*continued*)

T/°C	0	1	2	3	4	5	6	7	8	9	T/°C
					E/μV						
450	18516	18558	18601	18643	18686	18728	18771	18813	18856	18898	450
460	18941	18983	19026	19068	19111	19154	19196	19239	19281	19324	460
470	19366	19409	19451	19494	19537	19579	19622	19664	19707	19750	470
480	19792	19835	19877	19920	19962	20005	20048	20090	20133	20175	480
490	20218	20261	20303	20346	20389	20431	20474	20516	20559	20602	490
500	20644	20687	20730	20772	20815	20857	20900	20943	20985	21028	500
510	21071	21113	21156	21199	21241	21284	21326	21369	21412	21454	510
520	21497	21540	21582	21625	21668	21710	21753	21796	21838	21881	520
530	21924	21966	22009	22052	22094	22137	22179	22222	22265	22307	530
540	22350	22393	22435	22478	22521	22563	22606	22649	22691	22734	540
550	22776	22819	22862	22904	22947	22990	23032	23075	23117	23160	550
560	23203	23245	23288	23331	23373	23416	23458	23501	23544	23586	560
570	23629	23671	23714	23757	23799	23842	23884	23927	23970	24012	570
580	24055	24097	24140	24182	24225	24267	24310	24353	24395	24438	580
590	24480	24523	24565	24608	24650	24693	24735	24778	24820	24863	590
600	24905	24948	24990	25033	25075	25118	25160	25203	25245	25288	600
610	25330	25373	25415	25458	25500	25543	25585	25627	25670	25712	610
620	25755	25797	25840	25882	25924	25967	26009	26052	26094	26136	620
630	26179	26221	26263	26306	26348	26390	26433	26475	26517	26560	630
640	26602	26644	26687	26729	26771	26814	26856	26898	26940	26983	640
650	27025	27067	27109	27152	27194	27236	27278	27320	27363	27405	650
660	27447	27489	27531	27574	27616	27658	27700	27742	27784	27826	660
670	27869	27911	27953	27995	28037	28079	28121	28163	28205	28247	670
680	28289	28332	28374	28416	28458	28500	28542	28584	28626	28668	680
690	28710	28752	28794	28835	28877	28919	28961	29003	29045	29087	690
700	29129	29171	29213	29255	29297	29338	29380	29422	29464	29506	700
710	29548	29589	29631	29673	29715	29757	29798	29840	29882	29924	710
720	29965	30007	30049	30090	30132	30174	30216	30257	30299	30341	720
730	30382	30424	30466	30507	30549	30590	30632	30674	30715	30757	730
740	30798	30840	30881	30923	30964	31006	31047	31089	31130	31172	740
750	31213	31255	31296	31338	31379	31421	31462	31504	31545	31586	750
760	31628	31669	31710	31752	31793	31834	31876	31917	31958	32000	760
770	32041	32082	32124	32165	32206	32247	32289	32330	32371	32412	770
780	32453	32495	32536	32577	32618	32659	32700	32742	32783	32824	780
790	32865	32906	32947	32988	33029	33070	33111	33152	33193	33234	790
800	33275	33316	33357	33398	33439	33480	33521	33562	33603	33644	800
810	33685	33726	33767	33808	33848	33889	33930	33971	34012	34053	810
820	34093	34134	34175	34216	34257	34297	34338	34379	34420	34460	820
830	34501	34542	34582	34623	34664	34704	34745	34786	34826	34867	830
840	34908	34948	34989	35029	35070	35110	35151	35192	35232	35273	840
850	35313	35354	35394	35435	35475	35516	35556	35596	35637	35677	850
860	35718	35758	35798	35839	35879	35920	35960	36000	36041	36081	860
870	36121	36162	36202	36242	36282	36323	36363	36403	36443	36484	870
880	36524	36564	36604	36644	36685	36725	36765	36805	36845	36885	880
890	36925	36965	37006	37046	37086	37126	37166	37206	37246	37286	890

Type K thermocouple table   (*continued*)

T/°C	0	1	2	3	4	5	6	7	8	9	T/°C
					E/μV						
900	37326	37366	37406	37446	37486	37526	37566	37606	37646	37686	900
910	37725	37765	37805	37845	37885	37925	37965	38005	38044	38084	910
920	38124	38164	38204	38243	38283	38323	38363	38402	38442	38482	920
930	38522	38561	38601	38641	38680	38720	38760	38799	38839	38878	930
940	38918	38958	38997	39037	39076	39116	39155	39195	39235	39274	940
950	39314	39353	39393	39432	39471	39511	39550	39590	39629	39669	950
960	39708	39747	39787	39826	39866	39905	39944	39984	40023	40062	960
970	40101	40141	40180	40219	40259	40298	40337	40376	40415	40455	970
980	40494	40533	40572	40611	40651	40690	40729	40768	40807	40846	980
990	40885	40924	40963	41002	41042	41081	41120	41159	41198	41237	990
1000	41276	41315	41354	41393	41431	41470	41509	41548	41587	41626	1000
1010	41665	41704	41743	41781	41820	41859	41898	41937	41976	42014	1010
1020	42053	42092	42131	42169	42208	42247	42286	42324	42363	42402	1020
1030	42440	42479	42518	42556	42595	42633	42672	42711	42749	42788	1030
1040	42826	42865	42903	42942	42980	43019	43057	43096	43134	43173	1040
1050	43211	43250	43288	43327	43365	43403	43442	43480	43518	43557	1050
1060	43595	43633	43672	43710	43748	43787	43825	43863	43901	43940	1060
1070	43978	44016	44054	44092	44130	44169	44207	44245	44283	44321	1070
1080	44359	44397	44435	44473	44512	44550	44588	44626	44664	44702	1080
1090	44740	44778	44816	44853	44891	44929	44967	45005	45043	45081	1090
1100	45119	45157	45194	45232	45270	45308	45346	45383	45421	45459	1100
1110	45497	45534	45572	45610	45647	45685	45723	45760	45798	45836	1110
1120	45873	45911	45948	45986	46024	46061	46099	46136	46174	46211	1120
1130	46249	46286	46324	46361	46398	46436	46473	46511	46548	46585	1130
1140	46623	46660	46697	46735	46772	46809	46847	46884	46921	46958	1140
1150	46995	47033	47070	47107	47144	47181	47218	47256	47293	47330	1150
1160	47367	47404	47441	47478	47515	47552	47589	47626	47663	47700	1160
1170	47737	47774	47811	47848	47884	47921	47958	47995	48032	48069	1170
1180	48105	48142	48179	48216	48252	48289	48326	48363	48399	48436	1180
1190	48473	48509	48546	48582	48619	48656	48692	48729	48765	48802	1190

True/false questions

Introduction

Even in an advanced course of study where numerical calculations, design exercises and the like are expected to feature very centrally, there is the need for basic relevant *facts*. To present these in isolation from related numerical and design work would reduce a course to little more than a quiz game. Nevertheless, facts which are straightforward enough in themselves but previously unfamiliar to the student have to be learnt. The true/false questions are intended to help in this regard. They can be used to occupy a spare few minutes at the end of a lecture; in the author's experience, this often leads to very productive class discussion.

Questions

1. In a steam-powered device, heat where required is that released on condensation.

2. The Otto cycle is an example of a steam-driven cycle.

3. The first offshore oil/gas production was off the California coast in the 1940s.

4. 'Cracking' is the means by which molecular weights of hydrocarbons are adjusted.

5. Natural gas can be liquefied and transported long distances, in this form, by land or sea.

6. In single-phase flow, e.g., of natural gas through an orifice, critical conditions signify an independence of the flow rate on the upstream pressure.

7. The 'perfect gas in critical flow' model of discharge through an orifice assumes steady conditions.

8. The discharge coefficient in the equation for critical flow is an allowance for frictional effects.

9. γ for methane is 1.4.

10. In flow along a pipe, the Fanning friction factor is obtainable from the Reynolds number.

11. In Bernoulli's equation, flow work effects make no explicit appearance.

12. All petroleum distillates are less dense than water.

13. In two-phase flow, the 'mean density' of the two-phase hydrocarbon inventory is the simple arithmetic mean of the gas and liquid densities.

14. The drift distance for a cloud of hydrocarbon gas or vapour to drop below its lower flammability limit is a linear function of the wind speed.

15. A rapidly discharged quantity of gas or vapour tends to collapse from a tall cloud structure to a flatter one whilst retaining a horizontal top.

16. The length of a jet fire is, to a fair approximation, a function only of the fuel supply rate and not of orientation.

17. A jet fire is an example of a pre-mixed flame.

18. At an offshore installation any load-bearing member can be taken to fail instantly on contact with a jet fire.

19. A pool fire can itself, without 'escalation', cause death of persons.

20. Air is transparent to thermal radiation.

21. The Stefan–Boltzmann law is derivable from first principles by treating photons according to Bose–Einstein statistics.

22. The 'solid flame' model of thermal radiation from flames treats the flame as a black body at a single 'flame temperature'.

23. Typically, the proportion of combustion heat transferred as radiation from a hydrocarbon flame is in the range 0.2–0.4.

24. Pool fires have mass transfer rates from pool to flame which vary widely according to the combustion reactivity of the liquid.

25. Whether a pool fire is laminar or turbulent depends on the mass transfer rate from pool to flame.

26. The flame front in a pool fire has a low emissivity.

27. As well as substances which are liquids at room temperature, molten organic materials, e.g., molten plastics, will burn as pool fires.

28. The radiative flux from a flame depends strongly on the temperature of the surroundings.

29. All BLEVEs are fireballs, but vice versa is not true.

30. 'ALARP' means having a frequency of 10^{-6} per year or lower.

31. At an offshore installation, a fireball is the combustion behaviour expected when there is sudden leak of a large quantity of gas or two-phase inventory.

32. Overpressures accompany fireballs and jet fires only if there is significant confinement.

33. The thermal diffusivity can be viewed as the ratio of the capacity of a material to transmit heat to its capacity to store it.

34. Lewis and Prandtl numbers for gases tend to be of the order of ten.

35. The 1988 Piper Alpha disaster is believed to have resulted from hydrocarbon release at the gas compression module.

36. The blast energy due to a tonne of hydrocarbon in a v.c.e. is equivalent to that caused by about half a tonne of TNT.

37. LNG storage or transportation always requires refrigeration.

38. A hazard with LNG is the formation of layers of differing density through evaporation.

39. CNG is, in some parts of the world, used to power motor cars.

40. In troubleshooting plant utilising heavy fuel oil, the most likely cause of a fatal accident is electrocution.

41. Methane is the most reactive alkane.

42. Oxygenated hydrocarbons are of lower calorific value than the parent compound.

43. Oxygenated hydrocarbons tend to be of lower reactivity than the parent hydrocarbons.

44. The particles in smoke are not pure carbon but also contain some hydrogen.

45. The flash point of a pure hydrocarbon compound is calculable by stoichiometry from knowledge of its boiling point only.

46. An ignition source is always required for there to be a fire.

47. To be a fire hazard, a chemically reacting system must have a heat release of MJ (or tens of MJ) per kg of reactant.

48. Hydrocarbon liquids, including petroleum fractions, have heat values of about $25 \, MJ \, kg^{-1}$.

49. Crude oil contains significant amounts of alkenes (olefins).

50. Hydrocarbon leakages in industry frequently lead to detonations.

51. The 'adiabatic flame temperature' is that achieved when all heat released by the combustion goes into the products and remains there.

52. The value of the adiabatic flame temperature for a stoichiometric hydrocarbon–air system is typically 2000°C.

53. Excess fuel raises the adiabatic flame temperature.

54. The flash point of a pure hydrocarbon liquid occurs when the proportion of its vapour in the space above the liquid surface is approximately half-stoichiometric.

55. A leak of gas through an orifice is likely to result in a jet fire.

56. A flash fire involves significant overpressure.

57. A vapour cloud explosion may be either confined or unconfined.

58. At Flixborough both flash fire and vapour cloud explosion behaviour were observed.

59. Liquefied natural gas (LNG) can display BLEVE behaviour.

60. A BLEVE radiates as a black body.

61. A fatal BLEVE occurred in Sydney, Australia in 1989.

62. The 'semi-infinite solid' has one dimension much smaller than the other two.

63. The semi-infinite solid, as a means of representing non-steady conduction, has three possible sets of initial and boundary conditions.

64. Convection coefficients have the same units as thermal conductivity.

65. Probability and frequency are not interchangeable terms in risk assessment.

66. According to current philosophy, positive health benefits can accrue from employment with good H&S management.

67. All of the keywords in HAZOP will be relevant to every application.

68. A particular instruction might well involve several operations each needing its own HAZOP study.

69. The units $N\,mm^{-2}$, in which 'design stresses' are usually tabulated, are equivalent to MPa.

70. The same expressions for required vessel wall thickness apply whether or not there is a welded seam provided that any such seam is 'fully radiographed'.

71. The final answer in any calculation of pressure vessel wall thickness using approved equations involving internal pressure and design stress should always become, on implementation of the design, the actual wall thickness.

72. In a horizontal cylindrical tank, a domed end is less effective, in terms of withstanding pressure, than a flat end.

73. Liquids pumped along a pipeline in chemical processing usually have speeds on the order of $10\,m\,s^{-1}$.

74. Discharge pressures of pumps used in chemical plant are frequently of the order of tens of bar.

75. Pipes of the same schedule all have the same outer diameter.

76. In calculating the schedule number of a pipe, the material from which it is made has to be known.

77. In order to calculate the length of piping required to effect a particular degree of cooling of a fluid, the dynamic viscosity and the density are sufficient information with regard to the fluid properties.

78. At Flixborough, a gap between two reactors was bridged by a pipe having the same diameter as the vessel entry/exit.

79. At Flixborough, the pipe connecting the two reactors was deemed, as a straight pipe, able to withstand the pressure of its internal contents.

80. Mechanical support for the pipe which eventually failed is believed to have played a central part in the Flixborough accident.

81. Spent catalyst is a toxic, but not a thermal, hazard.

82. Hydrogenations of hydrocarbons are usually significantly endothermic.

83. Hydrogenations of hydrocarbons usually require a catalyst, e.g., palladium.

84. The odour threshold of elemental chlorine (Cl_2) is 5 p.p.m.

85. Reactions involving the nitration of hydrocarbons are significantly exothermic.

86. Chlorine is more toxic than ammonia in the same molar quantity.

87. Ammonia was present at Canvey Island at the time of the study.

88. Hydrogen fluoride (HF) has a critical temperature such that it cannot be stored as a liquid under its own vapour at ordinary temperatures.

89. Methyl isocyanate was the substance involved in the 1984 Bhopal accident.

90. Methyl isocyanate is an intermediate in the manufacture of certain synthetic fabrics.

91. Significant proportions of VCM are retained by PVC once the latter is produced in an autoclave.

92. There were once (circa 1960) serious proposals to use VCM as an anaesthetic for humans.

93. Acrylonitrile is used in the preparation of insecticides.

94. The odour threshold of acrylonitrile is 15–20 p.p.m.

95. Carbon tetrachloride and moist air are sufficient for the formation of phosgene.

96. Carbon tetrachloride currently finds wide use in the pharmaceutical and pesticide industries.

97. To paint the outside surface of a tank white is to ensure that solar radiation absorption is not excessive.

98. For heat balance purposes, the sky can be treated as a black body having a temperature 5–20 K below that of the local earth's surface.

99. The time taken for a pile of unstable material to explode can, with certain provisos, be taken to be proportional to the square of the dimension of the pile.

100. Stirring of a hydrocarbon liquid during processing often approximates to an adiabatic process.

101. Stirring of a hydrocarbon liquid requires about 10 kW per m^3 of liquid.

102. In the 1947 accident in the US when two ships in the same harbour, bearing ammonium nitrate, blew up within a few hours of each other, the first ship to blow up contained less ammonium nitrate than did the second.

103. Flaring at a production platform is usually intermittent.

104. Explosion of flares can result from air ingress to the pipes and valves.

105. At an oil production facility, on- or offshore, gas accompanying the oil is always collected and prepared for subsequent use.

106. The gas or vapour stream going to a flare might need to be supplemented with another flammable gas in order to maintain stable burning.

107. A gas needs to have a heat value of at least $1\,MJ\,m^{-3}$ in order to sustain a flame on a burner.

108. 'Recuperative burning' of waste hydrocarbon leads to economies in the amounts of supplementary gas required.

109. In operation of a heat exchanger, vibrations – either intrinsic or transmitted – during operation can cause structural damage.

110. Traditionally, stainless steels have been the most widely used materials for making heat exchangers.

111. Polymerisation of a hydrocarbon liquid during its residence time in a heat exchanger can lead to eventual blockage.

112. A previously unused heat exchanger can be taken to be free of significant corrosion.

113. In heat exchanger design, the simple mean temperature difference between the two ends of the exchanger is an adequate measure of the temperature difference for heat balance purposes.

114. An adsorbent carbon has an internal surface area of the order of $1000\,m^2\,g^{-1}$.

115. Refuse-derived fuels can be significantly improved by blending with waste hydrocarbon.

116. When hydrocarbon liquid is gasified with steam the dominant products are hydrogen and methane.

117. Ethylene can be made by dehydrogenation of ethane under fairly mild conditions using a catalyst.

118. Haber's rule, when applied to loss of consciousness through carbon monoxide inhalation, applies under all conditions of practical interest.

119. Carbon dioxide in post-combustion gas inhibits the inhalation of toxic gases including carbon monoxide.

120. The British Petroleum Company was originally called the Anglo–Persian Oil Company.

121. Ethylene in good yield can be made by cracking crude petroleum.

122. A 'dense gas' leak displays height decrease initially followed by height increase.

123. The Pasquil model for gas dispersion applies to dense gases.

124. The Gaussian model for gas dispersion is widely applied to hydrocarbon leaks.

125. Ethane when leaked is likely to display passive rather than 'dense gas' dispersion.

126. A brine, in the sense of the word in chemical engineering, is a refrigerant fluid.

127. Simple hydrocarbons such as propylene find wide application as refrigerants in hydrocarbon plant.

128. The ideal reversed Carnot cycle is an adequate approximation in the description of most actual refrigeration processes.

129. There are ten letter-designated thermocouple types.

130. A thermocouple obtained from a supplier and subjected to no further calibration has an intrinsic uncertainty of at least ±2–3°C, greater at higher temperatures.

131. Mineral-insulated, metal sheathed (MIMS) thermocouples are preferred to bare-wire thermocouples for many chemical processing applications.

132. In a MIMS thermocouple, the sheath provides total and indefinite protection of the thermoelements from chemical attack.

133. Resistance temperature detectors (RTDs) are intrinsically more accurate than thermocouples.

134. If an RTD is immersed in a liquid, protection of the platinum resistor is required.

135. When shale is retorted *in situ* it is necessary to regulate the temperature by means of a diluent gas.

136. High explosives are used in excavation of shale deposits.

137. Above-ground removal by hot water of tar sands from the porous rocks containing them is essentially the same process as the retorting of shale.

138. Above-ground and *in situ* separation of tar sands are, other things being equal, expected to yield the same product.

139. There has been some interest in nuclear means of fragmentation of the geological structure in the production both of shale and of tar sands.

140. LNG can be conveyed in pipes which operate essentially along the principles of the simple vacuum flask, having two concentric walls with the space between them evacuated.

141. Isopropyl nitrate (IPN) is an octane enhancer.

142. The Frank–Kamenetskii (FK) model of thermal ignition can be used to predict thermally critical sizes of assembly of unstable substances.

143. Cetane enhancers tend to have low activation energies.

144. Hot spots can be created by viscous heating.

145. Traditionally the place most associated with tar sand fuels is Alberta, Canada.

146. The FK model of thermal ignition is limited to spherical and cylindrical shapes of assembly.

147. Indonesia is the world's largest exporter of LNG.

148. Japan is, and has been throughout the entire industrial era, heavily dependent on imported fuels.

149. One natural gas reserve in China is known to have been in intermittent use since 211 BC.

150. Oil production in Vietnam is on an insignifcant scale.

151. Coal bed methane is rich in condensate.

152. The Gulf coast hurricanes led to unprecedented fluctuations in the price of oil.

153. Water is capable of displaying BLEVE behaviour.

154. The crude oil released into the sea during the 2010 Gulf Coast spill was of the order of 50 million barrels.

155. Biodiesels are less susceptible to ignition hazards from static electricity than petroleum derived fuels are.

156. The 2005 BP Texas City refinery accident originated in a cracking unit.

157. Natural gas, unlike crude oil, is marketed on a heat basis not a quantity basis.

158. The fire in an LNG production plant in Algeria in 2004 began at a gas turbine.

159. Maximum sea depths in offshore oil and gas production are about 1000 m

160. An alternative to refining crude oil is hydrocracking it.

161. The heat of polymerisation of ethylene is, on a kilogram basis, higher than that of vinyl chloride.

162. At least one of the deaths at Flixborough was from smoke inhalation.

163. Steam reforming of hydrocarbons is a form of partial oxidation.

164. Nitric acid is used as a catalyst in alkylation processes.

165. Cracking processes yield a range of products.

166. The largest LPG tank currently in existence can hold about 40 000 tonne.

167. Hydrocracking is of previously fractionated material only.

168. Natural gas can be used to generate hydrogen at a refinery.

169. The only drawing of the pipe used at Flixborough to bridge the two vessels was in chalk on the floor.

170. The thermal effect of the bromination of benzene to make bromobenzene is comparatively small

171. The fatal accident at a refinery in Delaware City in 2001 occurred at a hydrogen generation unit

172. LNG disperses as a passive gas.

173. To remove 1 p.p.m. by weight of hydrogen sulphide from natural gas by methyl diethanol amine treatment requires admittance of about 5 p.p.m. of the amine reagent.

174. Nitrogen in petroleum fractions is often a catalyst poison.

175. In the use of carbon dioxide fire extinguishers sublimation of any solid carbon dioxide present has a powerful effect.

176. The 2009 blowout at a jack-up rig in the Timor Sea had fatal consequences.

177. The fatal accident with hydroxylamine in PA in 1999 is have involved explosion of a tonne of the chemical.

178. Were a supertanker to explode catastrophically whilst holding a full payload, the blast would significantly exceed that of the nuclear weapon dropped at Hiroshima in 1945.

179. Stainless steel is the most common choice of material for pipe work within refineries.

180. Response by interested parties to the Proposed EU Regulation of Offshore Safety 2011 has been consistently positive

181. The laundering of overalls at offshore installations comes within the scope of Control of Substances Hazardous to Health (COSHH) Regulations 2002

182. A pellistor measures a gas concentration in units per cent molar.

Answers

1. True.

2. False. In the Otto cycle the working substance is air.

3. False. It was off the Louisiana coast.

4. True.

5. True.

6. False. Such conditions signify an independence on the downstream pressure.

7. True.

8. True.

9. False. It is 1.3.

10. True.

11. True.

12. True.

13. False.

14. False. It is an exponential function of the wind speed.

15. True.

16. True.

17. False.

18. False.

19. True.

20. True.

21. True.

22. True.

23. True.

24. False. The mass transfer rate depends only weakly on the nature of the liquid.

25. False. It depends on the pool diameter.

26. False. It is 'optically thick' and has a high emissivity.

27. True.

28. False. Such dependence is weak.

29. False. A BLEVE is basically a physical explosion.

30. False. ALARP is negotiated within certain limits.

31. True.

32. True.

33. True.

34. False. They are of the order of one.

35. True.

36. True.

37. False.

38. True.

39. True.

40. True.

41. False. It is the least reactive alkane.

42. True.

43. False. They tend to be of higher reactivity.

44. True.

45. False. The heat of vaporisation is also needed.

46. False. 'Spontaneous combustion' is a well-known and important phenomenon. Also, there is no ignition source in compression ignition (e.g., a diesel engine).

47. True.

48. False. They have heat values of about $45\,MJ\,kg^{-1}$.

49. False.

50. False.

51. True.

52. True.

53. False. It lowers it.

54. True.

55. True.

56. False.

57. True.

58. True.

59. False.

60. True.

61. False. A BLEVE did indeed occur in Sydney in 1989, but there were no deaths or injuries.

62. True.

63. True.

64. False. Convection coefficients have units $W\,m^{-2}\,K^{-1}$, whereas thermal conductivity has units $W\,m^{-1}\,K^{-1}$.

65. True.

66. True.

67. False. In the example in the main text, one of the keywords was deemed inapplicable.

68. True.

69. True.

70. True.

71. False. It is often necessary to build in a safety margin for corrosion or erosion.

72. False. It is more effective.

73. False. This figure is about an order of magnitude too high for liquids. Gases will travel at about that speed.

74. True.

75. False.

76. True.

77. False. Also need the heat capacity.

78. False. The pipe was narrower than the entry/exit 'stubs', and there were 'bellows' connecting them.

79. True.

80. True.

81. False. There are known cases of self-heating of spent catalyst.

82. False. They are significantly exothermic.

83. True.

84. False, it is 1 p.p.m.

85. True.

86. True.

87. True.

88. False. The critical temperature is 230°C and the compound frequently is stored as a liquid under its own vapour at ordinary temperatures.

89. True.

90. False. It is an intermediate in the manufacture of an insecticide.

91. True.

92. True.

93. False. It is used in the manufacture of synthetic fibres.

94. True.

95. False. There also has to be sunlight.

96. True.

97. False. Some white paints have high emissivities.

98. True.

99. True.

100. True.

101. False, it requires about $1\,kW$ per m^3 of liquid.

102. Probably true (see relevant calculation in main text) but documentary evidence is lacking.

103. False. It is usually continuous.

104. True.

105. False. If the gas is much less abundant than the oil, or if it contains high levels of sulphur, it will simply be flared.

106. True.

107. False. It needs to have at least $4\,MJ\,m^{-3}$ in order to sustain a flame on a burner.

108. True.

109. True.

110. False. Mild steel has been more widely used.

111. True.
112. False. It is known that heat exchangers have sometimes been stored out of doors before installation and have thus become corroded before use.
113. False. The log mean temperature difference should be used.
114. True.
115. True.
116. False. The dominant products are hydrogen and carbon monoxide.
117. True. This means of making ethylene is prevalent in North America: its preparation from the cracking of naphtha is more common in Europe.
118. False. It applies for high concentrations and short times, but underestimates times required for loss of consciousness at concentrations below about 1000 p.p.m.
119. False. The CO_2 actually stimulates inhalation of CO.
120. True.
121. True.
122. True.
123. False. It applies to passive dispersion where gravity effects are relatively unimportant.
124. False. It is most commonly applied to emissions from stacks.
125. True.
126. False. The brine transfers heat to the refrigerant fluid, but itself undergoes no refrigeration cycle.
127. True.
128. False. In a number of ways, including the assumption of reversibility along each part of the cycle, it is too simple, but it provides a bench-

mark against which the performance of practical refrigeration systems can be measured.
129. False. There are eight.
130. True.
131. True.
132. False. Any agent capable of attacking the thermoelements is likely to attack the sheath, impairing its protective function. Also, manganese present in small amounts in the sheath can, at high temperatures, migrate to the thermoelements and affect their calibration.
133. True.
134. True.
135. True.
136. True.
137. False. The removal of tar sands by hot water is a physical process – heating to lower the viscosity – only, whereas the retorting of shale is chemical decomposition.
138. False. In *in situ* methods there is likely also to be some chemical decomposition.
139. True.
140. True.
141. False. It is a cetane enhancer.
142. True.
143. False – they tend to have high activation energies.
144. True.
145. True.
146. False. It can be extended to very many shapes. See Chapter 9 for application to a cube.
147. False, since the first edition it has been surpassed by Malaysia.
148. True.
149. True. It is at Chi liu ching.

150. False. Vietnam is a very significant producer and exporter of oil.

151. False.

152. True.

153. True. That is what happens when an autoclave blows up.

154. False. It was about a tenth of this.

155. True.

156. False. It began at an isomerisation unit.

157. True

158. False. It began at a steam turbine.

159. False. They are at least twice this.

160. True.

161. True.

162. True.

163. True.

164. False. Hydrofluoric acid is.

165. True.

166. True. It was built by Samsung.

167. False. Crude oil is often so treated.

168. True.

169. The author has heard this from a number of sources.

170. True.

171. False.

172. False.

173. True.

174. True.

175. False.

176. False.

177. True.

178. True.

179. False. Carbon steel is.

180. False.

181. True.

182. False. It gives the measurement as a proportion of the LFL.

INDEX

acenaphthalene 35-36
adsorbent carbons 238–243
afterburning 235–237, 244
acetaldehyde
    adiabatic flame temperature 202–203
    low-temperature oxidation 38
acrylonitrile
    at Rijnmond 260
    leakage, case study 222
    toxic hazards of 220–221
ALARP (as low as reasonably practicable) 171–172, 182–183, 198–200
alkylation 110–111
ammonia
    at Canvey 256–257
    leakage of 259–261
    toxicity of 215–216
ammonium nitrate
    at Canvey 256–257
    explosive behaviour 207, 210, 211
Amuay refinery 76, 91
Antoine equation 131, 155, 229
API RP 75 178
azeotropic mixtures 130

barrel (US)
    definition of 10
benzene
    chemical structure of 2
    hydrogenation of 105–106
    nitration of 106–107
    occurrence of with toluene and xylenes 155, 217–219
    storage of 87

toxicity of 217–218, 229
vapour pressure of 229
Bernoulli's equation 15–16, 25, 137
Bhopal 217
Bourbon Dolphin 178
Bourdon gauge 141
brines 81
British thermal unit
conversion factor to calorie 9
Buncefield 61

carbon monoxide
toxicity of 224–225
cetane enhancers 69–73, 89
chlorination
of hydrocarbons 101–103, 113
chlorine
toxic hazards of 213–215
CNG (compressed natural gas) 197
coal
calorific value of 28, 187
hydrocarbons from 6, 103–104, 251–252
tars 31
compressibility factor (gases) 127
consequence analysis 163, 167–174
cool flames 38–39
COSSH Regulations (2012) 179–180
critical flow 12
critical temperature
propane and vinyl chloride monomer 63
hydrogen fluoride 216
crude oil
combustion of 184
leakage of 15-17, 25, 123–124
nature of 2, 73
refining of 73–77
separation from gas 160–162, 189
spillages of 186–187
stabilisation of 186

dense gas dispersion 20–21, 26, 261
design stress 117–120, 133,135
detonation
distinction from deflagration 47
possibility of with organic peroxides 206–207
Diesel cycle 1
dike 16, 257

Ekofisk 176
electrocution hazards 75
EXIT signs
   methods of illuminating 173
explosion times
   estimation of for unstable substances 207, 210

fire point 34
First Law of Thermodynamics 78–79, 129
flaring
   at an offshore platform 172–173
   hazards in 233–235
flash fires 46–47, 187
flash points
   apparatus 34–36
   background to 8, 30–33
   calculation of
      from the Factory Mutual Equation 33
      from vapour pressures 30–33, 53
   errors in 33
   values of for a selection of liquids 31
Fanning friction factor 14
Flixborough 18, 99, 124
fluidised beds 99
Frank-Kamenetskii model of thermal ignition
   applied to storage of unstable substances 69–73, 89
   extended to inert 'hot spots' 92

gas
   leakage of through an orifice 11–15, 25, 39–40, 52
   'perfect' 12
gasification
   hazards in 103–104
   of waste hydrocarbon 240–241
Gulf coast 4, 7, 159, 178

Haber's rule 224–225
Halons 223
   substitutes for 223
HAZOP 67–70, 78
heat exchangers
   basic designs of 79–81
   calculations of performance of 114
   for LNG 192, 208
   hazards with 80
   in refining of crude oil 73–74
   process integration with 77, 212

recuperative burning with 237–238, 245
Hoechst 7
'hot work'
    accident caused by 62
    need to regulate 93
hydrodesulphurisation 96–97, 114
hydrodenitrogenation 96–97
hydrogen fluoride
    at Canvey 256–258
    dispersion of 259
    toxicity of 216–217
    transportation of 229

ice
    malfunctioning of safety valves due to 76
'ideal reversed Carnot cycle' 81
Indonesia
    exporter of LNG 195
    offshore activity 160
inherent safety 161

Japan
    offshore activity 6, 160
    paucity of fuel reserves in 6, 196, 247

lightning
    as an ignition hazard 62
LNG (liquefied natural gas) 3, 191–197
    accidents with 193
    at Canvey Island 256, 260
    at Rijnmond 260 ??
    combustion behaviour 43, 196-197 (*see also* pool fires)
    exports and imports of 196
    hazards with 25, 191–192
    measurement of temperature of 152–153
    transportation of 195-196, 205
LPG (liquefied petroleum gas)
    at Canvey 256
    BLEVE behaviour with 43–46, 55, 200–201
    composition of 43, 197
    leakage of, case studies 49, 189, 200–201
    non-interchangeability of with natural gas 190
    production of, at refineries 73
    storage of 63–64, 117–119, 122–124,
    transportation of 197–201

manometer 137–138

Marcus Hook Refinery 76–77
Montara drilling rig 178
Morocco
  shale production in 249–250
'most economic diameter'
  pipes, formula for 121

natural gas
  afterburning with 235–238
  calorific value of 187
  composition of 2, 4
  compressed, *see* CNG
  condensate 3–4, 201–202
  early usage of 4, 187
  hydrates 188
  hydrogen sulphide removal by MDEA 115–116
  leakage of 11–15 (*see also* consequence analysis and leakage of gas
    through an orifice)
  hazards with 187–190
  liquefied, *see* LNG
  pricing of, in the US 9
  production of at offshore platforms 159–161
  steam reforming of 104
  synthesis gas from 188
naphtha
  gasification of 104
  production of by refining crude oil 73
nitrobenzene
  manufacture and uses of 88, 106–107
North Sea
  offshore activity at 159–160, 176–178 (*see also* Piper Alpha)
  Dutch sector of 74–75

offshore platforms
  accident case studies 174–175 (*see also* Piper Alpha)
  'Christmas tree' 12
  current activity 6
  early proliferation of 6
  features of 160–163
  hazards at 11, 160–161 (*see also* consequence analysis, risk assessment)
  legislation appertaining to 8
olefins
  accident in production of 82
  production of by cracking 6–7, 93–-96
  production of by hydrogenation 96
Otto cycle 1

overpressure
in offshore fires 161–163, 170–171
injuries due to 51, 57
introduction 47

partial molar volumes 126–127
Pasquil weather categories 19–20, 26
passive dispersion of gases 19–20, 26
pellistor 22–23
pipe schedule number 120–121, 133
Piper Alpha 8, 61, 176
polyesters
manufacture of 99
polymerisation
background to 1
hazards in 108–110, 115
unwanted
in heat exchangers 80
in storage of acrylonitrile 220
pool fires 41–43, 170, 196, 201, 209
probit tables 314
Proposed EU Regulation of Offshore Safety (2011) 179–181

Rankine cycle 1
Redwood viscosity 127
Refinery, layout of 82–86
refrigeration
hazards of 82
general principles 81–82
Reid vapour pressure 32–33
Re-refining 242–243
Reynolds analogy 14
Richmond CA refinery 76–77
risk assessment
at offshore platforms 11, 17–18, 165–176 (*see also* **ALARP**,
consequence analysis)
release of toxic substances 214–216
Rockefeller J.D. 4, 5
rotameter 139
RTD (resistance temperature detector) 149–153, 157
Russia
oil production in 5

semi-infinite solid 169, 183
shale oil
early usage of 4–5
production of by retorting 247–250, 254

slumping
    of a catastrophically leaked liquid 17
    of dense gases 20
smoke
    effects of in an offshore fire 172–175, 184
    toxicity of 223
solar flux
    hazards due to in hydrocarbon storage 65, 203–205
    role in the Ludwigshafen accidents 49
solid waste
    value-adding to by blending with oil waste 240, 245
static electricity
    as an ignition hazard 65, 192
stress
    in vessel support 122–124, 135
styrene
    flash point 54
    toxicity 229

tar sands 250
Tamaulipas, Mexico, natural gas explosion at 190
Thailand
    offshore activity 6
thermal radiation
    effects of on temperature measurement 100–101, 146–147
    from a BLEVE 55
    from a fireball 44–45
    from a jet fire 37, 39–41, 169–70
    from a pool fire 43
    general background 36–37
    non-gray body
        as a model for storage vessel surfaces 205, 210
    probit equation for injury by 164–165, 184
    to and from the sky 203–205 (*see also* solar flux)
thermocouple tables (Type K) 315–316
thermocouples
    background and principles 142–149
    radiation errors 100, 146–7
    use of in temperature control 56, 86-87, 89
tight gas 252–253
TNT equivalence 47–48, 55,
two-phase flow 17, 40

vapour cloud explosions 46–48, 197
venting
    of hydrocarbon to the atmosphere 239–240
venturi meter 137–139, 141

Vietnam
   oil production 6
vinyl acetate
   manufacture of 98
vinyl chloride monomer (VCM)
   BLEVE due to 44
   storage of 230
   toxic hazards of 219–220
   use of in PVC manufacture 108–110, 114

weir 139-140, 155
welding of pipes and vessels
   allowances for in design calculations 119–120, 133, 135 (*see also*
      'hot work')
wind
   effect of on dispersion of leaked hydrocarbon 17–18, 26
   effect of on smoke at an offshore fire 172–174